CAPE

Edited by
Luis Puigjaner and Georges Heyen

Related Titles

Kai Sundmacher, Achim Kienle, Andreas Seidel-Morgenstern (Eds.)

Integrated Chemical Processes
Synthesis, Operation, Analysis, and Control

2005
ISBN 3-527-30831-8

Kai Sundmacher, Achim Kienle (Eds.)

Reactive Distillation
Status and Future Directions

2002
ISBN 3-527-30579-3

Frerich Johannes Keil (Ed.)

Modeling of Process Intensification

2006
ISBN 3-527-31143-2

Ulrich Bröckel, Willi Meier, Gerhard Wagner

Best Practice in Product Design and Engineering

2007
ISBN 3-527-31529-2

Ullmann's Processes and Process Engineering

3 Volumes
2004
ISBN 3-527-31096-7

CAPE

Volume 2
Computer Aided Process and Product Engineering

Edited by
Luis Puigjaner and Georges Heyen

WILEY-VCH Verlag GmbH & Co. KGaA

The Editors

Professor Dr. Luis Puigjaner
Universitat Politèchnica de Catalunya
Chemical Engineering Department
ESTEIB, Av. Diagonal 647
08028 Barcelona
Spain

Professor Dr. Georges Heyen
Laboratoire d'Analyse et Synthèse des Systèmes
Chimiques
Université de Liège
Sart Tilman B6A
4000 Liège
Belgium

■ All books published by Wiley-VCH are carefully produced. Nevertheless, authors, editor, and publisher do not warrant the information contained in these books, including this book, to be free of errors. Readers are advised to keep in mind that statements, data, illustrations, procedural details or other items may inadvertently be inaccurate.

Library of Congress Card No.:
Applied for

British Library Cataloging-in-Publication Data:
A catalogue record for this book is available from the British Library.

Bibliographic information published by the Deutsche Nationalbibliothek
The Deutsche Nationalbibliothek lists this publication in the Deutsche Nationalbibliografie; detailed bibliographic data are available in the Internet at http://dnb.d-nb.de.

© 2006 WILEY-VCH Verlag GmbH & Co. KGaA, Weinheim

Typesetting   Mitterweger & Partner, Plankstadt
Printing   betz-druck GmbH, Darmstadt
Binding   Litges & Dopf GmbH, Heppenheim
Cover Design   4t Mattes + Traut, Werbeagentur GmbH, Darmstadt

Printed in the Federal Republic of Germany

Printed on acid-free paper

ISBN-13   978-3-527-30804-0
ISBN-10   3-527-30804-0

Table of Contents

Computer Aided Process and Product Engineering. Edited by Luis Puigjaner and Georges Heyen
Copyright © 2006 WILEY-VCH Verlag GmbH & Co. KGaA, Weinheim
ISBN: 3-527-30804-0

1
Preface

Computer Aided Process and Product Engineering (CAPE): Its Pivotal Role for the Future of Chemical and Process Engineering

Chemical and related industries are at the heart of the great number of scientific and technological challenges involving computer-aided processes and product engineering.

Chemical and related industries including process industries such as petroleum, pharmaceutical and health, agriculture and food, environment, textile, iron and steel, bitumous, building materials, glass, surfactants, cosmetics and perfume, and electronics are evolving considerably due to unprecedented market demands and constraints stemming from public concern over environmental and safety issues.

To respond to these demands, the following challenges faced by these process industries involve complex systems, both at the process-scale and at the product-scale:

1. Processes are no longer selected on a basis of economic exploitation alone. Rather, compensation resulting from the increased selectivity and savings linked to the process itself is sought after. Innovative processes for the production of commodity and intermediate products need to be researched where patents usually do not concern the products but the processes. The problem becomes more and more complex as factors such as safety, health, environment aspects including nonpolluting technologies, reduction of raw materials and energy losses, and product/by-product recyclability are considered. The industry, with large plants, must supply bulk products in large volumes and the customer will buy a process that is nonpolluting and perfectly safe, requiring computer-aided process engineering (CAPE).
2. New specialities, active material chemistry, and related industries involve the chemistry/biology interface of agriculture, food, and health industries. Similarly, it involves upgrading and conversion of petroleum feedstock and intermediates, conversion of coal-derived chemicals or synthesis gas into fuels, hydrocarbons or oxygenates. This progression from traditional chemistry is driven by the new market objectives where sales and competitiveness are dominated by the end-use properties of a product as well as its quality. It is important to underline that today, 60 % of all products sold by chemical companies are crystalline, polymer, or

Computer Aided Process and Product Engineering. Edited by Luis Puigjaner and Georges Heyen
Copyright © 2006 WILEY-VCH Verlag GmbH & Co. KGaA, Weinheim
ISBN: 3-527-30804-0

amorphous solids. These complex and structured materials have a clearly defined physical shape in order to meet the designed and the desired quality standards. This also applies to plastics, ceramics, soft solids, paste-like products, and emulsions. New developments require increasingly specialized materials, active compounds, and special effects chemicals. The chemicals are much more complex in terms of molecular structure than traditional, industrial chemicals. Control of the end-use property (size, shape, color, aesthetics, chemical and biological stability, degradability, therapeutic activity, solubility, touch, handling, cohesion, rugosity, taste, succulence, sensory properties, etc.), expertise in the design of the process, continual adjustments to meet changing demands, and speed to react to market conditions are the dominant elements. For these specialities and active materials the client buys the product that is the most efficient and first on the market. He will have to pay high prices and expect a large benefit from these short life-time and high-margin products, requiring most often computer-aided process and product engineering.

The triplet molecular processes-product-process engineering (3PE) approach requires the tools of CAPE.

Today, chemical and process engineering are concerned with understanding and developing systematic procedures for the design and optimal operation of process systems, ranging from nano and microsystems to industrial-scale continuous and batch processes; this is illustrated by the chemical supply chain concept. In the supply chain, it should be emphasized that product quality is determined at the micro and nanolevel and that a product with a desired property must be investigated for both structure and function. A comprehension of the structure-property relationship at the molecular (e.g., surface physics and chemistry) and microscopic level is required. The key to success is to obtain the desired end-use property of a product, and thus control product quality by controlling complexity in the microstructure formation. This will help to make the leap from the nanolevel to the process level. Moreover, most chemical and biological processes are nonlinear, belonging to the so-called complex systems for which multiscale structure is common nature. Therefore, an integrated system approach for a multidisciplinary and multiscale modeling of complex, simultaneous, and often coupled momentum, heat and mass transfer processes is required:

- Different time scales (10^{-15}–10^8 s) are used from femto and picoseconds for the motion of atoms in a molecule during a chemical reaction, nanoseconds for molecular vibrations, hours for operating industrial processes, and centuries for the destruction of pollutants in the environment.
- Different length scales (10^{-8}–10^6 m) are used from nanoscale for molecular kinetic processes; microscale for bubbles, droplets, particles, and eddies; mesoscale for unit operations dealing with reactors, columns and exchangers; macroscale for production units; and megascale for environment and dispersion of emissions (see the following figure):

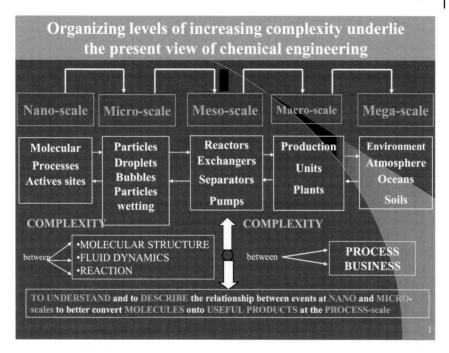

Therefore, organizing scales and complexity levels in process engineering is necessary in order to understand and describe the events at the nano and microscales, and to better convert molecules into useful products at the process scales.

This multiscale approach is now also encountered in biotechnology, bioprocesses, and product engineering, to manufacture products and to better understand and control biological tools such as enzymes and micro-organisms. In such cases, it is necessary to organize the levels of increasing complexity from the gene with known properties and structures, up to the product-process relation, by modeling coupled mechanisms and processes at different length scales: the nanoscale is used for molecular and genomic processes and metabolic transformations; pico and microscales are used for enzyme and integrated enzymatic systems, and biocatalyst and active aggregates; mesoscale is used for bioreactors, exchangers, separators; and macro and megascales are used for production units and interactions with the biosphere. Thus, organizing levels of complexity at different length scales, associated with an integrated approach to phenomena and simultaneous and coupled processes, are the heart of the new view of biochemical engineering (see next figure). Indeed this capability offers the opportunity to apply genetic-level controls to make better biocatalysts, novel products, or developing new drugs, new therapies, and biomimetic devices. Understanding an enzyme at the molecular level means that it may be tailored to produce a particular end-product. Also, the ability to think across length scales makes chemical engineers particularly well poised to elucidate the mechanistic understanding of molecular and cell biology and its large-scale manifestation, i.e., decoding communications between cells in the immune systems.

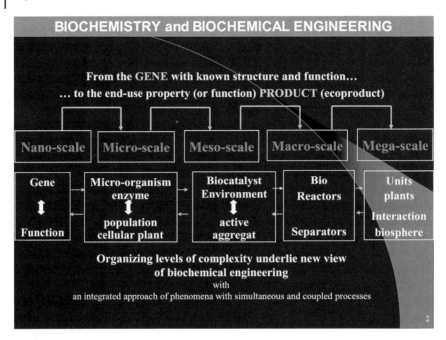

These examples are at the center of the new view of chemical and process engineering: organizing levels of complexity, by translating molecular processes into phenomenological macroscopic laws to create and control the required end-use properties and functionality of products manufactured by a continuous or batch process. I have defined this approach as the triplet molecular processes-product-process engineering (3PE): an integrated system approach of complex pluridisciplinary nonlinear and nonequilibrium processes and phenomena occurring on different length and time scales, involving a strong multidisciplinary collaboration between physicists, chemists, biologists, mathematicians, computer-aided specialists, and instrumentation specialists.

Today's tools are wide-ranging for the success of chemical and process engineering for modeling, complex systems at different scales encountered in the process and product engineering.

It's possible to understand and describe events on the nano and microscale in order to convert molecules into useful products on process and unit scales thanks to significant simultaneous breakthroughs in three areas: molecular modeling (both theory and computer simulation), scientific instrumentation and noninvasive measurement techniques, and powerful computational tools and capabilities for information collection and processing.

At the nanoscale, molecular modeling assists in maintaining better control of surface states of catalysts and activators, obtaining increased selectivity and facilitating asymmetrical synthesis, e.g., chiral technologies. Molecular modeling also assists in explaining the relationship between structure and activity at the molecular scale in order to control crystallisation, coating and agglomeration kinetics.

At the microscale, computational chemistry is very useful for understanding complex media and all systems whose properties are controlled by rheology and interfacial phenomena.

At the meso and macroscales, computer fluid dynamics (CFD) is required for scaling up new equipment, or for the design of new operation modes for existing equipment such as reversed flow, cyclic processes, and unsteady operations. It is especially useful when rendering multifunctional processes with higher yields in chemical or biological reactions coupled with separation or heat transfer. It also provides a considerable economic benefit.

At the production unit and multiproduct plant scale, dynamic simulation and computer tools for simulation of entire processes are needed more and more. These tools analyze the operating conditions of each piece of equipment in order to simulate the whole process in terms of time and energy costs. New performances (product quality and final cost) resulting from any change due to a blocking step or a bottleneck in the supply chain will be predicted in a few seconds. It is clear that such computer simulations enable the design of individual steps, the structure of the whole process at the megascale, and place individual processes in the overall context of production, emphasizing the role and the place of computer assistance in process and product engineering.

The previous considerations on the necessary multidisciplinary and multiscale integrated approach for managing complex systems encountered by chemical and related process industries in order to meet market demands led to the proposal of four main parallel objectives involving the tools of CAPE.

The first objective concerns a total multiscale control of the process to increase selectivity and productivity by the nanotailoring of materials. The nanotailoring can be produced with controlled structure, or by supplying the process with a local "informed" flux of energy and materials, or by increasing information transfer in the reverse direction, from process to man, requiring close computer control, relevant models, and arrays of local sensors and actuators.

The second objective concerns the process intensification by the design of novel equipment based on scientific principles, new operating modes, and new methods of production. Process intensification with multifunctional equipment that couples or uncouples elementary processes (transfer-reaction-separation), involving the reduction in the number of equipment units leads to reduced investment costs and significant energy recovery or savings. Cost reduction between 10 % and 20 % are obtained by optimizing the process. But the use of such hybrid technologies is limited by the resulting problems with control and simulation leading to interesting but challenging problems in dynamic modeling, design, operation and strong nonlinear control. Also, process intensification using microengineering and microtechnology will be used more and more for high-throughput and formulation screening. Indeed microengineered reactors have some unique characterictics that create the potential for high-performance chemicals and information processing on complex systems. Moreover, scale-up to production by replication of microreactor units used in the laboratory eliminates costly redesign and pilot plant experiments, thus accelerating the transfer from laboratory to commercial-scale production.

The third objective concerns the extension of chemical engineering methodology to product-focussed design and engineering in using the multiscale modeling of the above-mentioned approach, 3PE. Indeed to be able to design and control the product quality of structured materials, and make the leap from the nanolevel to the process level, chemical and process scientists and engineers face many challenges in fundamental concepts (structure-activity relationships on molecular level, interfacial phenomena, adhesive forces, molecular modeling, equilibria, kinetics, and product characterization techniques); in product design (nucleation growth, internal structure, stabilization, additive); in process integration (simulation and design tools based on population balance); and in process control (sensors and dynamic models). It should be underlined that much progress has been made in product-oriented engineering and in process control using the scientific methods of chemical engineering. The methods include examination of thermodynamic equilibrium states, analysis of transport processes and kinetics when they are separate and linked by means of models with or without the help of molecular simulation, and by means of computer tools of simulation, modeling and extrapolation at different scales for the whole supply chain up to the laboratory-scale. But how can operations be scaled up from laboratory to plant? Will the same product be obtained and will its properties be preserved? What is the role of the equipment design in determining product properties?

This leads to the fourth main objective, which is to implement the multiscale application of computational chemical engineering modeling and simulation to real-life situations from the molecular scale to the overall production scale in order to understand how phenomena at a smaller length scale relate to properties and behavior at a longer length scale. The long-term challenge is to combine the thermodynamics and physics of local structure-forming processes like network formation, phase separation, agglomeration, nucleation, crystallization, sintering, etc., with multiphase computer fluid dynamics. Indeed through the interplay of molecular theory, simulation and experimentation measurements a better quantitative understanding of structure-property relations evolves, which, when coupled with macroscopic chemical engineering science, form the basis for new materials and process design (CAPE). Turning to the macroscopic scale, dynamic process modeling and process synthesis are also increasingly developed. Moreover, integration and opening of modeling and event-driven simulation environments in response to the current demand for diverse and more complex models in process engineering is currently taking a more important place. The aim is to promote the adoption of a standard of communication between simulation systems at any time and length-scale level (thermodynamic, unit operations, numerical utilities for dynamic, static, batch simulations, fluid dynamics, process synthesis, energy integration, process control) in order to simulate processes and allow the customers to integrate the information from any simulator into another one. Thus expanding and developing interface specification standards to ensure interoperability of CAPE software components justifies the creation of a standardization body (CAPE-OPEN Laboratories Network, CO-LAN) to maintain and disseminate the software standards in the CAPE domain that have been developed in several international projects.

The CAPE Present Book: a Vade Mecum in Process Systems Engineering

It is clear and I have shown that chemical and process industries have to overcome challenges linked to the complexity. They need to master phenomena in order to produce products "first on the market" with "zero pollution, zero accident, and zero defects" processes. Therefore, never before have enterprises invested so much in information processing and computer-controlled production, which proved in many cases that it was capable of reducing the costs and greatly increasing the flexibility than any other technology in past decades. Moreover, information and communication technology offered a great number of standardized but also specific possibilities of applications and solutions as never before. Therefore, a strategy aiming at the strengthening of competitiveness of production should obviously incorporate CAPE as a guideline for the reunion of the flexible production, the technical and administrative data processing, and the complete penetration of the enterprise activities with date processing.

Within the European Federation of Chemical Engineering, the CAPE Working Party has been very active in this area since the end of the 1960s, as shown by the success of the ESCAPE series of symposia. The activity of the Working Party is also shown by the publication of this book, which aims to present and review the state-of-the-art and latest developments in process systems engineering. Its contents illustrates the modern day multidisciplinary and multiscale integrated approach for the integrated product-process design and decision support to the complex management of the entire enterprise. It also highlights the use of information technology tools in the development of new products and processes, in the optimal operation of complex and/or new equipment found in the chemical and process industries, and in the complex management of the supply chain.

Actually, this book, based on the competences of scientists and engineers confronted with industrial practice, is a reference tool. Its ensuing and clear objectives in the topic of process systems engineering are:

- the necessary multidisciplinary bases required for understanding and modeling the phenomena occurring at the different scales from molecular range up to the global supply chain and extended enterprise;
- the experimental and knowledge-based methods and tools available to assist in the conception of new processes and products design, and in the design of plants able to manufacture the products in a competitive and sustainable way;
- the presentation of needed advances to fight ever-increasing complexity involved within the product-process life cycle;
- some tutorial examples and cases studies aiming to the state-of-the-art computer-aided tools.

The theoretical and practical aspects of the computer-aided process engineering covered in this book, involving computer-aided modeling and simulation, computer-aided process and product design, computer-aided process operation, and computer integrated approaches in CAPE should find use in libraries and research facilities and make a direct impact in the chemical and related process industries.

This book judiciously titled "Computer Aided Process and Product Engineering – CAPE" is a vade mecum in process systems engineering. It is a valuable and indispensable reference to the scientific and industrial community and should be useful for engineers, scientists, professors and students engaged in the CAPE topic.

Bravo and many congratulations to our collegues Prof. Puigjaner and Prof. Heyen for this publication and to the many authors involved in the CAPE Working Party of the European Federation of Chemical Engineering that have made this vade mecum a reality.

Prof. Dr. Ing. *Jean-Claude Charpentier*
President of the European Federation of Chemical Engineering

Foreword

The European Working Party on Computer Aided Process Engineering has been an important and highly effective stimulus for and promoter of research and educational advances in process systems engineering for over 40 years. The 1991 redirection of the Working Party from the broad and all inclusive scope of embracing "the use of computers in chemical engineering", which was its theme for some thirty years, to its present emphasis on product and process engineering has had important consequences. It has reenergized the organization, sharpened its focus and promoted higher levels of technical achievement. Indeed over the past decade, the ESCAPE series of annual conferences sponsored by the Working Party has become a vital world forum for disseminating, discussing and analyzing progress in state-of-the-art methodologies to support product and process design, development, operation and management. This volume represents a well-directed effort by the Working Party to capture the current status of developments in this field and thus to give that field its current definition.

To be sure, the volume is an ambitious undertaking because the process systems engineering field has expanded enormously from its traditional primary focus on the design, control, and operations of continuously operated commodity chemical processes and its secondary concern with the design and control of batch unit operations. That expansion has included methodologies to support product design and development, increases in both scope and complexity of the processing systems under consideration, and approaches to quantify the risks resulting from technical and market uncertainties and incorporate risk-reward trade-offs in design and operational decisions. The scope of systems encompassed by process systems engineering now ranges from the molecular, biomolecular and nanoscale to the enterprise-wide arena. The levels of complexity include self-assembly processes at the nanoscale, self regulating processes at the cellular level, the combination of mechanical, electrical and surface energetic phenomena in heterogeneous particulate systems, the interplay between thermodynamic, reaction and transport phenomena in integrated reaction/separation operations, and even the decentralized and semiautonomous interactions of customers, suppliers, partners, competitors and government regulators at the enterprise level. Certainly, these developments have been greatly facilitated by remarkable advances in computing and information technologies. However, at least as important has been the expanded scope of the models that underpin design and operational decisions as well as key advances in the tools for creating, analyzing, solving and maintaining these models over the life cycle of the associate product/process. The models of interest are now defined not just in terms of the traditional alge-braic and differential equations, but also include systems of partial differential and integral equations, graphs/networks, logical relations/conditions, hybrids of logical conditions/relations and continuous equations, and even object-oriented representa-tions of information, decision and work flows.

Has this volume succeeded in addressing the expanded role of models, the thrusts in product design and development, the much enlarged scope and complexity of applications, and the innovative approaches to addressing uncertainty and risk? While it is impossible to address the full scope of these developments within the lim-ited pages of a single volume, the editors and authors, all active contributors of the Working Party, have indeed done remarkably well in capturing and highlighting many of the most important developments.

In Section 1, we find coverage of fundamental issues such as the development of modeling frameworks, model parameter estimation and verification methods, approaches to treatment of multiscale models, as well as numerical methods for solution of algebraic, differential and partial differential systems. The applications to particulate-based processes such as crystallization, grinding, and granulation are of continuing special interest. Computational fluid dynamics and molecular modeling tools, which have become integrated into the process systems engineering toolbox are reviewed and the state of methods for the modeling and analysis of micro-organisms are presented. The computational biology domain is receiving a high level of attention by the systems engineering community and will certainly receive even more extensive coverage in future reviews. The second section, which principally treats process design, reviews current developments in overall process synthesis as well as synthesis of reaction, separation and utility subsystems. The area of process intensification, which seeks to capture the potential synergies from exploiting the complex interactions of reaction-separation phenomena, is noted, discussed and rec-ognized as an important direction for further research in process systems engineer-ing. The third section on process operations covers important developments in the well-established functional levels of the process operations hierarchy: monitoring and data reconciliation, model based control, real-time optimization, scheduling, planning, and supply chain management. Additionally, issues related to the opera-tion of flexible batch plants are reviewed. Key to progress in developments in sched-uling, planning, supply chain and flexible batch plant operations have been advances in the formulation and solution of large-scale mixed integer optimization problems. The importance of these and need for continuing advances cannot be overempha-sized.

The fourth section treats three key integration issues as well as two supporting technology developments. While the basic features of product design are reviewed in Section 3, the progress in meeting the challenges of integrating product and process design are addressed in Section 4. As noted in that chapter, to date most of the prog-ress has been in applications such as structured or formulated products in which the linkage between product and process is very close but further developments are on the horizon. The modeling technology required to support the product/process life-cycle raised in one of the chapters is a key issue facing many industry sectors. This issue is not yet as intensively addressed in the process systems community as it should be, given its importance in capturing product/process knowledge and managing corporate risk. The chapter on integrated supply chain management at roots deals with strategic and tactical enterprise-wide decisions. Uncertainty and risk are critical components of such decisions requiring more attention and intense future study. The sections on physical property estimation and databases, as well as open standards for CAPE software, discuss components that comprise an essential infrastructure for CAPE developments. The importance of tools for physical property prediction/estimation is evident in domains such as pharmaceutical products, in which the absence of such predictive tools has significantly retarded CAPE efforts in product and process design. The volume concludes with several enlightening case studies spanning the technologies reviewed in the preceding sections that are well chosen to make these technologies, their strengths and weaknesses more concrete.

Given the limitations of a single volume, there necessarily are additional topics that will in the future require more intensive review and discussion. These include process intensification research at the micro and even nanoscales. While there have been research on microscale process design and control, the essential complexity of the phenomena that occurs at micro and nanoscales makes work in this area both challenging and of potential high impact. There has been progress in rigorous treatment of the full range of external and internal parameter uncertainties and promising computational methods for generating risk-reward frontiers that deserves notice, including the integration of discrete event simulation and discrete optimization methods. Algorithms and strategies for addressing multistage stochastic decision problems and incorporating the full valuation of the decision flexibility in multistage decision frameworks are receiving increased attention. Finally, large-scale optimization methods for attacking enterprise level decision problems of industrial scope are emerging and will become even more prominent in the near future.

In summary, this volume is remarkably thorough in capturing the current state of development of the process systems engineering field and representing its broad scope. It will serve this field well in stimulating further research and in encouraging students to learn and contribute to a vital and growing body of knowledge that has important applications in broad sectors of the chemical, petrochemical, specialty, pharmaceuticals, materials, electronics and consumer products industries. The editors and authors are to be congratulated for a job well done.

G. V. Reklaitis

List of Contributors

Jens Abildskov
CAPEC
Department of Chemical Engineering
Søltofts Plads, Building 229
2800 Kgs. Lyngby
Denmark

Alberto Alva-Argaez
Hankyong National University
Department Chemical Engineering
Kyonggi-do
Anseong 456-749
Korea

Jean-Pierre Belaud
National Polytechnic Institution of
Higher Learning at Toulouse INPT
Department of Process Systems
Engineering
118, route de Narbonne
31077 Toulouse Cedex 4
France

I. David L. Bogle
University College London
Centre for Process Systems
Engineering
Department of Chemical Engineering
Torrington Place
London WC1E 7JE
UK

Bertrand Braunschweig
Institut Française du Pétrole
Division Technologie, Informatique et
Mathématiques Appliquées
1 and 4 avenue de Bois Préau
92852 Rueil Malmaison Cédex
France

Ian T. Cameron
The University of Queensland
School of Engineering
Division of Chemical Engineering
St Lucia QLD 4072
Australia

Vivek Dua
University College London
Department of Chemical Engineering
Centre for Process Systems
Engineering
Torrington Place
London WC1E 7JE
UK

Sebastian Engell
University of Dortmund
Department of Biochemical and
Chemical Engineering
Process Control Laboratory
Emil-Figge-Str. 70
44221 Dortmund
Germany

Computer Aided Process and Product Engineering. Edited by Luis Puigjaner and Georges Heyen
Copyright © 2006 WILEY-VCH Verlag GmbH & Co. KGaA, Weinheim
ISBN: 3-527-30804-0

Antonio Espuña
Universitat Politècnica de Catalunya
Chemical Engineering Department
ESTEIB, Av. Diagonal 647
08028 Barcelona
Spain

Gregor Fernholz
Process Systems Enterprise Limited
Merlostrasse 12
50668 Köln
Germany

Guido Buzzi Ferraris
Politecnico di Milano
CMIC Department
Piazza Leonardo da Vinci, 32
20133 Milano
Italy

Rafiqul Gani
CAPEC
Department of Chemical Engineering
Søltofts Plads, Building 229
2800 Kgs. Lyngby
Denmark

Weihua Gao
GE Global Research Center
Real Time/Power Controls Laboratory
1800 Cailun Road
Zhangjiang High-tech Park, Pudong
New Area
201203 Shanghai
P. R. China

Michael C. Georgiadis
Imperial College London
Centre for Process Systems
Department of Chemical Engineering,
Engineering
Roderic Hill Building
South Kensington Campus
London SW7 2AZ
UK

Vincent Gerbaud
Laboratoire de Génie Chimique
BP 1301
5 rue Paulin Talabot
31106 Toulouse, Cedex 1
France

Krist V. Gernaey
Technical University of Denmark
Department of Chemical Engineering
Søltofts Plads, Building 229
2800 Kgs. Lyngby
Denmark

Johan Grievink
Delft University of Technology
Faculty of Applied Sciences
Department of Chemcial Technology
Julianalaan 136
2628 BL Delft
The Netherlands

Katalin M. Hangos
Hungarian Academy of Sciences
Computer and Automation Research
Institute
Process Control Research Group
Kende u. 11–13, PO Box 63
1518 Budapest
Hungary

Petra Heijnen
Delft University of Technology
Department of Technology, Policy
and Management
PO Box 5015
2600 GA Delft
The Netherlands

Georges Heyen
Laboratoire d'Analyse et Sythèse des
Systèmes Chimiques
Université de Liège
Alee de la Chimie 3–B6
4000 Liège
Belgium

Gordon D. Ingram
CSIRO
Land and Water
Private Bag 5
Wembley WA 6913
Australia

Sten Bay Jørgensen
Technical University of Denmark
Department of Chemical Engineering
Søltofts Plads, Building 229
2800 Kgs. Lyngby
Denmark

Xavier Joulia
ENSIACET –
Laboratoire de Génie Chimique
117 Route de Narbonne
310077 Toulouse, Cedex 4
France

Boris Kalitventzeff
BELSIM s.a.
Rue Georges Berotte 29A
4470 Saint-Georges-sur-Meuse
Belgium

Eustathois S. Kikkinides
University of West Macedoinia
School of Engineering and
Management of Energy Resources
Sialvera and Bakola Street
50100 Kozani
Greece

Antonis Kokossis
University of Surrey
Centre for Process and Information
Systems Engieering
Guildford, Surrey, GU2 7XH
UK

Margaritis Kostoglou
Aristotle University of Thessaloniki
Department of Chemical Technology,
School of Chemistry
Box 116
54124 Thessaloniki
Greece

Andrey Kraslawski
Lappeenranta University of Technology
Department of Chemical Technology
PO Box 20
53851 Lappeenranta
Finland

Rozalia Lakner
Department of Computer Science
Pannon University
Egyetem Street 10, PO Box 158
8200 Veszprem
Hungary

Young-il Lim
Natural Resources Canada
CETC – Varennes
Energy Technology and
Programs Sector
PO Box 27043
Calgary, Alberta T3L 2Y1
Canada

Morton Lind
Technical University of Denmark
Ørsted DTU, Automation
Elektrovej, Building 326
2800 Kgs. Lyngby
Denmark

Patrick Linke
University of Surrey
Centre for Process and Information
Systems Engineering
Guildford, Surrey, GU2 7XH
UK

Davide Manca
Politecnico di Milano
CMIC Department
Piazza Leonardo da Vinci, 32
20133 Milano
Italy

Francois Marechal
Ecole Polytechnique Fédérale de
Lausanne
Industrial Energy System Laboratory
LENI-ISE-STI-EPFL
Station 9
1015 Lausanne
Switzerland

Miguel Mateus
Belsim s.a.
Rue Georges Berotte 29A
4470 Saint-Georges-sur-Meuse
Belgium

Robert B. Newell
Daesim Technologies Pty Ltd
PO Box 309
Toowong QLD 4066
Australia

Lazaros G. Papageorgiou
University College London
Centre for Process Systems
Engineering
Department of Chemical Engineering
Torrington Place
London WC1E 7JE
UK

John D. Perkins
University of Manchester
Institute of Science and Technology
Sackville Street, PO Box 88
Manchester M60 IQD
UK

Efstratios N. Pistikopoulos
Imperial College London
Department of Chemical Engineering
Centre for Process Systems
Engineering
Roderic Hill Building
South Kensington Campus
London SW7 2AZ
UK

Petros Proios
Imperial College London
Department of Chemical Engineering
Centre for Process Systems
Engineering
Roderic Hill Building
South Kensington Campus
London SW7 2AZ
UK

Luis Puigjaner
Universitat Politècnica de Catalunya
Chemical Engineering Department
ESTEIB, Av. Diagonal 647
08028 Barcelona
Spain

Javier Romero
Universitat Politècnica de Catalunya
Chemical Engineering Department
ESTEIB, Av. Diagonal 647
08028 Barcelona
Spain

Richard Sass
DECHEMA e.V.
Department of Information Systems
and Databases
Theodor-Heuss-Allee 25
60486 Frankfurt-am-Main
Germany

Nilay Shah
Imperial College London
Department of Chemical Engineering
Centre for Process Systems
Engineering
Roderic Hill Building
South Kensington Campus
London SW7 2AZ
UK

Abdelaziz Toumi
Bayer Technology Services GmbH
PMT-AMS-APC, Bld. E41
51368 Leverkusen
Germany

Panagiotis Tsiakis
Process Systems Enterprise Ltd.
Bridge Studios,
107a Hammersmith Bridge Road
London W6 9DA
UK

B. Erik Ydstie
Carnegie Mellon University
Department of Chemical Engineering
Pittsburgh
5000 Forbes Avenue
Pennsylvania 15213-3890
USA

Section 3
Computer-aided Process Operation

Section 3 focuses on the application of computing technology to integrate and facilitate the key technical decision processes which arise in chemical manufacture. It comprises decision support systems at different levels of decision making. It discusses the problem of coordinated planning and scheduling of distributed plants at the top level, product sequencing, and precise allocation over time of detailed process operations resource constrained, coordination between units, process monitoring and regulatory control in a real-time environment extended to contemplate hybrid systems. An introduction to modeling the entire supply chain is also presented, which is further elaborated in Section 4.

Section 3 consists of seven chapters introducing the current problems facing process operations, the state of relevant methods and technology, and needed advances to combat ever-increasing complexity of computer-aided process operations in a business-wide context. Chapter 1 presents a comprehensive review of state-of-the-art models, algorithms, methodologies, and tools for the resource planning problem, covering a wide range of manufacturing activities and including a detailed critical discussion on the effect of uncertainty.

The emerging trend in the area of short-term scheduling is the development of efficient solution techniques and to render ever larger problems tractable. Chapter 2 deals with issues that must be resolved related mainly to problem scale. A sensible way forward is proposed by trying to capture the problem in all its complexity and then to explore rigorous or approximate solution procedures, rather than develop exact solutions to somewhat idealized problems. A final challenge relates to the seamless integration of the activities at different levels including data and functional fragmentation, inconsistencies between activities and datasets, and different tools being used for different activities, which are the subjects of subsequent chapters and sections.

The need for quality measurements to monitor process operations, evaluation of their efficiency and the equipment condition, thus avoiding equipment failure and any subsequent hazardous conditions is treated in Chapter 3. Recent progress in automatic data collection and current developments aiming at combining online data acquisition with data reconciliation are presented in detail. The measurement information can also be used in the control scheme in various ways The weakest form of feedback is to use the measurements for parameter adaptation only, which requires a structurally correct model. Chapter 4 introduces model-based control techniques and points out new trends. The techniques use the combination of first principles-based and black-box models the parameters of which are estimated from operational data as a way to obtain sufficiently accurate models without excessive effort.

Online optimization using measurement information in many cases is an attractive alternative to the tracking of pre-computed references because the process can be

operated much closer to its real optimum, while still meeting hard bounds on the specifications. A rigorous treatment of real-time optimization problem is given in Chapter 5.

The last two chapters expand the concept of computer-aided process operations to also
consider hybrid systems (Chapter 6) and the whole network of material procurement, material transformation to intermediates and final products, and distribution of these products
to customers (Chapter 7). The need for integrated solutions will be further explored in Section 4.

1
Resource Planning

Michael C. Georgiadis and Panagiotis Tsiakis

1.1
Introduction

Until recently, resources planning exercises in many companies were based on quantitative, managerial judgements about the future directions of the firms and the markets in which they compete. Complex interactions between the different decision-making levels were often ignored. In the past few years however, important planning decisions, such as those relating to capacity expansion, new product introduction, oil and chemical product distribution and energy planning, have been formally addressed based on recent developments in mixed-integer optimization. Today, most process and energy industries have turned to the use of optimization models in seeking efficient long-term planning use of their resources (Shapiro 2004).

In the last two decades, new techniques have developed to analyze large size planning models, while research in aggregation and decomposition techniques has multiplied. In addition, many managers have begun to recognize the major drawbacks of most current planning systems and the necessity for more intelligent and quantitative decisions tools instead of administrative routine procedures. This comes as a natural continuation of the pioneering work of F. W. Taylor and H. L. Gantt, who in the early 1900s identified the impact on productivity and other key performance indices, of general production planning systems based on scientific approaches (Wilson 2003). Today's computing offers more powerful techniques for modeling and solving planning problems, while Gantt charts still provide an excellent display tool for understanding and acceptance of plans in any type of environment, in addition to other available interfaces.

During the last year, companies have realized that in order to achieve significant competitive advantage within their sector they need to understand the operations hierarchy and solve their problems in a unified framework, a fact that is resulting in the development of corresponding tools. Towards that, the interest in planning and scheduling capabilities has given rise to the providers of solution systems designing, developing and implementing planning systems as part of general supply-chain sup-

Computer Aided Process and Product Engineering. Edited by Luis Puigjaner and Georges Heyen
Copyright © 2006 WILEY-VCH Verlag GmbH & Co. KGaA, Weinheim
ISBN: 3-527-30804-0

port systems or with open architecture to allow easy integration. Owing to the inherent complexity and the different scales of integration this has been accepted by the research community as a topic that needs urgent answers, since the planning software industry is in its infancy and under pressure to respond to the demand.

The objective of this chapter is to present a comprehensive review of state-of-the-art models, algorithms, methodologies and tools for the resource planning problem covering a wide range of manufacturing activities. For reasons of presentation, the remaining of this chapter is organized as follows. The long-range planning problem in the process industries is considered in Section 1.2 including a detailed discussion on the effect of uncertainty, the planning of refinery operations and offshore oil-fields, the campaign planning problem and the integration of scheduling and planning. Section 1.3 describes the planning problem for new product development with emphasis on pharmaceutical industries. Section 1.4 presents briefly the tactical planning problem, followed by a description of the resource planning problem in the power market and construction projects in Section 1.5. Section 1.6 is a review of recent computational solution approaches to the planning problem are reviewed, while available software tools are outlined in the Section 1.7. Finally, Section 1.8 will make some conclusions and propose future challenges in this area.

1.2
Planning in the Process Industries

1.2.1
Introduction

New environmental regulations, new processing technologies, increasing competition and fluctuating prices and demands in process industries have led to an increasing need for quantitative techniques for planning the selection of new processes, the expansion and shut down of existing processes, and the production of new products. Further decisions also include creation of production, distribution, sales and inventory plans. (Kallrath 2002). It has been recently realized that in a competitive and changing environment the need to plan new output levels and production mixes is likely to arise much more frequently than the need to design new batch plants.

Although the boundaries between planning and scheduling are not very clear we can distinguish the following basic features of the process planning problem:

- multipurpose equipment
- sequence-dependent set-up times and cleaning costs
- combined divergent, convergent and cyclic material flows
- multistage, batch and campaign production using shared intermediates
- multicomponent flow and nonlinear blending for the refinery operations
- finite intermediate storage, dedicated and variable tanks.

Structurally, these features often lead to allocation and sequencing problems and knapsack structures, or to the pooling problem for the petrochemical industries. In production planning we usually consider material flow and balance equations connecting sources and sinks of a production network. Time-indexed models using a relative coarse discretization of time, e.g., a year, quarter, months or weeks are usually accurate enough. linear programming (LP), mixed-integer linear programming (MILP) and mixed-integer nonlinear programming (MINLP) technologies are often appropriate and successful for problems with a clear quantitative objective function, as will become clear in the following sections.

Nowadays, it is possible to find the optimal way to meet business objectives and to fulfil all production, logistics, marketing, financial and customer constraints and especially:

- to accurately model single-site and multisite manufacturing networks;
- to perform capital planning and acquisition analysis, i.e., to have the possibility to change the structure of a manufacturing network through investment and to determine the best investment type, size and location based on user-defined rules related to business objectives and available resources; the results of such analysis can lead to nonintuitive solutions that provide management with scenarios that could dramatically increase profits;
- to produce integrated enterprise solutions and to enable a crossfunctional view of the planning process involving production, distribution and transport, sales, marketing and finance functions;
- to develop new product and introduction strategies along with capacity planning and investment strategies.

The following sections provide a comprehensive review of the above areas.

1.2.2
Long-Range Planning in the Process Industries

Chemical process industries are increasingly concerned with the development of planning techniques for their process operations. The incentive for doing so derives from the interaction of several factors (Reklaitis 1991, 1992). Recognizing the potential benefits of new resources when these are used in conjunction with existing processes is the first. Another major factor is the dynamic nature of the economic environment. Companies must assess the potential impact on their business of important changes in the external environment. Included are changes in product demand, prices, technology, capital market and competition. Hence, due to technology obsolescence, increasing competition, and fluctuating prices of and demands for chemicals, there is an increasing need to develop quantities techniques for planning the selection of new processes, and the production of chemicals (Sahinidis et al. 1989)

The long-range planning problem in process industries has received a lot of attention over the last 20 years and numerous sophisticated models exist in the literature. Sahinidis et al. (1989) consider the long-range planning problem for a chemical complex involving a network of chemical processes that are connected in a finite number of ways. The network also consists of chemicals: raw materials, intermediates and products that may be purchased from and/or sold to different markets. The objective function to be maximized is the net present value (NPV) of the planning problem over a long-range horizon of interest consisting of a number of *NT* time periods during which prices and demands of chemicals, and investment and operating costs of the processes can vary. The problem consists of determining the following items:

- capacity expansion and shut down policy
- selection of new processes and their capacity expansion and shut down policy
- production profiles
- sales and purchase of chemicals at each time period.

It is assumed that the material balance and the operating cost in each process can be expressed linearly in terms of the operating level of the plant. The investment costs of the processes and their expansions are considered to be linear expressions of the capacities with a fixed charge cost to account for economies of scale. This is a multi-product, multifacility, dynamic, location-allocation problem that has been formulated using MILP modes. Sahinidis and Grossmann (1991) extended the above model to account for production facilities that are flexible manufacturing systems operating in a continuous or in a batch mode. The suggested model provides a unified representation for the different types of processes.

Norton and Grossmann (1994) extended the original model of Sahinidis and Grossmann (1991) for dedicated and flexible processes by incorporating raw materials flexibility in addition to product flexibility. In their model, raw material flexibility is characterized by different chemicals as raw materials or different sources of the same raw material. The model was able to handle any combination of raw material and process flexibility, thus providing a truly unified representation for all types of process flexibility in the long-range planning problem.

The above industrial relevance of the chemical process planning problem motivated the need to develop more efficient solution techniques for large-scale problems. Liu and Sahinidis (1995) presented a comprehensive investigation of the effect of time discretization, data uncertainty and problem size, on the quality of the solution and computational requirements of the above MILP planning models. The importance of detailed time discretization was demonstrated and the effect of uncertainty was critically assessed. An exact branch-and-bound algorithm was also presented along with several heuristic approaches for the solution of larger problems. Extending this work, Liu and Sahinidis (1996a) investigated separation algorithms and cutting plane approaches that were demonstrated to be more robust and faster than conventional solution approaches for large-scale problems with long time-horizons.

Oxe (1997) considered a LP approach to choose an appropriate subset of existing production plants and lines and to optimize allocation, transportation paths and central stock profiles so that the overall costs are minimized while product delivery is ensured within some months (specified for each product) from the order.

McDonald and Karimi (1997) developed production planning and scheduling models for the case of semicontinuous processes, which are assumed to comprise several facilities in distinct geographical locations, each potentially containing multiple parallel lines. The models developed are deterministic in nature and are formulated as mixed-integer linear programs.

Oh and Karimi (2001a) presented a new methodology for determining the optimal campaign numbers for producing multiple products on a single machine with sequence-dependent set-ups. Their methodology is intended mainly for the purpose of capacity planning. In the second part of this work (Oh and Karimi 2001b) they addressed the problem of determining the sequence of these given product campaigns to obtained a detailed schedule of operation. Heuristic algorithms based on a decomposition scheme were investigated for the efficient solution of the underlying optimization problem.

1.2.3
Process Planning under Uncertainty

Decision making in the design and operation of industrial processes is frequently based on parameters of which the values are uncertain. Sources of uncertainty, which tend to imply the means for dealing with them, can be divided into:

- short-term uncertainties such as processing time variations, rush orders, failed batches, equipment breakdowns, etc.;
- long-term uncertainty such as market trends, technology changes, etc.

A detailed classification of different areas of uncertainty is suggested by Subrahmanyam et al. (1994) including uncertainty in prices and demand, equipment reliability and manufacturing uncertainty. An excellent review on the general subject of optimization under uncertainty has recently been presented by Sahinidis (2004).

In the area of process planning, uncertainty is usually associated with product demand fluctuations, which may lead to either unsatisfied customer demands or loss of market share or excessive inventory costs. A number of approaches have been proposed in the process systems engineering literature for the quantitative treatment of uncertainty in the design, planning and scheduling of batch process plants with an emphasis on the design. The most popular one so far has been the scenario-based approach, which attempts to forecast and account for all possible future outcomes through the use of scenarios. The scenario approach was suggested by Shah and Pantelides (1992) for the design of flexible multipurpose batch plants under uncertain production requirements, and was also used by Subrahmanyam et al. (1994). Scenario-based approaches provide a straightforward way to implicitly account for uncertainty (a comprehensive discussion is presented by Liu and Sahinidis (1996b)). Their main drawback is that they typically rely on either the *a priori* forecasting of all possible outcomes of the discretization of a continuous multivariable probability distribution, resulting in an exponential number of scenarios.

Liu and Sahinidis (1996a,b) and Iyer and Grossmann (1998) extended the MILP process and capacity planning model of Sahinidis and Grossmann (1991) to include multiple product demands in each period. They then propose efficient algorithms for the solution of the resulting stochastic programming problems (formulated as large deterministic equivalent models), either by projection (Liu and Sahinidis 1996a) or by decomposition and iteration. However, as pointed out by Shah (1998) a major assumption in their formulation is that product inventories are not carried from one period to the next. This has the advantage in ensuring that the problem size is of, $0(np \times ns)$, where np is the number of periods and ns is the number of demand scenarios, rather than $0(ns^{np+1})$. However, if the periods are too short, this compromise the solution from two perspectives:

- All products must be produced in all periods if demand exists for them – this may be suboptimal.
- Plant capacity must be designed for a peak demand period.

Clay and Grossmann (1994) addressed this issue. They considered the structure of both the two-period and multiperiod problem for LP models and derived an approximate model based on successive repartitioning of the uncertain space, with expectations being applied over partitions. This has the potential to generate solutions to a high degree of accuracy in a much faster time than the full deterministic equivalent model.

Liu and Sahinidis (1997) presented two different formulations for the planning in a fuzzy environment (the forecast model parameters are assumed to be fuzzy). The first considers uncertainty in demands and availabilities, whereas the second accounts for uncertainty of the right hand size of made constraints and objective function coefficient.

The approaches above mainly focus on relatively simple planning models of plant capacity. Petkov and Maranas (1997) considered the multiperiod planning model for multiproduct plants under demand uncertainty. Their planning model embeds the planning/scheduling formulation of Birewar and Grossmann (1990) and therefore calculates accurately the plant capacity. They do use discrete demand scenarios, but assume normal distributions and directly manipulate the functional forms to generate a problem which maximizes the expected profit and meets certain probabilistic bounds on demand satisfaction without the need for numerical integration. Ierapetritou and Pistikopoulos (1994) proposed a two-stage stochastic programming formulation for the long-range planning problem including capacity expansion options. Based on the Gaussian quadrature method for approximating multiple probability integrals, Ierapetritou et al. (1996) considered the operational and production planning problem under varying conditions and changing economic circumstances. The effect of uncertainty on future plant operation was investigated via the incorporation of explicit future plan feasibility constraints into a two-stage stochastic programming formulation, with the objective of maximizing an expected profit over a time horizon, and the use of the value of perfect information. The main drawback of this approach is its high computation cost. To address this issue Bernardo et al. (1999) investigated more efficient integration schemes for the solution of problems with many uncertain parameters. Recently, Ryu et al. (2004) addressed bilevel decision making problems under uncertainty in the context of enterprise-wide supply-chain optimization with one level corresponding to a plant planning problem, and the other to a distribution network problem. The bilevel problem was transformed into a family of single parametric optimization problems solved to global optimality.

Rodera et al. (2002) presented a methodology for addressing investments planning in the process industry using a mixed-integer multiobjective optimization approach. Romero et al. (2003) proposed a modeling framework integrating cash flow and budgeting aspects with an advanced scheduling and planning model. It was illustrated that potential budget limitation can significantly affect scheduling and planning decisions. Recently, Barbaro and Bagajewicz (2004) proposed a new mathematical formulation for problems dealing with planning under design uncertainty that allows management of financial risk according to the decision-maker's preferences.

Sanmarti et al. (1995) define a robust schedule as one which has a high probability of being performed, and it is readily adaptable to plant variations. They define an index of reliability for a unit scheduled in a campaign through its intrinsic reliability, the probability that a standby unit is available during the campaign, and the speed with which it can be repaired. An overall schedule reliability is then the product of the reliabilities of units scheduled in it, and solutions to the planning problem can be driven to achieve a high value of this indicator.

Ahmed and Sahinidis (1998) noted that the resulting two-stage stochastic optimization models in process planning under uncertainty minimize the sum of the costs of the first stage and the expected cost of the second stage. However, a limitation of this approach is that it does not account for the variability of the second-stage costs and might lead to solutions where the actual second-stage costs are unacceptably high. In order to resolve this difficulty they introduced a robustness measure that penalizes second-stage costs that are above the expected cost.

Pistikopoulos et al. (2001) presented a systems effectiveness optimization framework for multipurpose plants that involves a novel preventive maintenance model coupled with a multiperiod planning model. This provides the basis for simultaneously identifying production and maintenance policies, a problem of significant industrial interest. This framework was then extended by Goel et al. (2003) to incorporate the reliability allocation problem at the design stage. Li et al. (2003) employed probabilistic programming approach to plan operations under uncertainty and to identify the impact on profits based on reliability analysis. Recently, Suryadi and Papageorgiou (2004) presented an integrated framework for simultaneous maintenance planning and crew allocation in multipurpose plants.

1.2.4
Integration of Production Planning and Scheduling

The decisions made by planning, scheduling, and control functions have a large economic impact on process industry operations – estimated to be as high as US \$10 increased margin per ton of feed for many plants. The current process industry environment places even more of a premium on effective execution of these functions. In spite of these incentives, or perhaps because of them, there exists significant disagreement about the proper organization and integration of these functions, indeed even which decisions are properly considered by the planning, scheduling or control business processes. It has long been recognized that maintaining consistency among the decisions in most process companies continues to be difficult and the lack of consistency has real economic consequences. In their recent work Shobrys and White (2002) presented a critical and comprehensive analysis of several practical aspects that need to be carefully considered when challenges, associated with improving these functions and achieving integration, arise.

The planning and scheduling levels of the operations hierarchy are natural candidates for integration because the structure of these two decision problems is very similar. However, the direct merging of these two levels requires embedding the details of the scheduling level into a super-scheduling-problem defined over the entire planning horizon. The result is a problem that is extremely difficult to solve. Thus, in recent years research has been increasingly interested in the issues around the integration of production and scheduling, in order to provide greater consistency.

The most common approach for the simultaneous treatment of production planning and scheduling is a hierarchical decomposition scheme, where the overall production planning problem is decomposed into two levels (Bitran and Hax 1977). At the upper level, the planning problem, which usually involves a simplistic representation of the scheduling problem, is solved as a multiperiod LP problem in order to maximize the profit and set production targets. At the lower level, the scheduling problem is concerned with the sequencing of tasks that should meet the goals. An alternative integration approach is through the rolling schedule strategy (Hax 1978).

Production planning and scheduling are closely related activities. Ideally these two should be linked, in order that the production goals set at the production plan level should be implementable at the scheduling level. Birewar and Grossmann (1990), based on their initial LP flow-shop scheduling model, proposed aggregate methods that allow tackling longer time-horizons by reducing the combinatorial nature of the problem. The model accounts for inventory costs, sequence-dependent clean-up times and costs, and penalties for not meeting predefined product demands. Using a graph enumeration method, the production goals predicted by the planning model are applied to the actual schedule, with the key point that both problems are solved simultaneously, since the sequencing constraints can be accounted for at the planning level with very little error.

Bassett et al. (1996a), working in the same direction of model-based integrated applications and focused on integrating planning decisions with the actual schedule, proposed an aggregation/disaggregation technique that can be used to provide solutions to otherwise intractable mid-term planning models. The initiative is the exploitation of available enterprise information within the process operational hierarchy tree. A more formal approach to integration of production and scheduling is described based on the previous work of Subrahmanyam et al. (1996), where the planning model, based on an aggregate formulation, is modified to be consistent with detailed scheduling decisions.

Hierarchical production planning algorithms often make use of rolling horizon algorithms as a suboptimal to obtain feasible, but often good, solutions. The disadvantage of the method is reliance on the simplistic or rather poor representation of the scheduling problem within the aggregate part. Wilkinson (1996) derived an accurate aggregate formulation by applying formal aggregation operators to the resource-task network (RTN) formulation, and dividing the horizon into aggregate time periods (ATPs). This allows creating single MILP models that have varying time resolution. The first ATP is modeled in fine detail (scheduling) and the subsequent ATPs are modeled using the aggregate formulation (planning). The problem can then be solved as a single MILP, maintaining consistency between plan and schedule.

Rodrigues et al. (2000) presented a two-level decomposition procedure for integrating scheduling and planning decisions. At the planning level, demands are adjusted, a raw material plan is defined and a capacity analysis is performed. At the scheduling level an MILP model is proposed. Geddes and Kubera (2000) described a practical integration between planning and online optimization with application in olefins production. Das et al. (2000) developed a prototype system by integrating two higher-level hierarchical production planning application programs (aggregated production plan and master production schedule) using a common data model integration approach into an existing planning system for short-term scheduling and supervisory management, which was originally developed by Rickard et al. (1999). Bose and Pekny (2000) presented a similar approach to model predictive control for integrated planning and scheduling problems. Van den Heever and Grossmann (2003) addressed the integration of production planning and reactive scheduling for the optimization of a hydrogen supply network consisting of 5 plants, 4 interconnected pipelines and 20 customers. A multiperiod MINLP model was proposed for both the planning and scheduling levels, along with heuristic solution methods based on Lagrangean decomposition.

During the last Foundations on Computer Aided Process Operations (FOCAPO 2003) event, several contributions presented were on the integration between planning and scheduling decisions. Harjunkoski et al. (2003) provided a comprehensive analysis of different aspects needed for the integration of the planning, scheduling and control levels in the light of ABB's industrial initiative. They presented a framework introducing an approach to integrating all aspects relevant to decision making in a supportive way. An industrial case study was used to illustrate the benefits of the integrated framework. Yin and Liu (2003) developed a problem formulation and solution procedure for production planning and inventory management of systems under uncertainty. The production system is modeled by finite-event continuous-time Markov chains. Kabore (2003) presented a model predictive control formulation for the planning and scheduling problem in process industries. The main idea is to use moving-horizon techniques as well as a feedback control concept to continuously update production schedules. Wu and Ierapetritou (2003) proposed a method for simultaneously solving a planning and scheduling problem. The mathematical formulation of the planning problem involves scheduling decisions and results to a large MINLP problem, intractable to solve directly within reasonable computational time. A nonoptimal solution strategy is selected to provide near-optimal solutions within reasonable computational times.

Tsiakis et al. (2003) applied the algorithm of Wilkinson (1996) to obtain an integrated plan and schedule of the operations of a complex specialty oil refinery, focusing on the downstream products of the oil supply-chain. Operating in an uncertain environment, the company needed to schedule the refinery operations in detail over the next month, while producing plans for the next year that were both reasonably accurate and consistent with the short-term schedule.

1.2.5
Planning of Refinery Operations and Offshore Oilfields

The refinery industry is currently facing a rather difficult situation, typically characterized by decreasing profit margins, due to surplus refinery capacity, and increasing oil prices. Simultaneous market competition and stringent environmental regulations are forcing the industry to perform extensive modifications in its operations. As a result there is no refinery nowadays that does not use advanced process engineering tools to improve its business performance. Such tools range from advanced process control to long-range planning, passing through process optimization, scheduling and short-term planning. Despite their widespread use and the existence of quasi-standard technologies for these applications, their degree of commercial maturity varies greatly and there are many unresolved problems concerning their use. Moro (2003) presents a comprehensive discussion on current approaches to solving these problems and proposes directions for future development in this area.

Traditionally, planning and scheduling decisions in refinery plants have been addressed using LP techniques and several tools exist such as the Integrated System for Production Planning (SIPP) and the Refinery and Petrochemical Modelling System (RPMS). An excellent review has recently been presented by Pinto et al. (2000). These tools allow the development of general production plans of the whole refinery. As pointed out by Pelham and Pharris (1996), the planning technology in the refinery operations can be considered well-developed and the margins for further improvement are very tight. The major advances in this area should be expected in the form of more detailed and accurate modeling of the underlying processes, notably through the use of nonlinear programming (NLP) as illustrated by Moro et al. (1998) using a real-world application. Ballintjin (1993) compared continuous and mixed-integer linear formulations and emphasized the low applicability of models based solely on continuous variables.

In the literature, the first mathematical programming (MP) approaches utilizing advances in mixed-integer optimization are focused on specific applications such as gasoline blending (Rigby et al. 1995) and crude oil unloading. Shah (1996) presented a MP approach for scheduling the crude oil supply to a refinery, whereas Lee et al. (1996) developed a MILP model for short-term refinery scheduling of crude oil unloading with optimal inventory management decisions. Gothe-Lundgren (2002) proposed a planning and scheduling model which seems to be limited to the specific industrial problem to which it has been applied, whereas Jia and Ierapetritou (2004) addressed the optimal operation of gasoline blending and distribution, the transfer to product stock tanks and the delivery schedule to satisfy all of the orders.

Recent work by Pinto et al. (2000) is a key contribution in this area. A nonlinear planning model for refinery production was developed that is able to represent a general refinery topology. The model relies on a general representation for refinery processing units in which nonlinear equations are considered. The unit modes are composed of blending relations and process equations. Certain constraints are imposed to ensure product specifications, maximum and minimum unit feed flow rates, and limits on operating variables. Real-world industrial case studies for the planning of diesel production were used to illustrate the applicability and usefulness of the over-all approach. In the second part of their work scheduling problems in oil refineries were studied in detail. Discrete time representations were employed to model sched-uling decisions in important areas of the refinery such as crude oil inventory man-agement and fuel oil, asphalt, and liquefied petroleum gas (LPG) production. Several real-world refinery problems were presented and solved using the developed models.

Based on the above work, Neiro and Pinto (2004) proposed a general mathematical framework for modeling petroleum supply chains. A set of crude oil suppliers, refin-eries that can be interconnected by intermediate and final product streams and a set of distribution centres form the basis for this work.

The scheduling of well and facility operations is a very relevant problem in off-shore oil field development and represents a key subsystem of the petroleum supply-chain. The problem is characterized by long planning horizons (typically 10 years) and a large number of choices of platforms, wells, and fields and their interconnect-ing pipeline infrastructure. Resource constraints such as availability of the drilling rings make the requirement for proper scheduling more imperative to utilize resources efficiency. The sequencing of installation of well and production platforms is essential to ensure their availability before drilling wells. The operational design of the well and production platforms and the time of installation are critical, as they involve significant investment costs, these decisions must be optimized to maximize the return on investment. Thus, oil field development represents a complex and expensive undertaking in the oil industry. The process systems engineering commu-nity has recently made several key contributions in this area based on advances in mixed-integer optimization. Iyer et al. (1998) developed a multiperiod MILP formula-tion for the planning and scheduling of investments and operations in offshore oil field facilities. For a given time-horizon, the decision variables in their model are the choice of reservoir to develop, selection from among candidate well sites, and the well-drilling and platform installation planning, the capacities of well and production platforms and the fluid production rates from wells for each time period. The nonlin-ear reservoir behavior is handled with piecewise linear approximation functions.

Van den Heever and Grossmann (2000) presented a mixed-integer nonlinear model for oilfield infrastructure that involves design and planning decisions. The nonlinear reservoir behavior is directly incorporated into the formulation. For the solution of this model an iterative aggregation/disaggregation algorithm is proposed according to which time periods are aggregated for the design problem, and subse-quently disaggregated for the planning subproblem. Van den Heever et al. (2000) addressed the design and planning of offshore oilfield infrastructure focusing on business rules and complex economic objectives. A specialized heuristic algorithm that relies on the concept of Lagrangean decomposition was proposed by Van den Heever et al. (2001) for the efficient solution of this problem. Ierapetritou et al. (1999) studied the optimal location of vertical wells for a given reservoir property map. The problem is formulated as a large-scale MILP and solved by a decomposi-tion technique that relies on quality cut constraints. Kosmidis et al. (2002) described a MILP formulation for the well allocation and operation of integrated gas-oil sys-tems, whereas Barnes et al. (2002) focused on the production design of offshore platforms. Kosmidis (2003) presented a MINLP model for the daily well-scheduling, where the nonlinear reservoir behavior, the multiphase flow in the well, and constraints from the surface facilities are simultaneously considered. An efficient solution strategy is also proposed. Lin and Floudas (2003) presented a continuous-time modeling and optimization approach for the long-term planning problem for integrated gas-field development. They proposed a two-level formulation and solution framework taking into account complicated economic calculations. I.E. Goel, V. Grossmann Comput. Chem. Eng. 28 (2004), 1409 considered the optimal investment and operational planning of gas-field development under uncertainty in gas reserves. A novel stochastic programming model that incorporates the decision-dependence of the scenario was presented. Aseeri et al. (2004) discussed the financial risk management in the planning and scheduling of offshore oil infrastructures. They added budgeting constraints to the model of Iyer et al. (1998) by following the cash flow of the project, taking care of the distribution of proceeds and considering the possibility of taking loans.

1.2.6
The Campaign Planning Problem

The campaign planning problem has received rather limited attention in the past 20 years, yet it is considered a key problem in chemical batch production. If reliable long-term demand predictions are available, it is often preferable to partition the planning horizon into a smaller number of relatively long periods of time ("campaigns"), each dedicated to the production of single product. The campaign mode of operations may result in important benefits such as minimizing the number and costs of changeovers when switching production from one product to another. The complexity of management and control of the plant operation is further reduced by operating the plant in a more regular fashion, such as in a cyclic mode within each campaign, with the same pattern of operations being repeated at a constant frequency. Typical campaign lengths are from weeks to several months, with cycle times ranging from a few hours to a few days. The campaign mode of operations is often used for the manufacture of "generic" materials (e.g., base pharmaceuticals) which are produced in relatively large amounts and are then used as feedstocks for downstream processes producing various more specialized final products (Papageorgiou 1994, Grunow, et al. 2002).

Mauderli and Rippin (1979) studied the combined production planning and scheduling problem, developing a hierarchical procedure suitable for serial processing networks operated in a zero-wait mode. First, they consider each product individually, generating alternative production lines of a single product by assembling the available processing equipment in groups in order to achieve maximum path capacity. A LP-based screening procedure is used to determine a set of dominant campaigns. Finally, the production plan is generated by solving a LP or MILP problem, allocating the available production time to the various dominant campaigns for a given set of production requirements.

The generation of alternative production lines in the Mauderli and Rippin (1979) algorithm is based on an exhaustive enumeration procedure. A more efficient generation procedure is described by Wellons and Reklaitis (1989a), who formulated the optimal scheduling of a single-product production line as a MINLP model. However, this approach has several limitations, including high degeneracy, as many path assignments result in equivalent schedules. The elimination of this degeneracy was considered by Wellons and Reklaitis (1989b) who identified a set of dominant unique path sequences and hence improved the solution efficiency of the original formulation. A further improvement from the single-product production line scheduling to the single-product campaign formulation problem has been presented by Wellons and Reklaitis (1991a), including the automatic assignment of different equipment items to groups, and also the assignment of these groups to production stages. This work was extended by Wellons and Reklaitis (1991b) to the multiproduct campaign formulation problem for multipurpose batch plants. Finally, a multiperiod planning model is proposed, allocating the production time among the dominant campaigns while considering simultaneously profit from sales, changeover, inventory costs and campaign set-ups.

Papageorgiou and Pantelides (1993) presented a hierarchical approach attempting to exploit the inherent flexibility of multipurpose plants by removing various restrictions regarding the intermediate storage policies between successive processing steps, the utilization of multiple equipment items in parallel and also the use of the same item of equipment for more than one task within the same campaign. A three-step procedure was proposed. First, a feasible solution to the campaign planning problem is obtained to determine the number of campaigns and the active parts of the original processing network involved within each campaign. Secondly, the production rate in each campaign is improved by removing some assumptions and applying the cyclic scheduling algorithm of Shah et al. (1993). Finally, the timing of the campaigns is revised to take advantage of the improved production rates. An interesting feature of this approach is that any existing campaign planning algorithm can be used for its first step. However, this approach relies on several restricted assumptions, including limited flexibility in the utilization of processing equipment and limited operating modes, while multiple production routes or material recycles are not taken into account.

The algorithms described above are hierarchical in nature, and therefore relatively easy to implement given the reduction in the size of the problem solved at each step. On the other hand it is difficult to relate the exact objective for each individual step in the hierarchy to the overall campaign and planning objective function, and therefore it is very difficult to assess the quality of the final solution obtained.

Shah and Pantelides (1991) proposed a single-level mathematical (MILP) formulation for the simultaneous campaign formation and planning problem. Their algorithm simultaneously determines the number and the length of the campaigns and the products and/or stable intermediates manufactured within each campaign. They consider serial processing networks operating in a mixed Zero-Wait/Unlimited Intermediate Storage (ZW/UIS) mode, and nonidentical parallel equipment items operating in phase.

Voudouris and Grossmann (1993) extended the work originally presented by Birewar and Grossmann (1989a,b, 1990) to campaign planning problems for multiproduct plants. They introduced cyclic scheduling, location and sizing of intermediate storage, and inventory considerations along with novel linearization schemes transforming the resulting MINLP formulation.

Tsiroukis et al. (1993) considered the optimal operation of multipurpose plants operating in campaign mode to fulfil outstanding orders. Resource constraints are explicitly taken into account while the limited availability of resource levels affects the operation of the plant. To deal with the complexity, nonconvexity and nonlinearity of the MINLP formulation, more efficient formulations along with a problem-specific two-level decomposition strategy were proposed.

Papageorgiou and Pantelides (1996a) presented a general MP formulation for multiple campaigns planning/scheduling of multipurpose batch/semicontinuous plants. In contrast to hierarchical approaches presented above, a *single-level* formulation was developed, encompassing both overall planning considerations pertaining to the campaign structure and scheduling constraints describing the detailed operating of the plant during each campaign. The problem involves the simultaneous determination of the campaigns (i.e., duration and constituent products) and for every campaign the unit-task allocations, the tasks' timings and the flow of material through the plant. A cyclic operating schedule is repeated at a fixed frequency within each campaign, thus significantly simplifying the management and control of the plant operation. A rigorous decomposition approach to the solution of this problem is presented by Papageorgiou and Pantelides (1996b) and its effectiveness was demonstrated by applying it to a number of examples. Ways in which the special structure of the constituent mathematical models of the decomposition scheme can be exploited to reduce the size and associated integrality gaps are also considered.

1.3
Planning for New Product Development

Pharmaceutical industries are undergoing major changes to cope with the new challenges of the modern economy. The internationalization of the business, the diversity and complexity of new drugs, and the diminishing protection provided by patents are some of the factors driving these challenges. Market pressures are also forcing pharmaceutical industries to take a more holistic view of their product portfolio. The typical life cycles of new drugs are becoming shorter making it harder to recover the investments, especially with the expiry of short-life patents and the arrival of generic substitutes that can later appear in the market, reducing its profitability. It becomes necessary that the industry protects itself against these pressures while considering the limited physical and financial resources available. Several important issues and strategies for the solution of problems concerning pharmaceutical supply-chains are critically reviewed by Shah, (2004).

A large number of candidate new products in the agricultural and pharmaceutical industry must undergo a set of steps related to safety, efficacy, and environmental impact prior to commercialization. If a product fails any of the tests then all the remaining work of that product is halted and the investment in the previous tests is wasted. Depending on the nature of the products, testing may last up to 10 years and the problem of scheduling of tests should be made with the goal of minimizing the time-to-market and the expected cost of the testing. Another important challenge that the pharmaceutical and agrochemical industry faces today is how, then, to configure its product portfolio in order to obtain the highest possible profit, including any capacity investments, in a rapid and reliable way. These decisions have to be taken in the face of considerable uncertainty as demands, sales prices, outcomes of clinical tests, etc. may not turn out as expected.

These problems have recently received attention from the process systems engineering community utilizing advances from the process planning and scheduling area. The first approach appeared in the literature by Schmidt and Grossmann (1996), who considered the problem of optimal sequencing of testing tasks for new product development, assuming that unlimited resources are available. For a product involving a set of testing tasks with given costs, durations and probabilities of success, these authors formulated a MILP model based on a continuous-time representation to determine the sequence of those tasks. The objective of the model is to maximize the expected net present value (NPV) associated with a product, while a special case considers the minimization of cost, subject to a time completion constraint. Even though there may be a number of new products under consideration, the assumption of unlimited resources allows the problem, with either of the two objectives, to be decomposed by each product. Extending this work, Jain and Grossmann (1999) developed an MILP model that performs the sequencing and scheduling of testing tasks for new product development under resources constraints. It was shown that it is critical to incorporate resource constraints along with the sequencing of testing tasks to obtain a globally optimal solution. Blau et al. (2000) developed a simulation model for risk management in the new product development process andSubramanian et al. (2001) proposed a simulation-based framework for the management of the research and development (R&D) pipeline. The focus of these works, however, is the new products development processes and not the planning and design of manufacturing facilities. In most of these references it is assumed that there are no capacity limitations or that the production level of a new product is not affected by the production levels of other products. Furthermore, investments costs are not explicitly included in the calculation of the NPV of the projects.

The problem of simultaneous new product development and planning of manufacturing facilities has received rather limited attention. Papageorgiou et al. (2001) developed a novel optimization-based approach to selecting a product development and introduction strategy, as well as capacity planning and investment strategies. The overall problem is formulated as a MILP model that takes account of both the particular features of pharmaceutical active ingredient manufacturing and the global trading structures. Maravelias and Grossmann (2001) considered the simultaneous optimization of resource-constrained scheduling of testing tasks in new product development and design/planning of batch manufacturing facilities. A multiperiod MILP model was proposed that takes into account multiple tradeoffs and predicts which products should be tested, the detailed test schedule that satisfy design decisions for the process network, and production profiles for the different scenarios defined by the various testing outcomes. A heuristic algorithm based on Lagrange decomposition was investigated for the solution of larger problem instances. Roger et al. (2002) have addressed a similar problem.

In most of the above approaches it is assumed that the resources available for testing, such as laboratories and scientists, are constant throughout the testing horizon, and that all testing tasks have fixed costs, duration and resources requirements. Another common assumption in all the above approaches is that the cost of one test does not depend on the amount of resources allocated to one test. However, as noted in the recent contribution by Maravelias and Grossmann (2004) a company may decide to hire more scientists or build more laboratories to handle more efficiently a great number of potential new products in the R&D pipeline. As another option the company may have to outsource the tests, often at a high cost. All these issues have been addressed by proposing a MILP model that is efficiently solved with a heuristic decomposition algorithm.

In most of the above approaches uncertainty aspects have been neglected although clinical tests are highly uncertain in practice. The recent work by Gatica et al. (2003) explicitly considers uncertainty in clinical trial outcomes. A multistage, multiperiod stochastic problem was developed that was reformulated as a multiscenario MILP model. For this model, a performance measure that takes appropriate account of risk and potential returns has also been formulated. Levis and Papageorgiou (2004) extended the work of Papageorgiou et al. (2001) and proposed a two-stage multiscenario MILP model determining both the product portfolio and the multisite capacity planning in the face of uncertain clinical outcomes, while taking into account the trading structure of the company. They proposed a novel hierarchical algorithm to reduce the computational effort needed for the solution of the resulting large-scale MILP models.

1.4
Tactical Planning

Planning and scheduling is usually part of a company-wide logistics and supply-chain management platform. However, to distinguish between those topics, or even to distinguish further between planning and scheduling is often an artificial rather than a pragmatic approach. In reality, the borderline between all these areas is diffuse, due to the strong overlaps between scheduling and planning in production, distribution or supply-chain management and strategic planning.

Planning and scheduling considerations are very closely related and often confused. The most common distinction between the two concepts is based on the time horizon they consider. While scheduling considers problems that may be of some hours to a few weeks, planning problems may consider time horizons that are of a few weeks up to a few months, and in many applications can even be of years. Tactical planning aims to set the targets for the scheduling applications that will follow in order to determine the operational policy of the plant in the short term. Owing to its nature of involving longer time horizons, planning decisions are often subject to uncertainty that might arise from many sources.

The planning operation in the process industry is focused on analyzing the supply-chain operations as they are defined by strategic planning (see Fig. 1.1). Competitive environment and technological advances have resulted in enterprise resource planning (ERP) systems to be widely used within the process sector; they are considered to be software suites that help organizations to integrate their information flow and business processes (Abdinour-Helm et al. 2003).

The fundamental benefits of ERP systems do not in fact come from their planning capabilities but rather from their abilities to process transactions efficiently and to provide organized structured data bases (Jacobs 2003). Planning and decision support applications represent optional additions to this basic transactional, query and report capability. ERP has been designed to supersede the earlier concepts of material requirement planning (MRP) and manufacturing resource planning (MRP-II) that were designed to assist planners at a local level, by linking various pieces of information, especially in manufacturing. The advantage of a successful ERP implementation is the integration between different levels of the enterprise, such as financial, controlling, project management, human resources, plant maintenance and material flow logistics (Mandal and Gunasekaran 2003). The planning functions at a tactical level benefit from the existence in-place of an ERP system; the two systems do not replace each other but their relationship can be described as complementary. ERP systems play the role of an information highway that connects all planning levels and links various decision support systems to the same data.

MRP systems were designed to work backwards from the sales orders to determine the raw material required for production (Orlicky 1975). MRP-II was introduced as a follow-up to resolve obvious operational problems usually associated with the absence of capacity considerations from MRP that resulted in poor schedules (Wight 1984). The weakness of both approaches is that they were targeting and devel-

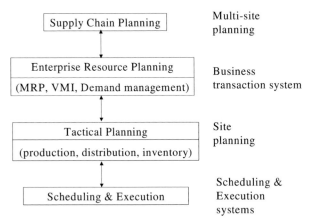

Fig. 1.1 Operations planning decision hierarchy.

oped for the manufacturing environment, and very often ignored the complexities of the process world. ERP on the other hand, is not limited only to manufacturing companies, but is useful for any company with the need to integrate their information across many functional areas.

Planning in the process industry is used to create production, distribution, order satisfaction and inventory plans, based on the information that can be extracted from ERP systems, while satisfying several constraints. In particular, operational plans have to be determined that are aimed to set targets for future production, plan the distribution of materials and allocate other related activities according to the business expectations. Business expectations are the product of strategic resource planning. A successful strategic resource planning, which can be performed either by activity-based cost (ABC), MP, resource-based view (RBV), or a combined approach, is sent to the ERP (Shapiro 1999). It is common practice that, based on these tactical plans, detailed schedules may be produced that define the exact sequence of operations, and determine the utilization of the available resources. Tactical planning is called on to address a number of decisions: the manufacturing policy (what shall we make?), the procurement policy (what do we need?), the inventory or stock policy (what stock already exists?), the resources utilization policy (what do I need to make it?).

Tactical planning supports different short- to medium-term objectives for the business by using different objective functions. By using different objective functions we can create several operational plans to support the various strategic supply-chain decisions. Its differentiation from other planning approaches is that it requires a more detailed representation of the resources in a system. These resources are tied with a number of constraints that might need to be satisfied.

A common approach to tactical planning in the process industry is to describe the problem using a MP model, and then to optimize towards a desired objective. The objective can be maximization of profit, customer order satisfaction, minimization of cost, minimization of tardiness, minimization of common resource utilization, etc. The production environment is a rather complex network and most standard heuristic production planning tools fail to address this complexity. This situation gave rise to the idea of employing MP-based models to provide planning systems with a higher degree of flexibility by considering both product demands as a function of the marketing and sales departments of an organization, and the plant capacity in terms of equipment, material, manpower and utility resources. The problem has been modeled using a number of approaches.

Bassett et al. (1996b) proposed a higher level planning model based on formal aggregation techniques and using uniform time discretization. The model contains aggregate material balance constraints and equipment allocation constraints similar to those of the state-task network (STN) description of a process. This planning model forms part of a decomposition strategy where production is allocated to different time zones, thus creating a set of scheduling problems that can then be solved independently.

Wilkinson (1995) presented a generic mathematical technique to derive aggregate planning models of high accuracy based on the resource-task network (RTN) representation. The proposed formulations allow a large number of the complicating features of multipurpose, multiproduct plant operation to be taken into account in a unified manner. Sequence-dependent changeovers, task utility requirements and limited intermediate storage are some of the additional features included. Also the use of linking variables allows the planning model to take into account inventory levels more accurately.

These two formulations are fairly generic and include most of the important features regarding the planning in process industries where fixed recipes are employed. Prior to them, most of the planning models contained complicated sets of constraints which had been tailored to a specific problem type.

1.5
Resource Planning in the Power Market and Construction Projects

The area of resources planning in the energy and power market and construction projects is worthy of a review in its own right; it will be considered somewhat briefly here, mainly due to its strong similarities with the process planning problem.

1.5.1
Resource Planning in the Power Market

In a traditional electric power system, a utility company is responsible for generating and delivering power to its industrial, commercial and residential customers in its service area. It owns generation facilities and transmission and distribution networks, and obtains necessary information for the economical and reliable operation of its system. For instance, an important problem faced daily by a traditional utility company is to determine which and when generating units should be committed, and how they should be dispatched to meet the system-wide demand and reserve requirements. The centralized resource planning problem involves discrete states (e.g., on/off of units) and continuous variables (e.g., units' generation levels), with the objective being to minimize the total generation costs. A 1% reduction in costs can result in more than US$10 million dollars savings per year for a large utility company. Various methods have been presented in the literature and impressive results have been obtained (Wang et al. 1995, Guan et al. 1997, Li et al. 1997).

Today, the deregulation and reconstruction of the electric power industry worldwide have raised many challenging issues for the economic and reliable operation of electric power systems. Traditional unit commitment of hydrothermal scheduling/ planning problems are integrated with resource bidding, and the development of optimization-based bidding strategies is a preliminary stage. Ordinal optimization approaches seek "good enough" binding strategies with high probabilities, and will turn out to be effective in handling market uncertainties with much reduced computational cost. Under this new structure, resource planning is intertwined with bidding in the market, and power suppliers and system operators are facing a new spectrum of issues and challenges (Guan and Luh 1999).

Many approaches have been presented in the literature to address resources planning in the deregulated power markets. In this context, modeling and solving the bid selection problem has recently received significant attention. In Hao et al. (1998), bids are selected to minimize the total system cost, and the energy clearing price is determining as the highest accepted price for each hour. In Alvey et al. (1998), a bid clearing system in New Zealand is presented. Detailed models are used, including network constraints, reserve constraints, and ramp-rate constraints, and LP is used to solve the problem.

Another very popular way to model the bidding process is to model the competitors' behavior as uncertainties. Therefore, the bidding problem can be converted to a stochastic optimization problem. One of the widely used approaches in stochastic optimization to address this problem is stochastic dynamic programming (Contaxis 1990, Li et al. 1990). The basic idea is to extend the backward dynamic programming procedure by having probabilistic input and probabilities state transmissions in place of determining input and transitions and by using expected costs to calculate deterministic costs-to-go. The direct consequence is increased computational cost due to the significant increase in the input space and the number of possible transitions. For example, when stochastic dynamic programming is used to solve a hydroscheduling problem with uncertain inflows, one more dimension is needed to consider probable inflows in addition to reservoir levels, which significantly worsens the dimension of the problem. Another approach is scenario analysis (Carpentier et al. 1998, Takriti et al. 1996). Each scenario (or a possible realization of random events) is associated with a weight representing the probability of its occurrence. The objective is to minimize the expected costs over all possible scenarios. Since the number of possible scenarios and consequently the computational requirements increase drastically as the number of uncertain factors and the number of possibilities per factor increase, this approach can only handle problems with a limited number of uncertainties. Recently, stochastic dynamic programming has been embedded within the Lagrangean relaxation framework for energy scheduling problems, where stochastic dynamic programming is used to solve uncertain subproblems after system-wide coupling constraints are relaxed. Since dynamic programming for each subproblem can be effectively solved without encountering the curse of dimensionality, good schedules are obtained without a major increase in computational requirements (Luh et al. 1999).

Among alternatives that are being investigated for the generation of electricity are a number of unconventional sources including solar energy and wind energy. In recent decades photovoltaic (PV) energy found its first commercial use in space. In many parts of the globe, PV systems are being considered as a viable alternative for generating electricity. Achieving this goal requires PV systems to enter the utility market, whereby electric utilities evaluate the potentials of each PV system corresponding to its impact on the electric utility expansion planning, and requirements for backup generating capacity to ensure a reliable supply of electricity. Abdul-Rahman (1996) presented a model for the short-term resource scheduling in power systems. An augmented Lagrangean relaxation was used to overcome difficulties with the solution convergence as realistic constraints were introduced (i.e., transmission flows, fuel emissions, ramp-rate limits, etc.) in the formulation of unit commitment. Marwali et al. (1998) presented an efficient approach to short-term resource planning for an integrated thermal and PV battery generation. The proposed model incorporates battery storage for peak loads. Several constraints including battery capacity, minimum up/down time and ramp-rates for thermal units as well as natural PV capacity are considered in the proposed model.

1.5.2
Resource Planning in Construction Projects

Traditionally, resource planning problems in construction projects have been solved either as resources-leveling or as a resource-constrained scheduling problem. The resources-constrained scheduling problem constitutes one of the most challenging facing the construction industry, due to the limited availability of skilled labor, and the increasing need for productivity and cost-effectiveness. These challenges have been discussed by many practitioners and have led researchers to investigate various avenues. One of the most promising solutions to the problem of the shortage of skilled labor has been to develop methods that optimize or better utilize the skilled workers already in the industry (Burleson et al. 1998). The resource-leveling problem arises when there are sufficient resources available and it is necessary to reduce the fluctuations in the resource usage over the project duration. These resource fluctuations are undesirable because they often require a short-term hiring and firing policy. The short-term hiring and firing presents labor, utilization, and financial difficulties because (a) the clerical costs for employee processing are increased, (b) top-notch journeymen are reluctant to join a company with a reputation of doing this and (c) new, less experienced employees require long periods of training. The scheduling objective of the resource-leveling problem is to make the resource requirements as uniform as possible or to make them match a particular nonuniform resource distribution in order to meet the needs of a given project. Resource usage usually varies over the project duration because different types of resources are needed in varying amounts over the life of the project. In construction projects, for example, operators are needed in the beginning of the project to dig the foundations, but they are not needed at the end of the project for the interior finish work. In resource-leveling, the project duration of the original critical path remains unchanged.

MILP models have been used to formulate the resource-constrained scheduling problem (Nutdtasomboon and Randhawa 1996). The efficiency of these models usually decreases due to the high combinatorial nature of the problem, and special algorithms have been developed as an attempt to reduce computational costs and improve the quality of the solution. Most of these algorithms rely on special branch-and-bound and implicit enumeration approaches (Sung and Lim 1996, Demeulemeester and Herroelen 1997). An alternative approach to improving the computational efficiency is the use of heuristic methods that produce feasible, but not necessarily optimal, solutions (Padilla and Carr 1991, Seibert and Evans 1991).

Savin et al. (1998) presented a neural network application for construction resource-leveling using an augmented Lagrangian multiplier. The formulation objective is to make the resource requirements as uniform as possible. Thus, the formulation does not consider the case of nonuniform resource usage. Also, it only allows for one precedence relationship (finish-start) and one resource type, and does not perform cost optimization.

Chan et al. (1996) proposed a resource scheduling method based on genetic algorithms (GAs). The method considers both resource-leveling and resource-constrained scheduling. It can minimize the project duration, but it does not consider the case of nonuniform resource usage, neither does it minimize the construction cost. Adeli and Karim (1997) presented a general mathematical formulation for project scheduling. Multiple crew strategies, work continuity considerations, and the effect of varying job conditions on the crew performance could be modeled. They developed an optimization framework for minimizing the direct construction cost. However, the resource-leveling and resource-constrained scheduling problems were not addressed. Recently, Senousi and Adeli (2001) presented a new formulation including project scheduling characteristics such as precedence relationships, multiple crew strategies, and time-cost tradeoff. The formulation can handle minimization of the total construction cost or duration while resource-leveling and resource-constrained scheduling are performed simultaneously.

An important problem that has received rather limited attention in the literature is related to the optimal allocation of multiskilled labor resources in construction projects. This strategy is commonly found in the manufacturing and process industries where some of the labor force is trained to be multiskilled. Various studies have demonstrated the benefits of multiskilled resources. Nilikari (1995) presented a study involving Finnish shipbuilding facilities, based on a multiskilled work team strategy and found savings of up to 50% in production time.

Burleson et al. (1998) explored several multiskill strategies such as a dual-skill strategy, a four-skill strategy and an unlimited-skill strategy. The study compared the economic benefits in a huge construction project to prove the benefits of multiskilling but did not develop a mechanism for selecting the best strategy for a given project. The work of Brusco and Johns (1998) presented an integer goal-programming model for investigating cross-training multiskilled resource policies to determine the number of employees in each skill category so as to satisfy the demand for labor while minimizing staff costs. The model was applied to the maintenance operations of a large paper mill in the USA. Hegazy et al. (2000) presented an approach for modifying existing resource scheduling heuristics that deal with limited resources, to incorporate the multiskills of available labor and accordingly to improve the schedule. The performance of the proposed approach was demonstrated using a case study and the solution is compared with that of a high-end software system that considers multiskilled resources.

1.6
Solution Approaches to the Planning Problem

Most of the planning problems in the process industry result in an LP/MILP or NLP/MINLP model. Planning problems are usually NP-hard and data-driven; no standard solution techniques are therefore available, and in many cases we are actually searching for a feasible solution to the problem rather than an optimal one. The solution approaches found in the literature may be categorized as:

- *exact* and *deterministic methods* such as mathematical optimization including MILP and MINLP, graph theory (GT) or constraint programming (CP), or *hybrid approaches* in which MILP and CP are integrated;
- *metaheuristics* (evolutionary strategies, tabu search, simulated annealing (SA), various decomposition schemes, etc.).

In this section we are going to focus on general solution approaches applied to planning problems, in addition to those that are mentioned in other sections and are problem-dependent. We are not going to describe extensively how these methods have been employed by a variety of authors, but we are going to describe the algorithms and the classes of problem to which they have been applied. Despite the extensive research work that exists for the solution of long-term planning and short-term scheduling problems, the interest in medium-term planning problem is limited. While the benefits of integrating tactical planning into strategic planning and production scheduling are becoming clear, interest in research into more effective methods has increased. Applequist et al. (1997) provide an excellent review on planning technology and the approaches available for solving planning and scheduling problems. Despite substantial efforts over the last 40 years, no algorithm, either exact or heuristic, has been found that can provide a solution to all planning problems.

1.6.1
Exact and Deterministic Methods

In real life applications we rarely see any NLP/MINLP planning models, except in pooling or refinery planning. The rest of the models proposed, despite their complexity, in term of features they include and mathematical terms, they remain or are transformed to be linear regarding their variables and constraints. Therefore, using state-of-the-art commercial solvers, such as, XPRESS-MP (Dash Optimization, http://www.dashoptimization.com<urle>), CPLEX (ILOG, http://www.ilog.com), or OSL (IBM, http://www.ibm.com), LP/MILP problems can be solved efficiently and at a reasonable computational cost.

In the case of NLP/MINLP, the solution efficiency depends strongly on the individual problem and the model formulation. Thus, in many cases the structure of the problem is exploited in order to provide valid cuts, or identify special structures in order to reduce computational times and increase the quality of the solution.

However, as both MILP and MINLP are NP-hard problems, it is recommended that the full mathematical structure of a problem to be exploited. Software packages may also differ with respect to their ability in presolving techniques, default strategies for the branch-and-bound algorithm, cut generation within the branch-and-cut algorithm, and last but not least, diagnosing and tracing infeasibilities, which is an important issue in practice. Kallrath (2000) provides an extensive review of mixed-integer optimization in the process industry by describing solution methods, algorithms, and applications.

Taking advantage of the special structure of mathematically formulated problems either as MILPs or MINLPs, several decomposition methods have been proposed and implemented in various types of problems.

Bassett et al. (1996a), focusing on chemical process industries, examined a number of time-based decomposition approaches along with their associated strengths and weaknesses. It is shown that the most promising of the approaches utilizes a reverse rolling window in conjunction with a disaggregation heuristic, applied to an aggregate production plan as part of their approach to integrate hierarchically related decisions. Resource- and task-based decompositions are also examined as possible approaches to reduce the problem to manageable proportions. To validate their proposed schemes a number of examples are presented.

Gupta, A. Maranas, C. D. Ind. Eng. chem. Res. 38 (1999) 1937 utilized an efficient decomposition procedure to solve mid-term planning problems based on Lagrangean relaxation. Having tried commercial MILP solvers, they found that the employed solution strategy is more efficient. The basic idea of the proposed solution technique is the successive partitioning of the original problem into smaller, more computationally tractable subproblems by hierarchical relaxation of key complicating constraints. Alongside with the hierarchical Lagrangean relaxation they employ a heuristic algorithm to obtain valid upper bounds. Two examples are used to demonstrate the capabilities of the proposed algorithm.

The size of the actual planning problem may be prohibitive for standard commercial solvers. Therefore, rigorous decomposition techniques that benefit from the special structure of MILP problems is exploited. Dimitriadis (2001), identified that block-diagonal MILP problems may be decomposed to simpler ones and introduced the concept of decomposable MILP (D-MILP). An algorithm based on the idea of "key variables", which break the problem down into a number of smaller partial MILPs that can be solved independently and in parallel, was implemented based on a standard branch-and-bound scheme. The decomposition branch-and-bound (dBB) as the algorithm is called, achieves better performance by obtaining quick upper bounds to the problem and assisting the solver to find an optimal solution within reasonable computational time. One of the advantages of the approach is that is can guarantee the optimality of the solution. Tsiakis et al. (2000) improved the algorithm by providing an automated method to decompose the problem and implementing a more generic solution scheme applicable to all MILP problems that have a similar structure.

1.6.2
Metaheuristics

In addition to the so called optimization methods we have techniques described as heuristics. These techniques differ in the sense that cannot guarantee an *optimal* solution; instead they aim to find reasonably good solutions in a relatively short time. Heuristics tend to be fairly generic and easily adaptable to a large variety of planning problems. There are a number of heuristic general-purpose approaches that can be applied to planning and scheduling problems (Pinedo 2003).

SA and *tabu search* are described as improvement algorithms. Algorithms of the improvement type are conceptually completely different from the constructive type algorithms. The algorithm starts by obtaining a complete plan that can be selected arbitrarily, and then tries to obtain a better plan by manipulating the current solution. The procedure is described as local search. A local search procedure does not guarantee an optimal solution, but aims to obtain a better solution in the neighborhood of the current one. They very often employ a probabilistic acceptance-rejection criterion with the hope it will lead to better solution. Reeves (1995) describes extensively the methods and applications in production planning systems.

GAs are more general than SA and tabu search and they can be classified as a generalization of the previously mentioned techniques. In this case a number of feasible solutions are initially found. Then, local search based on an evolution criterion is employed to select the most promising solution for further exploitation. The rest of the solutions are fathomed (Reeves 1995).

Heuristics are widely employed in industry to provide solutions to production planning problems. Stockton and Quinn (1995) describe how a GA based on aggregate planning techniques is used to develop a production plan that allows a strategic business objective to be implemented in short- and mid-term operational plans.

LeBlanc et al. (1999) utilize an extension of the multiresource generalized assignment problem (MRGAP) in order to provide an implementable solution to production planning problems. The model considers splitting of individual batches across multiple machines, while considering the effect of set-up times and set-up costs, features that the standard assignment problem (AP) fails to capture. The proposed formulations are solved using adaptations of a GA and SA.

A multiobjective GA (MOGA) approach was employed by Morad and Zalzada (1999) for the planning of multiple machines, taking into account their processing capabilities and the process costs incurred. The formulation is based on multiobjective weighted-sums optimization, which is to minimize makespan, to minimize total rejects produced and to minimize the total cost of production.

Tabu search is employed by Baykasoglu (2001) to solve multiobjective aggregate production planning (APP) problems based on a mathematically formulated problem. The model by Masud and Hawng was selected as the basis due to its extensibility characteristics.

1.7
Software Tools for the Resource Planning Problem

Enterprise resource planning (ERP) is a software-driven business management system which integrates all facets of the business, including planning, manufacturing, sales and marketing. Increasingly complex business environments require better and more accurate resource planning. Furthermore, management is under constant pressure to improve competitiveness by lowering operating costs and improving logistics, thus increasing the level of control within the operating environment. Organizations therefore have to be more responsive to the customer and competition. Resource planning as a business solution aims to help the management by setting up better business practices and equipping them with the right information to take timely decisions.

Production planning as a later business function is considered to be part of the supply-chain planning and scheduling suite, alongside other functions such as demand forecasting, supply-chain planning, production scheduling, distribution and transportation planning. Tactical production planning includes those software modules responsible for production planning within a single manufacturing facility. These solutions normally address tactical activities, although they may also be used to support both strategic and operational decisions and are very often integrated with them.

1.7.1
Enterprise Resource Planning

A big share of the software and services provided worldwide is targeting the integration of ERP and supply-chain operations. Most of the information needed by production planning software tools resides within ERP systems. Most of the ERP software providers already have developed their own fully integrated planning applications, have acquired smaller companies with production planning software or have been in partnership with such providers. The standard object-oriented approach to the implementation of ERP systems has contributed towards an easy integration. The leading suppliers and systems integrators to the worldwide ERP market across all industry sectors are alphabetically: Oracle (http://www.oracle.com), Manugistics (http://www.manugistics.com), PeopleSoft (http://www.peoplesoft.com), and SAP (http://www.sap.com) according to the latest market share studies. In small-medium enterprises (SMEs) the leading provider of ERP systems is Microsoft Business Solutions with its Navision system (http://www.microsoft.com/businessSolutions).

1.7.2
Production Planning

Production planning deals in medium-range time horizons, where decisions about incremental adjustments to the capacity or customer service levels are made.

Changes to supplier delivery dates, swings in raw materials purchases, and outsourcing agreements may require 3-5 months. Thus, production planning deals with what will be done, and when, in a factory over longer time frames. Tactical plans are updated frequently based on the operational plan and the actual schedule. This section provides the profiles of production planning software suppliers with main focus on the process industry, again in alphabetical order.

1.7.2.1

Advanced Process Combinatorics (http://www.combination.com)

The company's modular supply-chain product VirtECS contains a module, called Scheduler, with production planning capability. The package handles complex production planning models with multiple input/output bills of material, multiple routings, resource constraints and set-up times. Their algorithms used for production planning are based on a MILP formulation, with a number of techniques applied for their solution. Additionally, a set of Gantt-chart-based interactive tools provides the user with manipulating capabilities on the actual plan. A key strength of APC revolves around the research on optimization since the company was generated from an industrial research consortium at Purdue University.

1.7.2.2

Aspen Technologies (http://www.aspentech.com)

Aspen Technology's supply-chain capabilities are based largely on the company's acquisition of the Process Industry Modelling Systems (PIMS) from Bechtel and Manager for Interactive Modelling Interfaces (MIMI) of Chesapeake Decision Sciences. Aspen PIMS is a tactical level refinery planning package that is widely used in over 170 refineries worldwide. The Aspen MIMI production planner is focused on models that include material flows, set-up times, labor constraints and other resource restrictions. In addition to the standard heuristics and simulation employed for production planning and scheduling, the advanced planning offers LP-based optimization capabilities. Users can interact with the Gantt chart in order to develop "what-if" analysis cases and add constraints. Aspen has one of the larger installed bases of MIMI products for over 300 customers around the globe.

1.7.2.3

i2 Technologies (http://www.i2.com)

As part of its supply-chain platform, i2's Factory Planner manages material and capacity constraints to develop feasible operating plans for production plants. The tool aims to be a decision support system in the areas of production planning and scheduling, taking into account material and capacity requirements. It utilizes a number of heuristic algorithms and basic optimization to obtain feasible plans, and to answer capable-to-promise delivery-date quoting.

1.7.2.4
Manugistics (http://www.manugistics.com)
Manufacturing Planning and Scheduling, integrated within the Constraint-Based Master Planning supply-chain system of Manugistics, provides the detailed operational plan. It is based on a flow-oriented model, and uses the theory of constraints to solve the production planning problems. It takes into account throughput of equipment, determines the bill of materials, and allows what-if scenario analysis.

1.7.2.5
Process System Enterprise (PSE) (http://www.psenterprise.com)
PSE's ModelEnterprise has been designed as a modular supply-chain modeling platform that allows the construction and maintenance of complex enterprise models, and supports a wide range of tools applied to these models for solving different types of problem. The Optimal Single Site Planner and Scheduler (OSS Planner Scheduler) determines an optimal schedule for a plant producing multiple products. It is especially suited to multipurpose plant where products can be made on a selection of equipment units, via different routes and in different sizes. The plans produced are finite capacity and rigorously optimal. The objective of the optimization problem can be configured according to the economic requirements of the operation – for example, to deliver maximum profit, maximum output or on-time in-full. The OSS Planner Scheduler uses state-of-the-art MILP optimization algorithms that allow complex systems to be modeled. Utilizing comprehensive costing all costs may be accounted such as processing, storage, utilities, cleaning, supplies and penalties for late delivery. PSE originated at Imperial College, London, in the 1990s. and ModelEnterprise has been developed based on knowledge and research found there.

1.7.2.6
SAP AG (http://www.sap.com)
The APO Production Planning and Detailed Scheduling (PP/DS) tool comes under the umbrella of SAP APO supply-chain solutions. The software can be used to generate production plans and sequence schedules. A variety of approaches is included in this solution for theory of constraints and mathematical optimization, but in principle it is a heuristics-based tool, where the user-developed rules are employed. Other features of the tool include forward and backwards scheduling, simultaneous capacity and material planning in detail, what-if analysis to simulate effects of changes in constraints, and interactive scheduling via a Gantt chart interface.

1.8
Conclusions

The impact of accurate resource planning on the productivity and performance of both manufacturing and service organizations are tremendous. Researchers have found that organizations that had no resource planning information technology infrastructure in place performed poorly most of the time compared to those who had a specific plan. The successful implementation of planning capabilities means reduction in cost, increased productivity and improved customer services. The importance of resource planning models and systems therefore becomes significant. Moreover, the solution to the problems associated with that poses further challenges.

Despite many years of study in resource planning models, plus numerous examples of successful modeling systems implementations and industrial applications, there is still a great potential for applying them in a pervasive and enduring manner to a wider range of real-life industrial applications.

Several researchers have tackled the resource planning problem under uncertainty using different approaches. However, in most cases they have skirted around the problem of multiperiod, multiscenario planning with detailed production capacity models (i.e., embedding some scheduling information). Here, issues that must be addressed mainly relate to problem scale. Combined mixed-integer programming and stochastic optimization techniques seem to offer a promising solution alternative to this problem.

One of the major challenges will be to develop planning approaches that are consistent with detailed resource scheduling as part of the overall supply-chain integration. An obvious drawback is the problem size. This poses the need for rigorous decomposition algorithms and techniques that will enable handling problems of greater size without compromising the quality of the solution.

Over the last few years a trend has developed bringing MP and constrained programming techniques closer to each other. This results in hybrid approaches (i.e., in algorithms combining elements from both areas) that may have a great impact on reducing computational requirements for solving large-scale planning problems.

In addition to new techniques and solution approaches, advances in computational power in terms of hardware and software allow the exploitation of parallel algorithm optimization techniques. The tree structure in mixed-integer optimization, and the time- or scenario-dependent structures, indicates that more benefits are to be expected from parallelizing the combinatorial part. Dealing with large-scale NP-hard problems may lead to the implementation of distributed planning, where the computational effort and time is divided over a number of computers or clusters. Time-based or spatial decomposition methods will be exploited more and more.

Resource planning is a fundamental business process that exists in every production environment. It has long been recognized that in the process industries there are very large financial incentives for planning, scheduling and control decisions to function in a coordinated fashion. Nevertheless, many companies have not achieved integration in spite of multiple initiatives. An important challenge thus relates to the development of efficient theoretical methodologies, algorithms and tools to achieve this integration in a formal way, allowing process industries to take steps to practically improve the integration activities at different levels.

The planning problem of refinery operations and offshore oilfields has been recently attacked by several researchers. However, the practical implementation of most of the developed approaches is usually limited to subsystems of a plant with considerable simplifications. Here, the trend is to expand the planning process to include larger systems, such as a group of refineries instead of a single one. Another area that deserves further attention is the inclusion of scheduling decisions in planning processes. Furthermore, there is a lot of scope for developing commercial tools to serve refineries to cope with daily operational problems.

In the area of product planning, the integration of development management and capacity and production planning seems to be very important. Currently, capacity issues are often not considered at the development stage. The development of integrated models of the life cycle, from the discovery through to consumption would greatly facilitate strategic decision making.

Demand for advanced planning systems (APS) is expected to grow with the solutions being increasingly industry- and supply-chain-specific. The standards are specified by the large software suppliers, such as i2 and Manugistics. The scope for smaller suppliers is to have a more specific focus in segments of the industry. There is a clear trend for industry-specific solutions, this being due to the different operating environments and the detail required in order to generate a meaningful plan. The development of resource planning systems very much depends on the industry segment (industrial or nonindustrial) and the manufacturing type (process or discrete industries). While segmentation based on the type of industry is common, it is important to be able to segment the operational environment based on the supply-chain type. In this case we have distribution, manufacturing or source intensive supply chains, each one with their own needs. Many companies are competing as software providers for planning systems. However, they have realized that they need to be able to communicate with other libraries and software modules as part of supply-chain solutions, and at minimum cost. Systems with open architecture and ease of integration are in demand. Initiatives such as CAPE-OPEN aim to define industry-wide standards (CO-LaN 2001).

References

1 *Abdinour-Helm S. Lengnick-Hall M. L. Lengnick-Hill C. A. Eur. J. Oper. Res. 146 (2003) p. 258*

2 *Abdul-Rahman K. H. Shahidehpour S. M. Aganagic M. Mokhtari S. IEEE Trans. Power Syst. 11 (1996) p. 254*

3 *Adeli H. Karim A.J. Constr. Eng. Manage. 123 (1997) p. 450*

4 *Ahmed S. Sahinidis N. V. Ind. Eng. Chem. Res. 37 (1998) p. 1883*

5 *Alvey T. Goodwin D. Ma X. Streiffert D. Sun D. IEEE Trans. Power Syst. 13 (1998) p. 986*

6 *Applequist G. Samikoglu O. Pekny J. Reklaitis G. ISA Trans. 36 (1997) p. 81*

7 *Aseeri A. Gorman P. Bagajewicz M. J. Ind. Eng. chem. Res. 43 2004, 3063*

8 *Ballintjin K. 1993 In: Ciriani T. A. Leachman R. C. (eds.), Vol. 3, Wiley, New York*

9 *Barbaro A. Bagajewicz AICH Journal, 50 (2004) 963*

10 *Barnes R. Linke P. Kokossis A. 2002 In: Proceedings of ESCAPE-12, The Hague, The Netherlands, p 631*

11 *Bassett M. H. Dave P. Doyle F. J. Kudva G. K. Pekny J. F. Reklaitis G. V. Subrahmanyam S. Miller D. L. Zentner M. G. Comput. Chem. Eng. 20 (1996a) p 821*

12 *Bassett M. H. Pekny J. F. Reklaitis G. V. AICHE J. 42 (1996b) p 3373*

13 *Bassett M. H. Pekny J. F. Reklaitis G. V. Comput. Chem. Eng. 21 (1997) p. S1203*

14 *Baykasoglou A. Int. J. Prod. Res. 39 (2001) p. 3685*

15 *Bernardo F. P. Pistikopoulos E. N. Saraiva P. Ind. Eng. Chem. Res. 38 (1999) p. 3056*

16 *Birewar D. B. Grossmann I. E. Comput. ChemM. Eng. 13 (1989a) p. 141*

17 *Birewar D. B. Grossmann I. E.* Ind. Eng. Chem. Process Des. Dev. 28 (1989b) p. 1333

18 *Birewar D. B. Grossmann I. E.* Ind. Eng. Chem. Res. 29 (1990) p. 570

19 *Bitran G. R. Hax A. C.* Decision Sci. 8 (1977) p. 28

20 *Blau G. Metha B. Bose S. Pekny J. Sinclair G. Keunker K. Bunch P.* Comput. Chem. Eng. 24 (2000) p. 659

21 *Bose S. Pekny J. F.* Comput. Chem. Eng. 24 (2000) p. 329

22 *Brusco M. J. Johns T. R.* Decision Sci. J. 29 (1998) p. 499

23 *Burleson R. C. Hass C. T. Tucher R. L. Stanley A.* J. Constr. Eng. Manage. 124 (1998) p. 480

24 *Carpentier P. Woodwin D. Ma X. Streiffert D. Sun D.* IEEE Trans. Power Syst. 13 (1998) p. 986

25 *Chan W. T. Chua D. K. H. Kannan G.* J. Constr. Eng. Manage. 122 (1996) p. 125

26 *Clay R.L. Grossmann I. E.* Chem. Eng. Res. Des. 72 (1994) p. 415

27 *CO-LaN*: The CAPE-OPEN Laboratories Network (http://zann.informatik.rwth-aachen.de:8080/opencms/opencms/COLAN-gamma/index.html)

28 *Contaxis G. Kavantza S.* IEEE Trans. Power Syst. 5 (1990) p. 766

29 *Das B. P. Rickard J. G. Shah N. Macchietto S.* Comput. Chem. Eng. 24 (2000) p. 1625

30 *Demeulemeester E. Herroelen W.* Manage. Sci. 43 (1997) p. 1485

31 *A. D. Dimitriadis 2001 PhD thesis Title: Algorithms for the solution of Large-scale Scheduling problems. Imperial College of science, Technology and Medicine*

32 *Gatica G. Papageorgiou L. G. Shah N.* Chem. Eng. Res. Des. 81 (2003) p. 665

33 *Geddes D. Kubera T.* Comput. Chem. Eng. 24 (2000) p. 1645

34 *Goel H. D. Grievink J. Weijnen M. P. C.* Comput. Chem. Eng. 27 (2003) p. 1543

35 *Gothe-Lundgren M. Lundgren J. T. Persson J. A.* Int. J. Prod. Econ. 78 (2002) p. 255

36 *Grunow M. Gunther H. Lehmann M.* OR Spectrum 24 (2002) p. 281

37 *Guan X.. Luh P. B.* Discrete Event Dyn. Syst. Theor Appl. 9 (1999) p. 331

38 *Guan X. Ni E. Li R. Luh P. B.* IEEE Trans. Power Syst. 12 (1997) p. 1775

39 *Hao S. Angelidis G. A. Singh H. Papalexopoulos A. D.* IEEE Trans. Power Syst. 13 (1998) p. 986

40 *Harjunkoski I. Grossmann I. E. Friedrich M. Holmes R.* 2003. FOCAPO 2003 p. 315

41 *Hax A. C.* 1978 Handbook of Operations Research Models and Applications

42 *Hegazy B. T. Shabeeb A. K. Elbeltagi E. Cheema T.* J. Constr. Eng. Manage. 126 (2000) p. 414

43 *Ierapetritou M. G. Floudas C. A. Vasantharajan S. Cullick A. S.* AIChE J. 45 (1999) p. 844

44 *Ierapetritou M. G. Pistikopoulos E. N.* Ind. Eng. Chem. Res. 33 (1994) p. 1930

45 *Ierapetritou M. G. Pistikopoulos E. N. Floudas C. A.* Comput. Chem. Eng. 20 (1996) p. 1499

46 *Iyer R. R. Grossmann I. E.* Ind. Eng. Chem. Res. 37 (1998) p. 474

47 *Iyer R. R. Grossmann I. E. Vasantharajan S. Cullick A. S.* Ind. Eng. Chem. Res. 37 (1998) p. 1380

48 *Jacobs R. F. Bendoly E.* Eur. J. Oper. Res. 146 (2003) p. 233

49 *Jain V. Grossmann I. E.* Ind. Eng. Chem. Res. 38 (1999) p. 3013

50 *Jia Z. Ierapetritou M. G.* Ind. Eng. Chem. Res. 43 (2004), p. 3782

51 *Kabore P.* FOCAPO 2003 p. 285

52 *Kallrath J.* Trans. IChemE 78 (2000) p. 809

53 *Kallrath J.* OR Spectrum 24 (2002) p. 219

54 *Kosmidis V. D.* 2003 Ph.D. Thesis University of London U.K

55 *Kosmidis V. D. Perkins J. D. Pistikopoulos E. N.* 2002 In: Proceedings of ESCAPE-12 The Hague The Netherlands p 327.

56 *LeBlanc L. J. Shtub A. Anandnalingham G.* Eur. J. Oper. Res. 112 (1999) p. 54

57 *Lee H. Pinto J. M. Grossmann I. E. Park S.* Ind. Eng. Chem. Res. 35 (1996) p. 1630

58 *Levis A. A. Papageorgiou L. G.* 28 (2004), p. 707 Comput. Chem. Eng.

59 *Li C. Johnson R. B. Svaboda A. J.* IEEE Trans. Power Syst. 12 (1997) p. 1834

60 *Li C. Yan R. Zhou J.* IEEE Trans. Power Syst. 5 (1990) p. 1487

61 *Li P. Wendt M. Wozny G.* 2003 FOCAPO 2003 p. 289

62 *Lin X. Floudas C. A.* Optimisation Eng. 4 (2003) p. 65

63 *Liu M.L. Sahinidis N. V.* Ind. Eng. Chem. Res. 34 (1995) p. 1662

64 *Liu M.L. Sahinidis N. V.* Comput. Oper. Res. 23 (1996a) p. 237

65 *Liu M.L. Sahinidis N. V.* Ind. Eng. Chem. Res. 35 (1996b) p.4154

66 *Liu L. Sahinidis N. V.* Eur. J. Oper. Res. 100 (1997) p. 142

67 *Luh P. B. Chen D. Tahur L. S.* IEEE Trans. Robot. Autom. 15 (1999) p. 328

68 *Mandal P. Gunasekaran A.* Eur. J. Oper. Res. 146 (2003) p. 274

69 *Maravelias C. T. Grossmann I. E.* Ind. Eng. Chem. Res. 40 (2001) p. 6147

70 *Maravelias C. T. Grossmann I. E.* 28 (2004), p. 1921 Comput. Chem. Eng. in press

71 *Marwali M. K. C. Ma H. Shahidehpour S. M. Abdul-Rahman H.* IEEE Trans. Power Syst. 13 (1998) p. 1057

72 *Mauderli A. Rippin D. W. T.* Comput. Chem. Eng. 3 (1979) p. 199

73 *McDonald C. M. Karimi I. A.* Ind. Eng. Chem. Res. 36 (1997) p. 2691

74 *Morad N. Zalzada A.* J. Intell Manuf. 10 (1999) p. 169

75 *Moro L. F. L.* Comput. Chem. Eng. 27 (2003) p. 1303

76 *Moro L. F. L. Zanin A. C. Pinto J. M.* Comput. Chem. Eng. S22 (1998) p. S1039

77 *Neiro S. M. S. Pinto J. M.* Comput. Chem. Eng. 28 (2004), p. 871

78 *Nilikari M. J.* Ship Prod. 11 (1995) p. 239

79 *Norton L. C. Grossmann I. E.* Ind. Eng. Chem. Res. 33 (1994) p. 69

80 *Nutdtasomboon N. Randhawa S.* Comput. Ind. Eng. 32 (1996) p. 227

81 *Oh H. Karimi I. A.* Comput. Chem. Eng. 25 (2001a) p. 1021

82 *Oh H. Karimi I. A.* Comput. Chem. Eng. 25 (2001b) p. 1031

83 *Orlicky J.* 1975 Material Requirements Planning. McGraw Hill. New York

84 *Oxe G.* Eur. J. Oper. Res. 97 (1997) p. 337

85 *Padilla E. M. Carr R. L.* J. Constr. Eng. Manage. 117 (1991) p. 279

86 *Papageorgiou L. G.* 1994; Ph.D. Thesis University of London

87 *Papageorgiou L. G. Pantelides C. C.* Comput. Chem. Eng. 17S (1993) p. S27

88 *Papageorgiou L. G. Pantelides C. C.* Ind. Eng. Chem. Res. 35 (1996a) p. 488

89 *Papageorgiou L. G. Pantelides C. C.* Ind. Eng. Chem. Res. 35 (1996b) p. 510

90 *Papageorgiou L. G. Rotstein G. E. Shah N.* Ind. Eng. Chem. Res. 40 (2001) p. 275

91 *Pelham R. Pharris C.* Hydrocarb. Process. 75 (1996) p. 89

92 *Petkov S. B. Maranas C. D.* Ind. Eng. Chem. Res. 36 (1997) p. 4864

93 *Pinedo M.* 2003 Scheduling: Theory Algorithms and Systems 2nd Edn. Prentice Hall, New Jersey

94 *Pinto J. M. Joly M. Moro L. F. L.* Comput. Chem. Eng. 24 (2000) p. 2259

95 *Pistikopoulos E. N. Vassiliadis C. G. Arvela J. A. Papageorgiou L. G.* Ind. Eng. Chem. Res. 40 (2001) p. 3195

96 *Reeves C. R.* 1995 Modern Heuristic Techniques for Combinatorial Problems Wiley

97 *Reklaitis G. V.* 1991 Proceedings of the 4th International Symposium on Process Systems Engineering Montebello Canada

98 *Reklaitis G. V.* 1992 NATO Advanced Study Institute on Batch Processing Systems Eng. Antalya Turkey

99 *Rickard J. G. Macchietto S. Shah N.* Comput. Chem. Eng. S23 (1999) p. S539

100 *Rigby B. Lasdon L. S. Waren A. D.* Interfaces 25 (1995) p. 64

101 *Rodera H. Bagajewicz M. J. Tsafalis T. B.* Ind. Eng. Chem. Res. 41 (2002) p. 4075

102 *Rodrigues M. M. Latre L. G. Rodrigues L. A.* Comput. Chem. Eng. 24 (2000) p. 2247

103 *Roger M. J. Gupta A. Maranas C. D.* Ind. Eng. Chem. Res. 41 (2002) p. 6607

104 *Romero J. Badell M. Bagajewicz M. Puigjaner L.* Ind. Eng. Chem. Res. 42 (2003) p. 6125

105 *Ryu J.-H. Dua V. Pistikopoulos E. N.* Comput. Chem. Eng. 28 (2004), p. 1121

106 *Sahinidis N. V.* Comput. Chem. Eng. 28 (2004), p. 971

107 *Sahinidis N. V. Grossmann I. E.* Ind. Eng. Chem. Res. 30 (1991) p. 1165

108 *Sahinidis N. V. Grossmann I. E. Fornari R. E. Chathrathi C.* Comput. Chem. Eng. 13 (1989) p. 1049

109 *Sanmarti E. Espuna A. Puigjaner L.* Comput. Chem. Eng. S19 (1995) p. S565

110 *Savin D. Alkass S. Fazio P.* J. Comput. Civ. Eng. 12 (1998) p. 241

111 *Schmidt C.W. Grossmann I. E.* Ind. Eng. Chem. Res. 35 (1996) p. 3498

112 *Seibert J. E. Evans G. W.* J. Constr. Eng. Manage. 117 (1991) p. 503

113 *Senousi A. B. Adeli H.* J. Constr. Eng. Manage. 127 (2001) p. 28

114 *Shah N.* Comput. Chem. Eng. S20 (1996) p. S1227

115 *Shah N.* AIChE Symp Ser. 1998 No. 320 94 75

116 *Shah N.* Comput. Chem. Eng. 28 (2004), p. 929

117 *Shah N. Pantelides C. C.* Ind. Eng. Chem. Res. 30 (1991) p. 2308

118 *Shah N. Pantelides C. C.* Ind. Eng. Chem. Res. 31 (1992) p. 1325

119 *Shah N. Pantelides C. C. Sargent R. W. H.* Ann. Oper. Res. 42 (1993) p. 193

120 *Shapiro J.F.* Eur. J. Oper. Res. 118 (1999) p. 295

121 *Shapiro J. F.* Comput. Chem. Eng. 28 (2004), p. 855

122 *Shobrys D. E. White D. C.* Comput. Chem. Eng. 26 (2002) p. 149

123 *Stockton D. J. Quinn L.* Proc. Inst. Mech. Eng. 209(3) (1995) p. 201

124 *Subrahmanyam S. Pekny J. F. Reklaitis G. V.* Ind. Eng. Chem. Res. 33 (1994) p. 2668

125 *Subrahmanyam S. Pekny J. F. Reklaitis G. V.* Ind. Eng. Chem. Res. 35 (1996) p. 1866

126 *Subramanian D. Pekny J. F. Reklaitis G. V.* AIChE J. 47 (2001) p. 2226

127 *Sung C. S. Lim S. K.* Comput. Ind. Eng. 12 (1996) p. 227

128 *Suryadi H. Papageorgiou L. G.* Int. J. Prod. Res. 42 (2004) p. 355

129 *Takriti S. Birge J. Long E.* IEEE Trans. Power Syst. 11 (1996) p. 1497

130 *Tsiakis P. Rickard J. G. Shah N. Pantelides C. C.* 2003 AIChE Annual Conference San Francisco California USA

131 *Tsiakis P. Dimitriadis A. D. Shah N. Pantelides C. C.* 2000 AIChE Annual Conference Los Angeles California USA

132 *Tsiroukis A.G. Papageorgaki S. Reklaitis G. V.* Eng. Chem. Res. 32 (1993) p. 3037

133 *van den Heever S. A. Grossmann I. E.* Ind. Eng. Chem. Res. 39 (2000) p. 1955

134 *van den Heever S. A. Grossmann I. E.* Comput. Chem. Eng. 27 (2003) p. 1813

135 *van den Heever S. A. Grossmann I. E. Vasantharajan S. Edwards K.* Comput. Chem. Eng. 24 (2000) p. 1049

136 *van den Heever S. A. Grossmann I. E. Vasantharajan S. Edwards K.* Ind. Eng. Chem. Res. 40 (2001) p. 2857

137 *Voudouris T. V. Grossmann I. E.* Ind. Eng. Chem. Res. 32 (1993) p. 1962

138 *Wang S. J. Shahidehpour S. M. Kirschem D. S. Mokhtari S. Irissari G. D.* IEEE Trans. Power Syst. 10 (1995) p. 1294

139 *Wellons M. C. Reklaitis G. V.* Comput. Chem. Eng. 13 (1989a) p. 201

140 *Wellons M. C. Reklaitis G. V.* Comput. Chem. Eng. 13 (1989b) p. 213

141 *Wellons M. C. Reklaitis G. V.* Ind. Eng. Chem. Res. 30 (1991a) p. 671

142 *Wellons M. C. Reklaitis G. V.* Ind. Eng. Chem. Res. 30 (1991b) p. 688

143 *Wight O.* 1984 Manufacturing Resource Planning: MRP-II Wight Williston

144 *Wilkinson S. J.* 1996 PhD Thesis University of London

145 *Wilkinson S. J. Shah N. Pantelides C. C.* Comput. Chem. Eng. 19 (1995) p. S583

146 *Wilson J. M.* Eur. J. Oper. Res. 149 (2003) p. 430

147 *Wu D. Ierapetritou M. G.* 2003. FOCAPO 2003 p. 257

148 *Yin K. K. Liu H.* 2003. FOCAPO 2003 p. 261

2
Production Scheduling

Nilay Shah

2.1
Introduction

The theme of production planning and scheduling has been the subject of great attention in the recent past. Initially, especially from the early 1980s to the early 1990s, this was due to the resurgence in interest in flexible processing either as a means of ensuring responsiveness or of adapting to the trends in chemical processing towards lower volume, higher-value-added materials in the developed economies (Reklaitis 1991, Rippin 1993, Hampel 1997). More recently, the topic has received a new impetus as enterprises attempt to optimize their overall supply chains in response to competitive pressures or to take advantage of recent relaxations in restrictions on global trade, as well as the information storage and retrieval capabilities provided by ERP systems.

It is widely recognized that the complex problem of what to produce and where and how to produce it is best considered through an integrated, hierarchical approach that also acknowledges typical corporate structures and business processes. This type of structure is illustrated in Figure 2.1. In the most general case, the extended supply chain is taken to mean the multienterprise network of manufacturing facilities and distribution points that perform the functions of materials procurement, transformation into intermediate and finished materials and distribution of the finished products to customers.

The most common context for planning at the supply-chain level is the coordination of manufacturing and distribution activities across multiple sites operated by a single enterprise (enterprise-wide or multisite planning). Here, the aim is to make the best use of geographically distributed resources over a certain time period.

The result of the multisite planning problem is typically a set of production targets for each of the individual sites, and rough transportation plans for the network as a whole. The production scheduling activity at each individual site seeks to determine precisely how these targets can be met (or indeed how best to compromise them if they cannot be met in whole). This involves determining the precise details of resource allocation over time.

Computer Aided Process and Product Engineering. Edited by Luis Puigjaner and Georges Heyen
Copyright © 2006 WILEY-VCH Verlag GmbH & Co. KGaA, Weinheim
ISBN: 3-527-30804-0

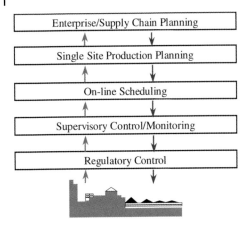

Figure 2.1   Process operations hierarchy

Once a series of activities has been determined, these must be implemented in the plant. The role of the supervisory control system is to initiate the correct sequences of control logic with the correct parameters at the correct time, making sure that conflicts for plant resources are resolved in an orderly manner. It is also useful at this level to create a schedule of planned operations over a short future interval using a model detailed enough to ensure that there are no anticipated resource conflicts. This "online" scheduling allows current estimates of the starting and finishing times of each operation to be known at any time. Although this capability is not essential for the execution of operations in the plant, it is vital if the hierarchical levels are to be integrated so that production scheduling is performed in response to deviations in expected plant operation ("reactive scheduling").

Finally, the lowest levels of the hierarchy relate to execution of individual control phases and ensuring safe and economic operation of the plant.

2.1.1
Why Is Scheduling Important?

The planning function aims to optimize the economic performance of the enterprise as it matches production to demand in the best possible way.

The production scheduling component is of vital importance as it is the layer which translates the economic imperatives of the plan into a sequence of actions to be executed on the plant, so as to deliver the optimized economic performance predicted by the higher-level plan.

2.1.2
Challenges in Scheduling

There is clearly a need for research and development in all the levels of the operations hierarchy. Four previous reviews in this area (Reklaitis 1991, Rippin 1993, Shah 1998, Kallrath 2002a) summarized some of the main challenges as:

- the development of efficient general-purpose solution methods for the mixed-integer optimization problems that arise in planning and scheduling;
- the design of tailored techniques for the solution of specific problem structures, which either arise out of specific types of scheduling problems or are embedded substructures in more general problems;
- the design of algorithms for efficient solution of general resource constrained problems, especially those based on a continuous representation of time;
- the development of hybrid methods based on optimization and constraint propagation methods;
- the development of commercially available software packages for optimization-based scheduling (as distinct from planning);
- the systematic treatment of uncertainty;
- the advancement of online techniques for rapid adaptation of operations;
- the development of methods for the integrated planning and scheduling of multi-site systems.

Multisite and supply-chain planning and scheduling is dealt with in section 5.7, and the focus here is on scheduling at a single site. Progress towards these challenges will be described. The remainder of this chapter is organized as follows: Section 2.2 describes the problem in more detail. Sections 2.3–2.9 review research into alternative solution methods for scheduling problems, both with deterministic and uncertain data, and Section 2.10 describes some successful industrial applications of advanced scheduling methods. The remaining sections list some new application domains and describe conclusions drawn.

2.2
The Single-Site Production Scheduling Problem

The scheduling problem at a single site is usually concerned with meeting fairly specific production requirements. Customer orders, stock imperatives or higher-level supply chain or long-term planning would usually set these requirements, as described in subsequent sections. It is concerned with the allocation over time of scarce resources between competing activities to meet these requirements in an efficient fashion. The data required to describe the scheduling problem typically include:

- Production recipes: details of how each product is to be produced, including details of production of intermediates. This will include material balance information, resource requirements, processing times/rates of the process tasks, etc.

- Resource data: for process equipment, storage equipment, utilities (capacities, capabilities, availabilities, costs, etc.)
- Material data: stability, opening inventories, anticipated receipts of raw materials/intermediates.
- Demand data: time horizon of interest, firm orders, forecasted demands, sales prices.

The key components of the scheduling problem are resources, tasks and time. The resources need not be limited to processing equipment items, but may include material storage equipment, transportation equipment (intra- and interplant), operators, utilities (e.g., steam, electricity, cooling water), auxiliary devices and so on.

The tasks typically comprise processing operations (e.g., reaction, separation, blending, packaging) as well as other activities that change the nature of materials and other resources such as transportation, quality control, cleaning, changeovers, etc.

There are both external and internal elements to the time component. The external element arises out of the need to coordinate manufacturing and inventory with expected product liftings or demands, as well as scheduled raw material receipts and even service outages. The internal element relates to executing the tasks in an appropriate sequence and at the right times, taking account of the external time events and resource availabilities.

Overall, this arrangement of tasks over time and the assignment of appropriate resources to the tasks in a resource-constrained framework must be performed in an efficient fashion, which implies the optimization, as far as possible, of some objective. Typical objectives include the minimization of cost or maximization of profit, maximization of customer satisfaction, minimization of deviation from target performance, etc. Generally speaking, depending on raw material lead times, production lead times, forecast accuracy and other similar factors, production scheduling is driven either by firm customer orders ("make-to-order") or forecasted demands ("make-to-stock").

As noted by Gabow (1983), all but the most trivial scheduling problems belong to the class of NP hard (Non-deterministic Polynomial-time hard) problems; there are no known solution algorithms that are of polynomial complexity in the problem size. This has posed a great challenge to the research community, and a large body of work aiming to develop either tailored algorithms for specific problem instances or efficient general-purpose methods has arisen.

Solving the scheduling problem requires methods that search through the decision space of possible solutions. The search processes can be classified as follows:

- Heuristic: a series of rules (e.g., the sequence of production should be based on order due-dates) are used to generate alternative schedules.
- Metaheuristic: higher level generic search algorithms (e.g., simulated annealing, genetic algorithms) are used to explore the decision space.
- Mathematical programming: the scheduling problem is posed as a formal mathematical optimization problem and solved using general-purpose or tailored methods (see section 4.2).

The research into production scheduling techniques may be further subdivided into specific and general application domains. The latter division is intended to reflect the scope of the technique (in terms of plant structure and process recipes). Rippin (1993) classified different flexible plant structures as follows:

- Multiproduct plants, where each product has the same processing network, i.e., each product requires the same sequence of processing tasks (often known as "stages"). Owing to the historic association between the work on batch plant scheduling and that on discrete parts manufacturing, these plants are sometimes called "flowshops".
- Multipurpose plants ("jobshops"), where the products are manufactured via different processing networks, and there may be more than one way in which to manufacture the same product. In general, a number of products undergo manufacture at any given time.

In addition to the process structure, the storage policies for intermediate materials are critical in production scheduling, especially for batch plants. Any intermediate material can usually be classified as being subject to one of five intermediate storage policies:

- Zero-wait (ZW): the material is not stable and must be processed further upon production.
- No intermediate storage (NIS): the material is stable, but no storage vessels are provided. However, it may reside temporarily in the processing equipment that produced it before being processed further.
- Shared intermediate storage (SIS): the material is stable, and may be stored in one or more storage vessels that may also be used to store other materials (though not at the same time).
- Finite intermediate storage (FIS): the material is stable, and one or more dedicated storage vessels are available.
- Unlimited intermediate storage (UIS): the material is stable, and one or more dedicated storage vessels are available, the total capacity of which is effectively unlimited.

The importance of these is clearly evident from the following example. Consider a process whereby a material C, is made from raw material A, via the following reactions:

$A \rightarrow B$ (reaction 1, duration 3 h)
$B \rightarrow C$ (reaction 2, duration 1 h)

Reaction 1 takes place in reactor 1 (capacity 10.000 kg) and reaction 2 takes place in reactor 2 (capacity 5000 kg). The average production rate of C depends strongly on the storage policy for B. The rates for the ZW, NIS and UIS cases are calculated below.

2.2.1
ZW Case

Here, only 5000 kg of A can be loaded into reactor 1, because once this batch is complete, it must be immediately transferred to reactor 2, which limits the size of the batch. A sample operating schedule is shown in Figure 2.2.

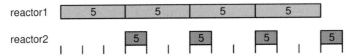

Figure 2.2   Sample schedule for the zero-wait (ZW) case

According to the schedule, 5000 kg of C is produced every 3 h, so the average production rate is 1667 kg h^{-1}.

2.2.2
NIS Case

Here, 10.000 kg of A can be loaded into reactor 1. After the reaction is complete, 5000 kg of B can be transferred to reactor 2, and 5000 is held in reactor 1 for an extra hour before being transferred to reactor 2. A sample operating schedule is shown in Figure 2.3.

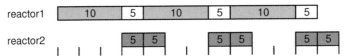

Figure 2.3   Sample schedule for the no intermediate storage (NIS) case

In this case, 10.000 kg of C is produced every 4 h, so the average production rate is 2500 kg h^{-1}.

2.2.3
UIS Case

In this case, there is sufficient storage (e.g., 10.000 kg) to decouple the operation of reactor 1 and reactor 2 completely. The production rate is then limited by the bottleneck stage; in this case reactor 1. A sample operating schedule is shown in Figure 2.4.

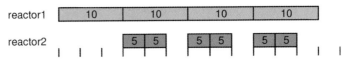

Figure 2.4   Sample schedule for the unlimited intermediate storage (UIS) case

Here, 10.000 kg of C is produced every 3 h; the production rate is 3333 kg h^{-1}.

The above discussion serves to define categories for scheduling techniques and categories for process structures.

The next sections review developments in the solution of scheduling problems and are organized along the categories listed above.

2.3
Heuristics/Metaheuristics: Specific Processes

Most scheduling heuristics are concerned with formulating rules for determining sequences of activities. They are therefore best suited to processes where the production of a product involves a prespecified sequence of tasks with fixed batch sizes; in other words, variants of multiproduct processes. Often, it is assumed that fixing the front-end product sequence will fix the sequence of activities in the plant (the so-called permutation schedule assumption; see Figure 2.5). Generally, the processing of a product is broken down into a sequence of jobs that queue for machines, and the rules dictate the priority order of the jobs.

Dannebring (1977), Kuriyan and Reklaitis (1985, 1989) and Pinedo (1995) give a good exposition on the kinds of heuristics (dispatching rules) that may be used for different plant structures. Typical rules involve ordering products (see, e.g., Hasebe et al. 1991) by processing time (either shortest or longest), due dates and so on.

Most of the heuristic methods originated in the discrete manufacturing industries, and might be expected not to perform as well in process industry problems, because in the latter material is infinitely divisible and batch sizes are variable (unlike discrete "jobs"). Furthermore, batch splitting and mixing are allowed and are becoming increasingly popular as a means of effecting late product differentiation.

Stochastic search approaches ("metaheuristics") are based on continual improvement of trial solutions by the application of an evolutionary algorithm which modifies solutions and prioritizes solutions from a list for further consideration. The two main evolutionary algorithms applied to this area are simulated annealing and genetic algorithms. An early application of simulated annealing to batch process scheduling problems was undertaken by Ku and Karimi (1991), where they applied the algorithm to multiproduct plant scheduling. They concluded that such algorithms are easy to implement and tended to perform better than conventional heuristics, but often required significant computational effort.

Xia and Macchietto (1997) described the application of simulated annealing and genetic algorithm techniques to the scheduling of multiproduct plants with complex material transfer policies. More recently, Murakami et al. (1997) described a repetitive simulated annealing procedure which avoids local minima by using many starting points with fewer evolutionary iterations per starting point.

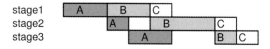

Figure 2.5 Permutation schedule

Sunol et al. (1992) described the application of a genetic algorithm approach to a simple flowshop sequencing problem, and found the technique to be superior to explicit enumeration. As noted by Hasebe et al. (1996), the performance of a genetic algorithm depends on the operators used to modify trial solutions. They applied a technique that selects appropriate operators during the solution procedure for the scheduling of a parallel-unit process.

Overall, the stochastic search processes are best applied to problems of an entirely discrete nature, where an objective function can be evaluated quickly. The classic example is the sequencing and timing of batches in a multiproduct plant, where the decision variables are the sequence of product batches, and the completion time of any candidate solution is easily evaluated through recurrence relations or minimax algebra. The main disadvantages are that it is difficult to consider general processes, and inequality constraints and continuous decisions, although some recent work (e.g., Wang et al. 2000) aims at addressing this.

2.4
Heuristics/Metaheuristics: General Processes

The problem of scheduling in general multipurpose plants is complicated by the additional decisions (beyond the sequencing of product batches) of assignment of equipment items to processing tasks, task batch sizes and intermediate storage utilization. It is difficult to devise a series of rules to resolve these, and there are therefore few heuristic approaches reported for the solution of this problem.

Kudva et al. (1994) consider the special case of "linear" multipurpose plants, where products flow through the plant in a similar fashion, but potentially using different stages and with no recycling of material. A rule-based constructive heuristic is used, which requires the maintenance of a status sheet on each unit and material type for each time instance on a discrete-time grid. The algorithm uses this status sheet with a sorted list of orders and develops a schedule for each order by backwards recursive propagation. The schedule derived depends strongly on the order sequence. Solutions were found to be within acceptable bounds of optimality when compared with those derived through formal optimization procedures.

Graells et al. (1996) presented a heuristic strategy for the scheduling of multipurpose batch plants with mixed intermediate storage policies. A decomposition procedure is employed where subschedules are generated for the production of intermediate materials. Each subschedule consists of a mini production path determined through a branch-and-cut enumeration of possible unit-to-task allocations. The mini-paths are then combined to form the overall schedule. The overall schedule is checked for feasibility with respect to material balances and storage capacities. Improvements to the schedules may be effected manually through an electronic Gantt chart or through a simulated annealing procedure.

Lee and Malone (2000) describe the application of a simulated annealing metaheuristic to a variety of batch process planning problems. Here, intermediate products, inventory costs and a variety of process flow networks can be represented.

As mentioned earlier, the application of heuristics to such problems is not straightforward. Although this effectively represents current industrial practice, most academic research has been directed towards the development of mathematical programming approaches for multipurpose plant scheduling. As will be described later, these approaches are capable of representing all the complex interactions present.

2.5
Mathematical Programming: Specific Processes

Here, we shall first outline some of the features of mathematical programming approaches in general, and then consider their application to processes other than the general multipurpose one. The latter will be considered in the Section 2.6. Mathematical programming approaches to production scheduling in the process industries have received a large amount of attention recently. This is because they bring the promise of generality (i.e., ability to deal with a wide variety of problems), rigor (the avoidance of approximations) and the possibility of achieving optimal or near-optimal solutions.

The application of mathematical programming approaches implies the development of a mathematical model and an optimization algorithm. Most approaches aim to develop models that are of a standard form (from linear programming (LP) models for refinery planning to mixed-integer nonlinear programming (MINLP) models for multipurpose batch plant scheduling). These may then be solved by standard software or specialized algorithms that take account of problem structure.

The variables of the mathematical models will tend to include some or all of the following choices, depending on the complexity considered:

- sequence of products or individual tasks
- timing of individual tasks in the process
- selection of resources to execute tasks at the appropriate times
- amounts processed in each task
- inventory levels of all materials over time.

The discrete nature of some of the variables (sequencing and resource selection) implies that binary or integer-valued variables will be required.

The selection of values for all the variables will be subject to some or all of the following constraints:

- nonpreemptive processing: once started, processing activities must proceed until completion;
- resource constraints: at any time, the utilization of a resource must not exceed its availability;
- material balances;
- capacity constraints: processing and storage;
- orders being met in full by their due dates.

Finally, optimization methods dictate that an objective function be defined. This is usually of an economic form, involving terms such as production, transition and inventory costs and possibly revenues from product sales.

A very good review of mathematical programming techniques applied to scheduling and an associated classification of methods and models is provided by Pinto and Grossmann (1998).

A critical feature of mathematical programming approaches is the representation of the time horizon. This is important because activities interact through the use of resources; therefore, the discontinuities in the overall resource utilization profiles must be tracked with time, to be compared with resource availabilities to ensure feasibility. The complexity arises because these discontinuities (unlike discontinuities in availabilities) are functions of any schedule proposed and are not known in advance. The two approaches for dealing with this are:

- Discrete-time (or "uniform discretization"): the horizon is divided into a number of equally spaced intervals so that any event that introduces such discontinuities (e.g., the starting of a task or a due date for an order) can only take place at an interval boundary. This implies a relatively fine division of the time grid, so as to capture all the possible event times, and in the solution to the problem it is likely that many grid points will not actually exhibit resource utilization discontinuities.
- Continuous time (or "nonuniform discretization"): here, the horizon is divided into fewer intervals, the spacing of which will be determined as part of the solution to the problem. The number of intervals will correspond more closely to the number of resource utilization discontinuities in the solution.

In addition to the above, another attribute of time representation is whether the same grid is used for all major equipment items in the plant (the "common grid" approach) or whether each major equipment item operates on its own grid (the "individual resource grid", only used with continuous-time models). Generally speaking, the former approach is more suitable for processes in which activities on the major equipment items also interact with common resources (materials, services, etc.) and the latter where activities on the major equipment items are quite independent in their interactions with common resources. These distinctions will become clearer when individual pieces of research are discussed. The distinctions between these representations are shown in Figures 2.6 and 2.7.

The simplest specific scheduling process is probably a single production line which produces one product at time in a continuous fashion. Work in this area has been directed towards deriving cyclic schedules (where the production pattern is

Figure 2.6 Discrete-time representation

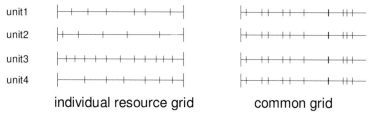

unit1

unit2

unit3

unit4

individual resource grid   common grid

Figure 2.7   Continuous-time representations

repeated at a fixed frequency) that balance inventory and transition costs by determining the best sequence of products and their associated run-lengths or lot-sizes. A review of this so-called economic lot scheduling problem is given by Elmaghraby (1978).

Sahinidis and Grossmann (1991) consider the more general problem of the cyclic scheduling of a number of parallel multiproduct lines, where each product may in principle be produced on more than one line and production rates and costs vary between lines. They utilize a continuous-time individual resource grid model, which turns out to be a MINLP. This includes an objective function that includes combined production, product transition and inventory costs for a constant demand rate for all products. Their work was extended by Pinto and Grossmann (1994), who considered the case of multiple production lines, each consisting of a series of stages decoupled by intermediate storage and operating in a cyclic mode. Each product is processed through all stages, and each product is processed only once at each stage. The model again uses a continuous-time model, and it is possible to use the independent grid approach despite the fact that stages interact through material balances; this is due to the special structure of the problem.

A number of mathematical programming approaches have been developed for the scheduling of multiproduct batch plants. All are based (either explicitly or implicitly) on a continuous representation of time.

Pekny et al. (1988) considered the special case of a multiproduct plant with no storage (zero wait (ZW)) between operations. They show that the scheduling problem has the same structure as the asymmetric traveling salesman problem, and apply an exact parallel computation technique employing a tailored branch-and-bound procedure which uses an assignment problem to provide problem relaxations. The work was extended to cover the case of product transition costs, where the problem structure is equivalent to the prize-collecting traveling salesman problem (Pekny et al. 1990), and LP relaxations are used. For both cases, problems of very large magnitude were solved to optimality with modest computational effort. Gooding et al. (1994) augmented this work to cover the case of multiple units at each stage (the so-called "parallel flowshop" stage).

A more complete overview of the development of algorithms for classes of problems ("algorithm engineering") is given by Applequist et al. (1997) and a commercial development in this area is described by Bunch (1997).

Birewar and Grossmann (1989) developed a mixed-integer programming model for a similar type of plant. They show that through careful modeling of slack times, and by exploiting the fact that relatively large numbers of batches of relatively few products will be produced (which allows end-effects to be ignored), a straightforward LP model can be used to minimize the makespan. The result is a family of schedules from which an individual schedule may be extracted. They extend the work to cover simultaneous long-term planning and scheduling, where the planning function takes account of scheduling limitations (Birewar and Grossmann 1990).

Pinto and Grossmann (1995) describe a mixed-integer linear programming (MILP) model for the minimization of earliness of orders for a multiproduct plant with multiple equipment items at each stage. The only resources required for production are the processing units. Pinto and Grossmann (1997) then augmented the model to take account of interactions between processing stages and common resources (e.g., steam). Rather than utilize a common grid, they retained individual grids, and accounted for the resource discontinuities through complex mixed-integer constraints which weakened the model and resulted in large computational times. They therefore proposed a hybrid logic-based/MILP algorithm where the disjunctions relate to the relative timing of orders. This dramatically reduces the computational effort expended.

Moon et al. (1996) also developed a MILP model for ZW multiproduct plants. The objective was to assign tasks to sequence positions so as to minimize the makespan, with nonzero transfer and set-up times being included.

The extension of the work to more general intermediate storage policies was described by Kim et al. (1996), who proposed several MINLP formulations based on completion time relations.

The case of single-stage processes with multiple units per stage has been considered by Cerda et al. (1997) and McDonald and Karimi (McDonald and Karimi, 1997; Karimi and McDonald, 1997). Both describe continuous-time-based MILP models. Cerda et al. focus on changeovers and order fulfilment, while Karimi and McDonald focus on semicontinuous processes and total cost (transition, shortage and inventory) with the complication of minimum run lengths. A characteristic of both approaches is that discrete demands must be captured on the continuous-time grid.

Méndez and Cerdá (2000) developed a MILP model for a process with a single production stage with parallel units followed by a storage stage with multiple units, with restricted connectivity between the stages. This was extended to the multistage case with general production resources by Méndez et al. (2001). In common with other models, there are no explicit time slots in the model; the key variables are allocations of activities to units and the relative orderings of activities.

The work described above all relates to special process structures, which means that mathematical models can be designed specifically for the problem class. This ensures that, despite the typical concerns about computational complexity of discrete optimization problems, solutions are available with reasonable effort. The drawback of the work is its limited applicability. Nevertheless, several models appear to have been developed with specific industrial applications in mind (e.g., Sahinidis and Grossmann (1991a), Pinto and Grossmann (1995) and Karimi and McDonald (1997)).

2.6
Mathematical Programming: Multipurpose Plants

A large portion of the most recent research in planning and scheduling undertaken by the process systems community relates to the development of mathematical programming approaches applied to multipurpose plants. As intimated earlier, in this case the application domain tends to imply the solution approach: mathematical models are the best way of representing the complex interactions between resource allocations, task timings, material flows and equipment capacities. Much of the recent work reported in the literature deals with this class of problem.

The work in this are can be characterized by three different assumptions about plant operation:

- The unique assignment case: each task can only be performed by a unique piece of equipment, and there are no optional tasks in the process recipe and batch sizes are usually fixed.
- The campaign mode of operation: the horizon is divided into relatively long campaigns, and each campaign is dedicated to one or a few products.
- Short-term operation: products are produced as required and no particular scheduling pattern may be assumed.

The first assumption is particularly restrictive. The second relates to a mode of operation that is becoming relatively scarce, as it implies a low level of responsiveness. One sector in which campaign operation is still prevalent is in the manufacture of active ingredients for pharmaceuticals and agrochemicals. The short-term mode of operation is tending to become the most prevalent elsewhere, as it best exploits operational flexibility to meet changing external circumstances.

Mauderli and Rippin (1979) developed a procedure for campaign planning which attempts to optimize the allocation of equipment to tasks. An enumerative procedure (based on different equipment-to-task allocations) is used to generate possible single-product campaigns which are then screened by LP techniques to select the dominant ones. A production plan is then developed by the solution of a MILP that sequences the dominant campaigns and fixes their lengths. The disadvantages of this work are the inefficiency of the generation procedure and the lower level of resource utilization implied by single-product campaigns. Wellons and Reklaitis (1991a,b) addressed this through a formal MINLP method to generate campaigns and production plans in a two-stage procedure, as did Shah and Pantelides (1991) who solved a simultaneous campaign generation and production planning problem.

An early application of mathematical programming techniques for short-term multipurpose plant scheduling was the MILP approach of Kondili et al. (1988). They used a discrete representation of time, and introduced the state-task network (STN) representation of the process (see Figure 2.8).

The STN representation has three main advantages:

- It distinguishes the process operations from the resources that may be used to execute them, and therefore provides a conceptual platform from which to relax the unique assignment assumption and optimize unit-to-task allocation.

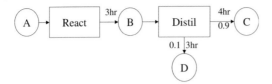

Figure 2.8 Example of a state-task network. *Circles*: material states, *rectangles*: tasks

- It avoids the use of task precedence relations, which become very complicated in multipurpose plants: a task can be scheduled to start if its input materials are available in the correct amounts and other resources (processing equipment and utilities) are also available, regardless of the plant history.
- It provides a means of describing very general process recipes, involving batch splitting and mixing and material recycles, and storage policies including ZW, NIS, SIS and so on.

The formulation of Kondili et al. (1988) (described in more detail in Kondili et al. (1993)) is based on the definition of binary variables that indicate whether tasks start in specific pieces of equipment at the start of each time period, together with associated continuous batch sizes. Other key variables are the amount of material in each state held in dedicated storage over each time interval, and the amount of each utility required for processing tasks over each time interval.

Their key constraints related to equipment and utility usage, material balances and capacity constraints. The common, discrete-time grid captures all the plant resource utilizations in a straightforward manner; discontinuities in these are forced to occur at the predefined interval boundaries. Their approach was hindered in its ability to handle large problems by the weakness of the allocation constraints and the general limitations of discrete-time approaches, such as the need for relatively large numbers of grid points to represent activities with significantly different durations.

Their work formed the basis of several other pieces of research aiming to take advantage of the representational capabilities of the formulation while improving its numerical performance. Sahinidis and Grossmann (1991b) disaggregated the allocation constraints and also exploited the embedded lot-sizing nature of the model where relatively small demands are distributed throughout the horizon. They disaggregate the model in a fashion similar to that of Krarup and Bilde (1977), who were able to improve the solution efficiency despite the larger nature of the disaggregated model. This was due to a feature particular to mixed-integer problems: other things being equal, the computational effort for problem solution through standard procedures is dictated mainly by the difference between the optimal objective function and the value of the objective function obtained by solving the continuous relaxation where bound constraints rather than integrality restrictions are imposed on the integer variables (the so-called "integrality gap"). The formulation of Sahinidis and Grossmann (1991b) was demonstrated to have a much smaller integrality gap than the original.

Shah et al. (1993a) modified the allocation constraints even further to generate the smallest possible integrality gap for the type of formulation. They also devised a tailored branch-and-bound solution procedure which utilizes a much smaller LP relaxation and solution processing to improve integrality at each node. The same authors (Shah et al. 1993b) considered the extension to cyclic scheduling, where the same schedule is repeated at a frequency to be determined as part of the optimization. This was augmented by Papageorgiou and Pantelides (1996a,b) to cover the case of multiple campaigns, each with a cyclic schedule to be determined.

Elkamel (1993) also proposed a number of measures to improve the performance of the STN-based discrete-time scheduling model. A heuristic decomposition method was proposed, which solves separate scheduling problems for parts of the overall scheduling problem. The decomposition may be based on the resources ("longitudinal decomposition") or on time ("axial decomposition"). In the former, the recipes and suitable equipment for each task are examined for the possible formation of unique task-unit subgroups which can be scheduled separately. Axial decomposition is based on grouping products by due dates and decomposing the horizon into a series of smaller time periods, each concerned with the satisfaction of demands falling due within it. He also described a perturbation heuristic, which is a form of local search around the relaxation. Elkamel and Al-Enezi (1998) describe valid inequalities that tighten the MILP relaxations of this class of model.

Yee and Shah (1997, 1998) and Yee (1998) also considered various manipulations to improve the performance of general discrete-time scheduling models. A major feature of their work is variable elimination. They recognize that in such models, only about 5–15 % of the variables reflecting task-to-unit allocations are active at the integer solution, and it would be beneficial to identify as far as possible inactive variables prior to solution. They describe a LP-based heuristic, a flexibility and sequence reduction technique and a formal branch-and-price method. They also recognize that some problem instances result in poor relaxations and propose valid inequalities and a disaggregation procedure similar to that of Sahinidis and Grossmann (1991b) for particular data instances (Romero and Puigjaner (2004)). Bassett et al. (1996) and Dimitriadis et al. (1997a, 1997b) describe decompostion-based approaches which solve the problems in stages, eventually generating a complete solution.

Blömer and Günther (1998) also introduced a series of LP-based heuristics that can reduce solution times considerably, without compromising the quality of the solution obtained.

Grunow et al. (2002) show how the STN tasks can be aggregated into higher level processes for the purposes of longer-term campaign planning.

Gooding (1994) considers a special case of the problem with firm demands and dedicated storage only. The scheduling model is described in a digraph form where nodes correspond to possible task-unit-time allocations and arcs the possible sequences of the activities. The explicit description of the sequence in this form addresses one of the weaknesses of the discrete-time formulation of Kondili et al. (1988, 1993), which was that it did not model sequence-dependent changeovers very well. Gooding's (1994) model therefore performed relatively well in problems with a strong sequencing component, but suffers from model complexity in that all possible sequences must be accounted for directly.

Pantelides et al. (1995) reported a STN-based approach to the scheduling of pipe-less plants, where material is conveyed between processing stations in movable vessels. This requires the simultaneous scheduling of the movement and processing operations.

Pantelides (1994) presented a critique of the STN and associated scheduling formulations. He argued that despite its advantages, it suffers from a number of drawbacks:

- The model of plant operation is somewhat restricted: each operation is assumed to use exactly one major item of equipment throughout its operation.
- Tasks are always assumed to be processing activities which change material states: changeovers or transportation activities have to be treated as special cases.
- Each item of equipment is treated as a distinct entity: this introduces solution degeneracy if multiple equivalent items exist.
- Different resources (materials, units, utilities) are treated differently, giving rise to many different types of constraints, each of which must be formulated carefully to avoid unnecessarily increasing the integrality gap.

He then proposed an alternative representation, the resource-task network (RTN), based on a uniform description of all resources (Figure 2.9). In contrast to the STN approach, where a task consumes and produces materials while using equipment and utilities during its execution, in this representation, a task is assumed only to consume and produce resources. Processing items are treated as though consumed at the start of a task and produced at the end. Furthermore, processing equipment in different conditions (e.g., "clean" or "dirty") can be treated as different resources, with different activities (e.g., "processing" or "cleaning") consuming and generating them › this enables a simple representation of changeover activities.

Pantelides (1994) also proposed a discrete-time scheduling formulation based on the RTN that, due to the uniform treatment of resources, only requires the description of three types of constraint, and does not distinguish between identical equipment items (which results in more compact and less degenerate optimization

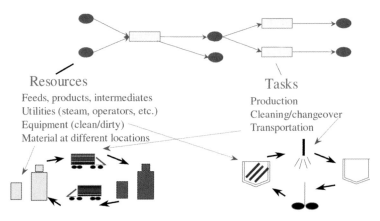

Figure 2.9 The resource-task network representation

models). He illustrated that the integrality gap could not be worse than the most efficient form of STN formulation, but that the ability to capture additional problem features in a straightforward fashion made it an ideal framework for future research.

The review above has mainly considered the development of discrete-time models. As argued by Schilling (1997), discrete-time models have been able to solve a large number of industrially relevant problems (see, e.g., Tahmassebi 1996), but suffer from a number of inherent drawbacks:

- The discretization interval must be fine enough to capture all significant events; this may result in a very large model.
- It is difficult to model operations where the processing time is dependent on the batch size.
- The modeling of continuous and semicontinuous operations must be approximated, and minimum run-lengths give rise to complicated constraints.

A number of researchers have therefore attempted to develop scheduling models for multipurpose plants which are based on a continuous representation of time, where fewer grid points are required as they will be placed at the appropriate resource utilization discontinuities during problem solution.

Zentner and Reklaitis (1992) described a formulation based on the unique assignment case and fixed batch sizes. The sequence of activities as well as any external effects can be used to infer the discontinuities and therefore the interval boundaries. A MILP optimization is then used to determine the exact task starting times.

Reklaitis and Mockus (1995) detailed a continuous-time formulation based on the STN formulation, and exploiting its generality. A common resource grid is used, with the timing of the grid points ("event orders" in their terminology) determined by the optimization. The model is a MINLP, which may be simplified to a mixed-integer bilinear problem by linearizing terms involving binary variables. This is solved using an outer-approximation algorithm. Only very preliminary findings were reported, but the promise of such models is evident.

Mockus and Reklaitis (1996) then reported an alternative solution procedure. They introduce the concept of Bayesian heuristics, which are heuristics that can be described through parameterized functions. The Bayesian technique iteratively modifies the parameters to develop a heuristic that is expected to perform well across a class of problem parameters. They illustrate the procedure using a material requirements planning (MRP) backward-scheduling heuristic which outperforms a standard discrete-time MILP formulation solved using branch-and-bound. They extend this work (Mockus and Reklaitis 1999a,b) to the case where a variety of heuristics are used in combination with optimization).

Zhang and Sargent (1994, 1996) presented a continuous-time formulation based on the RTN representation for both batch and continuous operations, with the possibility of batch-size-dependent processing times for batch operations. Again, the interval durations are determined as part of the optimization. A MINLP model ensues; this is solved using a local linearization procedure combined with what is effectively a column generation algorithm.

A problem with continuous-time models of the form described above arises from the inclusion of products of binary variables and interval durations or absolute starting times in the constraints. The linearization of these products gives rise to terms involving products of binary variables and maximum predicted interval durations or starting times. The looser these upper bounds, the worse the integrality gap of the formulation and, in general, the more difficult it becomes to solve the scheduling problem. Furthermore, it is difficult to predict good duration bounds *a priori*. The poor relaxation performance of the continuous-time models is the main obstacle to their more widespread application.

Schilling and Pantelides (1996) and Schilling (1997) attempted to address this deficiency. They developed a continuous-time scheduling model based on the RTN. They proposed a number of modifications to the formulation of Zhang and Sargent (1996) ,which simplify the model and improve its general solution characteristics. A global linearization gives rise to a MILP. They then developed a hybrid branch-and-bound solution procedure which branches in the space of the interval durations as well as in the space of the integer variables. For a given problem instance, this can be viewed as generating a number of problem instances, each with tighter interval duration bounds. The independence of these new instances was recognized by Schilling (1997), who implemented a parallel solution procedure based on a distributed computing environment. The combination of the hybrid and parallel aspects of the solution procedure resulted in a much improved computational performance on a wide class of problems. Their model and solution procedure was then extended to the cyclic scheduling case (Schilling and Pantelides 1999). Castro et al. (2001) made some adjustments to the model to account for stable materials that can be held temporarily in processing units; these improve the computational performance considerably.

Ierapetritou and Floudas (1998a–c) and Ierapetritou et al. (1999) introduced a new continuous-time model where the task and unit events are not directly coordinated against the same grid, but have their own grids (i.e., individual resource grids). Sequencing and timing constraints are then introduced to ensure feasibility. This has the effect of reducing the model size and reducing the associated computational effort required to find a solution. Their model is able to deal with semicontinuous processes and products with due dates falling arbitrarily within the horizon. Lin and Floudas (2001) extended this work to cover simultaneous design and scheduling.

Wu and Ierapetritou (2003) employed time-, recipe- and resource-based decomposition procedures to this class of model. They indicated that near-optimal solutions may be obtained with modest effort. The work was generalized further by Janak et al. (2004) who included mixed storage policies, batch-size-dependent processing times, general resource constraints and sequence-dependent changeover constraints.

Lee et al. (2001) extended this body of work with a formulation that uses binary variables to represent the start, process and end components of a process task. The computational performance of their model is similar to that of Castro et al. (2001).

Orcun et al. (2001) used the concept of operations through which batches flow to develop a continuous-time model. Each batch has a prespecified alternative set of recipes for its manufacture; each recipe defines the flow through the operations. One of the major decisions is then the choice of recipe for each batch. The complicating constraints are those that ensure the timing and sequence of batches and operations are feasible.

Majozi and Zhu (2001) modified the STN concept by removing tasks and units; thereby generating a state-sequence network (SSN); essentially a state-transition network. They developed a continuous-time model based on this, which relies on specialized sequencing constraints to ensure feasibility. Resources other than processing units cannot be treated.

Giannelos and Georgiadis (2002) described a very straightforward continuous-time model for multipurpose plant scheduling based on the STN process representation. In their work, they introduced "buffer times", which means that although all tasks must start at event points, they do not need to finish on event points. This reduced synchronization improves the computational performance considerably when compared to similar mathematical models.

Giannelos and Georgiadis (2003) applied their continuous-time model to the scheduling of consumer goods factories. The latter are characterized by sequence-dependent changeovers and flexible intermediate storage. By preprocessing the data, good upper bounds on the number of changeover tasks can be estimated and used to tighten the MILP model. Good solutions were found for the case of a medium-size industrial study.

Maravelias and Grossmann (2003) used a mixed time representation, where tasks which produce ZW states must start and finish at slot boundaries, while others are only anchored at the start, as per Giannelos and Georgiadis (2002). A different set of binary variables and assignment constraints is used from other works in this area. General resource constraints, sequence-dependent cleaning, and variable processing times are also included. Good computational performance is observed.

Castro et al. (2003) investigated both discrete- and continuous-time RTN models for periodic scheduling applied to a pulp and paper case study. A coarse and then a fine grid is used with the discrete-time model to optimize the cycle time, and an iterative search on the number of event points is used in the continuous-time model. The discrete-time model allowed more flexibility in the statement of the objective function and is easier to solve, while the continuous-time model in principle allows more accurate modeling of the operations.

Castro et al. (2004) presented a simple model for both batch and continuous processes which has an improved LP relaxation compared to that of Castro et al. (2001), due to a different set of timing constraints; this model generally outperforms other continuous-time models proposed in the literature.

Overall, considerable progress has been made towards the development of general-purpose mathematical programming-based methods for process scheduling. At least two commercial packages ModelEnterprise (see http://www.psenterprise.com/products-me.html) and Verdict (see http://www.combination.com) have resulted from these academic endeavours.

2.7
Hybrid Solution Approaches

So far, we have described individual solution approaches. A new class of solution methods has arisen out of the recognition that mathematical programming approaches are very effective when the scheduling problems are dominated by flow-type decisions, but often struggle when sequencing decisions dominate. Hybrid approaches decompose the problem into flow and sequence-based components and then apply different algorithms to these components. An example of this is described by Neumann et al. (2002), who solve separate batching and scheduling problems. The batching problem is a mixed-integer optimization problem which determines the number of instances of each task on each unit. The scheduling problem then determines the sequence and timing of the batches. A tailored algorithm based on project scheduling concepts is used for the scheduling/sequencing problem (Schwindt and Trautmann 2000).

Most of the other hybrid methods are based on the same decomposition principle, but recognize that constrained logic programming (CLP) (also called constraint programming) is a powerful technique for the solution of sequencing problems. It is based on the concept of domain reduction and constraint propagation (van Hentenryck 1989).

Jain and Grossmann (2001) use a single, hybrid model where different degrees of freedom are determined by the two solvers. Effectively, this requires an iteration between MILP and CLP solvers which proceeds until an optimal and feasible solution is found. A specific (parallel flowshop), rather than general process is studied.

Harjunkoski et al. (2000) extend this work by developing a solution procedure which starts with the relaxed MILP and then iterates through different CLP solutions, each with a different objective function target. This approach performed better on a trim-loss problem than a traditional jobshop scheduling problem as tackled by Jain and Grossmann (2001).

Huang and Chung (2000) explained how CLP can be used on its own (along with some dispatching rules) for the scheduling of a simple pipeless batch plant.

Maravelias and Grossmann (2004) and Roe et al. (2003, 2004) also recognized that mixed-integer optimization techniques are appropriate for the batching problem, and constrained logic programming (CLP) approaches may be appropriate for the scheduling problem.

Roe et al. (2003, 2005) developed an algorithm appropriate to all types of processes. A STN-based description is used. A "batching" optimization is solved to determine the number of allocations of STN tasks to units and the average batch sizes, and then a tailored algorithm based on the ECLiPse framework (Wallace et al. 1997) is used to derive the schedule which aims to ensure completion of all tasks in the minimum possible time. A series of constraints are introduced in the batching problem (as per Maravelias and Grossmann 2004) to try to ensure schedule feasibility. Their approach has a single pass, while that of Maravelias and Grossmann (2004) is more sophisticated in that the algorithm iterates between MILP and CLP levels, and the two solution methods deal with different parts of a single model (essentially that of

Maravelias and Grossmann 2003), which allows for variable batch sizes. The CLP level, rather than just adding simple integer cuts, adds a more general type of cut that excludes similar permutations of the solution to be excluded.

Romero et al. (2004) used a graph theoretical framework to tackle scheduling problems which involve complex shared intermediate storage systems. Their representation is based on two types of graph: the recipe graph which depicts possible material flow routes, and the schedule graph which shows a unique solution to the scheduling problem. A branch-and-bound algorithm is used to search for optimal solutions.

2.8
Combined Scheduling and Process Operation

A feature of scheduling problems is that the representation of the production process depends on the gross margin of the business. Businesses with reasonable to large gross margins (e.g., consumer goods, specialties) tend to use "recipe-based" representations, where processes are operated at fixed conditions and to fixed recipes. Recipes may also be fixed by regulation (e.g., pharmaceuticals) or because of poor process knowledge (e.g., food processing). On the other hand, businesses with slimmer margins (e.g., refining, petrochemicals) are moving towards "property-based" representations, where process conditions and (crude) process models are used in the process representation, and stream properties are inferred from process conditions and mixing rules. Hence some degrees of freedom associated with process operation are optimized during production scheduling. Some examples of this type of process are described below.

Castro et al. (2002) described the use of both dynamic simulation and scheduling for the improved operation of the digester part of a pulp mill. A dynamic model is used to determine task durations and an RTN-based model is used for scheduling. The schedule optimization indicated that steam availability limits throughput. Alternative task combinations based on different steam sharing options were generated through the detailed modeling, and these were made available to the scheduling model. This then (approximately) enables the scheduling model to optimize the details of process operation.

Glismann and Gruhn (2001) describe a model which combines scheduling with nonlinear recipe blend optimization. Here, a long-range planning problem using nonlinear programming identifies products to be produced and different blending recipes to produce them. A short-term RTN-based scheduling model then schedules the blending activities in detail. Deviations between plan and schedule can then be reduced in a further step.

Alle and Pinto (2002) developed a model for the cyclic scheduling of continuous plants including operational variables such as processing rates and yields. This enables the optimization procedure to trade off time and material efficiencies. A tailored algorithm is used to find the global optimum of this mixed-integer nonconvex optimization problem.

A different type of operational consideration is performance degradation over time. Jain and Grossmann (1998) describe the cyclic scheduling of an ethylene cracking process. Here, the conversion falls with time, until a cleaning activity is undertaken to restore the cracker to peak performance. There is a tradeoff between frequent cleaning (and high downtime) and high average performance and infrequent cleaning (and lower downtime) and lower average performance. The problem is complicated by the presence of multiple furnaces and model nonlinearity.

Joly and Pinto (2003) describe a discrete-time MINLP model for the scheduling of fuel oil and asphalt production at a large Brazilian refinery. The nonlinear operational component comes from the calculation of the viscosity through variable flow rates rather than through fixed recipes. Because the viscosity specifications are fixed, a linear model can be derived from the MINLP and solved to global optimality.

Pinto et al. (2000) and Joly et al. (2002) described a refinery planning model with nonlinear process models and blending relations. They demonstrated that industrial scale problems can in principle be solved using commercially available MINLP solvers.

Neiro and Pinto (2003) extended this work to a set of refinery complexes, and also added scenarios to account for uncertainty in product prices. To ensure a robust solution, the decision variables are chosen "here and now". They demonstrate that nonlinear models reflecting process unit conditions and mixture property prediction can be used in multisite planning models. They also show that there are significant cost benefits in solving for the complex together rather than for the individual refineries separately.

Moro (2003), in his review of technology in the refining industry, indicates that scheduling tools that include details of process operation are still not available, but their application should results in benefits of US$10 million per year for a typical refinery.

2.9
Uncertainty in Planning and Scheduling

As with any other industrially relevant optimization problem, production scheduling requires a considerable amount of data. This is often subject to uncertainty (e.g., task processing times, yields, market demands, etc.). Sources of uncertainty (which tend to imply the means for dealing with them) can crudely be divided into:

- short-term uncertainties such as processing-time variations, rush orders, failed batches, equipment breakdowns, etc.;
- long-term uncertainties such as market trends, technology changes, etc.

Traditionally, short-term uncertainties have been treated through online or reactive scheduling, where schedules are adjusted to take account of new information. Longer-term uncertainties have been tackled through the solution of some form of stochastic programming problem. These two areas are considered below.

2.9.1
Reactive Scheduling

A major requirement of reactive scheduling systems is the ability to generate feasible updated schedules relatively quickly. A secondary objective is often to minimize deviations from the original schedule. As plants become more automated, this may become less important.

Cott (1989) presented some schedule modification algorithms to be used in conjunction with online monitoring, in particular to deal with processing-time variations and batch-size variations.

Kanakamedala et al. (1994) presented a least-impact heuristic beam search for reactive schedule modification in the face of unexpected deviations in processing times and resource availability. This is based on evaluating possible product reroutings and selecting that which has least overall impact on the schedule.

Rodrigues et al. (1996) modified the discrete-time STN formulation to take account of due-date changes and equipment unavailability. They use a rolling horizon (rolling out a predefined schedule) approach which aims to look ahead for a short time to resolve infeasibilities. This implies a very small problem size and fast solution times.

Schilling (1997) adapted his RTN-based continuous-time formulation to create a hierarchical family of MILP-based reactive scheduling formulations. At the lowest level, the sequence of operations is fixed as in the original schedule and only the timing can vary. At the topmost level, a full original scheduling problem is solved. The intermediate levels all trade off degrees-of-freedom with computational effort. This allows the best solution in the time available to be implemented on the plant.

Bael (1999) combined constraint satisfaction and iterative improvement (based on local perturbations) in his rescheduling strategy for jobshop environments. A trade-off between computational time and solution quality was identified.

Castillo and Roberts (2001) described a real-time scheduling approach for batch plants, based on model predictive control methods. The future allocation of orders to machines can be investigated using a fast tree-search algorithm, and robust solutions are generated in real time. This works due to the assumption of batch integrity throughout the process (i.e., there is no mixing or splitting of material).

Wang et al. (2000) described a genetic algorithm for online scheduling of a complex multiproduct polymer plant with many conflicting constraints. They describe how this technique may be successfully applied to scheduling problems and give guidance on appropriate mutation and crossover operations.

Henning and Cerdá (2000) described a knowledge-based framework which aims to support a human scheduler performing both offline and reactive scheduling. They argue that purely automated scheduling is difficult because plant circumstances change regularly. An object-oriented, knowledge-based framework is used to capture problem information and a scheduling support system developed (within which a variety of scheduling algorithms can be encoded) that enhances the capabilities of the human domain expert via an interactive front end. Because schedules can be generated very quickly using this approach, it is suitable for reactive scheduling.

One reason for reactive scheduling is the need to rework batches when quality criteria are not met. Flapper et al. (2002) provide a review of methods for planning and control of rework.

2.9.2
Planning and Scheduling under Uncertainty

Most of the work in this area is based on models in which product demands are assumed to be uncertain and to differ between a number of time periods. Usually, a simple representation of the plant capacity is assumed, and the sophistication of the work relates to the implementation of stochastic planning algorithms to select amounts for production in the first period (here and now) and potential production amounts in different possible demand realizations in different periods (see, e.g., Ierapetritou et al. (1996)).

In relatively long-term planning, it is reasonable to introduce additional degrees of freedom associated with potential capacity expansions. Liu and Sahinidis (1996a,b) and Iyer and Grossmann (1998) extended the MILP process and capacity planning model of Sahinidis and Grossmann (1991b) to include multiple product-demand scenarios in each period. They then proposed efficient algorithms for the solution of the resulting stochastic programming problems (formulated as large deterministic equivalent models), either by projection (Liu and Sahinidis 1996a) or by decomposition and iteration (Iyer and Grossmann 1998). A major assumption in their formulation is that product inventories are not carried over from one period to the next.

Clay and Grossmann (1994) also addressed this issue. They considered the structure of both the two-period and multiperiod problem for LP models and derived an approximation method based on successive repartitioning of the uncertain space with expectations being applied over partitions. This has the potential to generate solutions to a high degree of accuracy in a much faster time than the full-scale deterministic equivalent model.

The approaches above are based on relatively simple models of plant capacity. Petkov and Maranas (1997) treat the multiperiod planning model for multiproduct plants under demand uncertainty. Their planning model embeds the planning/scheduling formulation of Birewar and Grossmann (1990) and therefore accurately calculates the plant capacity. They do not use discrete demand scenarios but assume normal distributions and directly manipulate the functional forms to generate a problem which maximizes expected profit and meets certain probabilistic bounds on demand satisfaction without the need for numerical integration. They also make the no-inventory-carryover assumption, but show how this can be remedied to a certain extent at the lower level scheduling stage. Sand and Engell (2004) use a rolling horizon, two-stage stochastic programming approach to schedule an expandable polystyrene plant that is subject to uncertainty in processing times, yields, capacities and demands. The former two sources of uncertainty are considered short-term and the latter two medium-term. A hierarchical scheduling technique is used where a master schedule deals with medium-term uncertainties and a detailed schedule with the short-term ones. The uncertainties are represented through discrete scenarios and the two-stage problem solved using a decomposition technique.

Alternative approaches have attempted to characterize the effects of some sources of uncertainty on detailed schedules.

Rotstein et al. (1994) defined flexibility and reliability indices for detailed schedules. These are based on data for equipment reliability and demand distributions. Given a schedule (described in network flow form), these indices can be calculated to assess its performance.

Dedopoulos and Shah (1995) used a multistage stochastic programming formulation to solve short-term scheduling problems with possibilities of equipment failure at each discrete time instant. The technique can be used to assess the impact of different failure characteristics of the equipment on expected profit, but suffers from the very large computational effort required even for small problems.

Sanmarti et al. (1995) define a robust schedule as one which has a high probability of being performed, and is readily adaptable to plant variations. They define an index of reliability for a unit scheduled in a campaign through its intrinsic reliability, the probability that a standby unit is available during the campaign, and the speed with which it can be repaired. An overall schedule reliability is then the product of the reliabilities of units scheduled in it, and solutions to the planning problem can be driven to achieve a high value of this indicator.

Mignon et al. (1995) assess schedules obtained from deterministic data for performance under variability by Monte Carlo simulation. Although a number of parameters may be uncertain, they focus on processing time. Performance and robustness (predictability) metrics are defined and features of schedules with good indicators are summarized (e.g., introducing an element of conservatism when fixing due dates).

Honkomp et al. (1997) build on this to compare schedules generated by discrete-time and continuous-time algorithms and two means of ensuring robustness in the face of processing time uncertainties, namely increasing the processing times of bottleneck stages and increasing all processing times at the deterministic scheduling level. They found that the latter heuristic was better, and that the rounding effect of the discrete-time model results in marginally better robustness. Robustness is defined with respect to variance in the objective function. Strictly speaking, penalizing the variance of a metric to ensure robustness assumes that the metric is two-sided (i.e., "the closer to nominal the better" in the Taguchi sense). Since economic objective functions are one-sided ("the more the better"), robustness indicators such as these should be used with caution. This has been noted recently by Ahmed and Sahinidis (1998).

Gonzalez and Realff (1998a) analyze MILP solutions for pipeless plants that generated by assuming lower level controls for detailed vehicle movements and fixed, nominal transfer times. The analysis performed using stochastic simulation with variabilities in the transfer times. The system performance was found not to degrade considerably from its nominal value. They extended the work (Gonzalez and Realff 1998b) to consider the development of dispatching rules based on both general flexible manufacturing principles and properties of the MILP solutions. They found that rules abstracted from the MILP solutions were superior, and could be used in realtime.

A similar "multimodel" technique that combines optimization, expert systems and discrete-event simulation is described by Artiba and Riane (1998), although their focus is on a robust package for an industrial environment rather than uncertainty *per se*.

Bassett et al. (1997) contrasted aggregate planning and detailed scheduling under uncertainties in processing times and equipment failure. They argue that aggregate models that take these into account miss critical interactions due to the complex short-term interactions. They therefore propose the use of detailed scheduling to study the effects of such uncertainties on aggregate indicators such as average probabilities in meeting due dates and makespans. They also use Monte Carlo simulation, but use each set of sampled data to generate a detailed scheduling problem instance, solved using a reverse rolling horizon algorithm. Once enough instances have been solved for statistical significance, a number of comparisons can be made. For example, they conclude that long, infrequent breakdowns are more desirable, with obvious implications for maintenance policies.

Lee and Malone (2001a, 2001b) developed a hybrid Monte Carlo simulation-simulated annealing approach to planning and scheduling under uncertainty. They treat uncertainties in demands, due dates, processing times, product prices and raw material costs. An expected NPV objective is chosen; this is calculated through simultaneous Monte Carlo simulation and simulated annealing. This can also be used to devise strategies to ensure flexibility and robustness, for example by including enforced idle times in the schedule to allow for adjustments or rush orders.

Ivanescu et al. (2002) describe an approach for makespan estimation and order acceptance in multipurpose plants with uncertain task processing times (following an Erlang distribution). Instead of using a large mathematical model, regression analysis is used instead, based on a family of problem classes.

Balasubramanian and Grossmann (2003) presented an alternative approach to scheduling under uncertainty, arguing that probabilistic data on the uncertainties (e.g., in processing times) are unlikely to be available, and instead proposing a fuzzy set and interval theory approach. A rigorous MILP that can provide bounds on the makespan is developed for the flowshop case, based on the evaluation of a fuzzy, rather than crisp, makespan, and rules for comparing alternative makespans in order to determine optimality.

Kuroda et al. (2002) use the simple concept of due-date buffers to allow for flexibility in adjusting schedules in an operational environment. Here, orders further out in the horizon are allowed to move around within the buffer, while those near the current time remain fixed. This facilitates responsiveness to unforeseen orders.

2.10
Industrial Applications of Planning and Scheduling

Honkomp et al. (2000) give a list of reasons why the practical implementation of scheduling tools based on optimization is fraught with difficulty. These include:

- The large amount of user-defined input for testing purposes.

- The difficulty in capturing all the different types of operational constraints within a general framework, and the associated difficulty in defining an appropriate objective function.
- The large amounts of data required; Book and Bhatnagar (2000) list some of the issues that must be faced if generic data models are to be developed for planning/scheduling applications.
- Computational difficulties associated with the large problem sizes found in practice.
- Optimality gaps arising out of many shared resources.
- Intermediate storage and material stability constraints.
- Nonproductive activities (e.g., set-up times, cleaning, etc.)
- Effective treatment of uncertainties in demands and equipment effectiveness.

Nevertheless, there have been several success stories in the application of state-of-the-art scheduling methods in industry.

Schnelle (2000) applied MILP-based scheduling and design techniques to an agrochemical facility. The results indicated that sharing of equipment items between different products was a good idea, and the process reduced the number of alternatives to consider to a manageable number.

Berning et al. (2002) describe a large-scale planning-scheduling application which uses genetic algorithms for detailed scheduling at each site and a collaborative planning tool to coordinate plans across sites. The plants all operate batchwise, and may supply each other with intermediates, creating interdependencies in the plan. The scale of the problem is large, involving on the order of 600 different process recipes, and 1000 resources.

Kallrath (2002b) presented a successful application of MILP methods for planning and scheduling in BASF. He describes a tool for simultaneous strategic and operational planning in a multisite production network. The aim was to optimize the total net profit of a global network, where key decisions include: operating modes of equipment in each time period, production and supply of products, minor changes to the infrastructure (e.g., addition and removal of equipment from sites), and raw material purchases and contracts. A multiperiod model is formulated where equipment may undergo one mode change per period. The standard material balance equations are adjusted to account for the fact that transportation times are much shorter than the period durations. Counterintuitive but credible plans were developed that resulted in cost savings of several millions of dollars. Sensitivity analyses showed that the key decisions were not too sensitive to demand uncertainty.

Keskinocak et al. (2002) describe the application of a combined agent- and optimization-based framework for the scheduling of paper products manufacturing. The framework solves the problems of order allocation, run formation, trimming and trim-loss minimization and load planning. The deployment of the system is claimed to save millions of dollars per year. The "asynchronous agent-based team" approach uses constructor and improver agents to generate candidate solutions that are evaluated against multiple criteria.

2.11
New Application Domains

The scheduling techniques described above have in the main been applied to batch chemical production, particularly fine and specialty chemicals and pharmaceuticals. They are also appropriate to the wider and emerging process industries, and have started to find application in other domains, some of which are reviewed below.

Kim et al. (2001) tackle the problem of semiconductor wafer fabrication scheduling involving multiple products with different due dates. A series of dispatching ("lot release") and lot scheduling rules are evaluated.

Bhushan and Karimi (2003) describe the application of scheduling techniques to the wet-etching component of a semiconductor manufacturing process. This process is complicated by its "re-entrant" nature, where a product revisits stages of manufacture, and therefore does not fit the classical flowshop structure. A MILP formulation combined with a heuristic is used to minimize the makespan required to complete an outstanding set of jobs.

Pearn et al. (2004) describe the challenges associated with the scheduling of the final testing stage of integrated circuit manufacture and compare the performance of alternative heuristic algorithms.

El-Halwagi et al. (2003) describe a system for efficient design and scheduling of recovery of nutrients from plant wastes and reuse of the nutrients, with a view to developing a strategy for future planetary habitation.

Lee and Malone (2000b) show how planning can be useful in waste minimization. They combine scheduling with the main process and scheduling of the solvent recovery system. They show that such simultaneous scheduling can reduce waste disposal costs significantly.

Pilot plant facilities can become very scarce resources in the modern chemical industry, with many more short-run processes. Mockus et al. (2002) describe the integrated planning and scheduling of such a facility. The long-term planning problem is primarily concerned with skilled human resource allocation while the short-term scheduling problem deals with production operations.

Röslof et al. (2001) describe the application of production scheduling techniques to the paper manufacturing industry. In this sector, large numbers of orders, some of which are for custom products, are the norm.

Van den Heever and Grossmann (2003) describe the production planning and scheduling of a complex of plants producing hydrogen. This requires the description of the behavior of a pipeline and its associated compressors, which adds complexity and nonlinearity. The combination of longer-term planning and short-term reactive scheduling enables the decision-makers to deal effectively with uncertainty.

2.12
Conclusions and Future Challenges

Production scheduling has been a fertile area for CAPE research and the development of technology. Revisiting the challenges posed by Reklaitis (1991), Rippin (1993), Shah (1998) and Kallrath (2002a), it is clear that considerable progress has been made towards meeting them.

Overall, the emerging trend in the area of short-term scheduling is the development of techniques for the solution of the general, resource-constrained multipurpose plant scheduling problem. The recent research is all about solution efficiency and techniques to render ever-larger problems tractable. There remains work to be done on both model enhancements and improvements in solution algorithms if industrially relevant problems are to be tackled routinely, and software based on these are to be used on a regular basis by practitioners in the field.

Many algorithms have been developed to exploit the tight relaxation characteristics of discrete-time formulations. There remains work to be done in this area, in particular to exploit the sparsity of the solutions. Direct intervention at the LP level during branch-and-bound procedures (e.g., column generation and branch-and-price) seems a promising way of solving very large problems without ever considering the full variable space. Decomposition techniques (e.g., rolling horizon methods) will also find application here.

Much of the more recent research has focussed on continuous-time formulations, but little technology has been developed based on these. The main challenge here is in continual improvement in problem formulation and preprocessing to improve relaxation characteristics, and tailored solution procedures (e.g., branch-and-cut, and hybrid logic-continuous variable-integer variable branching) for problems with relatively large integrality gaps.

Probably the most promising recent development is the implementation of hybrid MILP/CLP solution methods which recognize that different algorithms are suitable for different components of the scheduling problem.

An important contrast between early and recent work is that the early algorithms tended to be tested on "motivating" examples (e.g., to find the best sequence of a few products), while recent algorithms are almost always tested on (and often motivated by) industrial or industrially based studies.

The multisite problem has received relatively little attention, and is likely to be a candidate for significant research in the near future. A major challenge is to develop planning approaches that are consistent with detailed production scheduling at each site and distribution scheduling across sites. An obvious stumbling block is problem size, and a resource-task-based decomposition based on identifying weak connections should find promise here as the problems tend to be highly structured. As scheduling and planning become integrated, the financial aspects will require more rigorous treatment. For example, Romero and Puigjaner (2004) describe the integration of cash flow modeling with production scheduling. This facilitates an accurate forward prediction of cash flow and even allows the enterprise to optimize treasury management simultaneously with decisions on purchasing, production and sales, and can be used to enforce upper and lower bounds on cash balances.

Researchers have attacked the problem of planning and scheduling under uncertainty from a number of angles, but have tended to skirt around the fundamental problem of multiperiod, multiscenario planning with realistic production capacity models (i.e., embedding some scheduling information) in the case of longer-term uncertainties. Issues that must be resolved relate mainly to problem scale. A sensible way forward is to try to capture the problem in all its complexity and then to explore rigorous or approximate solution procedures, rather than develop exact solutions to somewhat idealized problems. Process industry models are complicated by having multiple stages (periods) and integer variables in the second and subsequent stages, so most of the classical algorithms devised for large scale stochastic planning problems are not readily applicable.

The treatment of short-term uncertainties through the determination of characteristics of resilient schedules and then to use online monitoring and rescheduling seems eminently sensible. Further work is required in such characterization and in the design of rescheduling algorithms with guaranteed real-time performance.

A final challenge relates to the seamless integration of the activities at different levels – this is of a much broader and more interdisciplinary nature. Shobrys and White (2002) describe some of the difficult challenges to be faced here, including data and functional fragmentation, inconsistencies between activities and datasets, different tools being used for different activities, time and material buffers at each function for protection, slow responses and information flow.

References

1 *Ahmed S. Sahinidis N. V.* Robust process planning under uncertainty, Ind. Eng. Chem. Res. 37 (1998) p. 1883–1892

2 *Alle A. Pinto J. M.* A general framework for simultaneous cyclic scheduling and operational optimization of multiproduct continuous plants, Braz. J. Chem. Eng. 19:4 (2002) p. 457–466

3 *Artiba A. Riane F.* An application of a planning and scheduling multi-model approach in the chemical industry, Comput. Ind. 36 (1998) p. 209–229

4 *Applequist G. O. Samikoglu J. Pekny G. V. Reklaitis* Issues in the use, design and evolution of process scheduling and planning systems, ISA Trans. 36 (1997) p. 81–121

5 *Bael P.* A study of rescheduling strategies and abstraction levels for a chemical process scheduling problem, Prod. Planning Control 10(4) (1999) p. 359–364

6 *J. Balasubramanian Grossmann I. E.* Scheduling optimization under uncertainty – an

alternative approach, Comput. Chem. Eng. 27 (2003) 469–490

7 *Bassett M. H. Pekny J. F. Reklaitis G. V.* Decomposition techniques for the solution of large-scale scheduling problems, AIChE J. 42 (1996) p. 3373–3387

8 *Bassett M. H Pekny J. F. Reklaitis G. V.* Using detailed scheduling to obtain realistic operating policies for a batch processing facility, Ind. Eng. Chem. Res. 36 (1997) p. 1717–1726

9 *Berning G. Brandenburg M. Gursoy K. Mehta V. Tölle F.-J.* An integrated system for supply chain optimisation in the chemical process industry, OR Spectrum 24 (2002) p. 371–401

10 *Bhushan S. Karimi I. A.* An MILP approach to automated wet-etch station scheduling, Ind. Eng. Chem. Res. 42 (2003) p. 1391–1399

11 *Birewar D. B. Grossmann I. E.* Efficient optimization algorithms for zero-wait scheduling of multiproduct batch plants, Ind. Eng. Chem. Process Des. Dev. 28 (1989) p. 1333–1345

12 *Birewar D. B. Grossmann I. E.* Simultaneous production planning and scheduling in multiproduct batch plants, Ind. Eng. Chem. Res. 29 (1990) p. 570–580

13 *Blömer F. Günther H.-O.* Scheduling of a multi-product batch process in the chemical industry, Comput. Ind. 36 (1998) p. 245–259

14 *Book N. L. Bhatnagar V.* Comput. Chem. Eng. 24 (2000) p. 1641–1644

15 *Bunch P.* A simplex-based primal-dual algorithm for the perfect B-matching problem – a study in combinatorial optimisation, Phd Thesis, Purdue University (1997)

16 *Castillo I Roberts C. A.* Real-time control/scheduling for multipurpose batch plants, Comp. Ind. Eng. 41 (2001) p. 211–225

17 *Castro P. Barbosa-Póvoa A. P. F. D. Matos H.* An improved RTN continuous-time formulation for the short-term scheduling of multipurpose batch plants, Ind. Eng. Chem. Res. 40 (2001) p. 2059–2068

18 *Castro P. Barbosa-Póvoa A. P. F. D. Matos H.* Ind. Eng. Chem. Res. 43 (2004) p. 105–118

19 *Castro P. M. Barbosa-Póvoa A. P. Matos H. A.* Optimal periodic scheduling of batch plants using RTN-Based discrete and continuous-time formulations: a case study approach, Ind. Eng. Chem. Res. 42 (2003) p. 3346–3360

20 *Castro P. M. Barbosa-Póvoa A. P. Matos H. A. Novais A. Q.* Ind. Eng. Chem. Res. 43 (2004) p. 105–118

21 *Castro P. M. Matos H. Barbosa-Póvoa A. P. F. D.* Dynamic modelling and scheduling of an industrial batch system, Comput. Chem. Eng. 26 (2002) 671–686

22 *Cerda J. Henning G. P. Grossmann I. E.* A mixed integer linear programming model for short-term scheduling of single-stage multiproduct batch plants with parallel lines, Ind. Eng. Chem. Res. 36 (1997) p. 1695–1707

23 *Clay R. L. Grossmann I. E.* Optimization of stochastic planning-models, Chem. Eng. Res. Des. 72 (1994) p. 415–419

24 *Cott B. J.* An integrated computer-aided production management system for batch chemical processes, PhD Thesis, University of London (1989)

25 *Dannebring D. G.* An evaluation of flowshop sequencing heuristics, Manage. Sci. 23 (1977) p. 1174–1182

26 *Dedopoulos I. T. Shah N.* Preventive maintenance policy optimisation for multipurpose plant equipment, Comput. Chem. Eng, S19 (1995) p. S693–S698

27 *Dimitriadis A. D. Shah N. Pantelides C. C.* RTN-based rolling horizon algorithms for medium-term scheduling of multipurpose plants, Comput. Chem. Eng. S21 (1997a) p. S1061–S1066

28 *Dimitriadis A. D. Shah N. Pantelides C. C.* A rigorous decomposition algorithm for solution of large-scale planning and scheduling problems, paper presented at AIChE Annual Meeting, Nov. 16–21, Los Angeles (1997b)

29 *El-Halwagi M. Williams L. Hall J. Aglan H. Mortley D. Trotman A.* Mass integration and scheduling strategies for resource recovery in planetary habitation, Chem. Eng. Res. Des. 81 (2003) p. 243–250

30 *Elkamel A.* Scheduling of process operations using mathematical programming techniques, PhD Thesis, Purdue University (1993)

31 *Elkamel A. Al-Enezi G.* Structured valid inequalities and separation in optimal scheduling of the resource-constrained batch chemical plant, Math. Eng. Ind. 6 (1998) p. 291–318

32 *Elmaghraby S.* The economic lot scheduling problem. review and extensions, Manage. Sci. 24 (1978) p. 587–598

33 *Flapper S. D. P. Fransoo J. C. Broekmeulen R. A. C. M. Inderfurth K.* Planning and control of rework in the process industries: a review, Prod. Planning Control, 13 (2002) p. 26–34

34 *Gabow H. N.* On the design and analysis of efficient algorithms for deterministic scheduling, Proceedings of the 2nd International Conference Foundations of Computer-Aided Process Design, Michigan, June 19–24, (1983) USA, Cache Publications, pp. 473–528 (1983)

35 *Giannelos N. F. Georgiadis M. C.* A simple new continuous time formulation for short tern scheduling of multipurpose batch processes, Ind. Eng. Chem. Res. 41 (2002) p. 2178–2184

36 *Giannelos N. F. Georgiadis M. C.* Efficient scheduling of consumer goods manufacturing processes in the continuous time domain Comput. Oper. Res. 30 (2003) p. 1367–1381

37 *Glismann K. Gruhn G.* Short-term scheduling and recipe optimization of blending processes, Comput. Chem. Eng. 25 (2001). p. 627–634

38 *Gonzalez R. Realff M. J.* Operation of pipeless batch plants – I. MILP schedules, Comput. Chem. Eng. 22 (1998a) p. 841–855

39 *Gonzalez R. Realff M. J.* Operation of pipeless batch plants – II. Vessel dispatch rules, Comput. Chem. Eng. 22 (1998b) p. 857–866

40 *Gooding W. B.* Specially structured formulations and solution methods for optimisation problems important to process scheduling, PhD Thesis Purdue University (1994)

41 *Gooding W. B. Pekny J. F. McCroskey P. S.* Enumerative approaches to parallel flowshop scheduling via problem transformation Comput. Chem. Eng. 18 (1994) p. 909–927

42 *Graells M. Espuña A. Puigjaner L.* Sequencing intermediate products: a practical solution for multipurpose production scheduling, Comput. Chem. Eng. S20 (1996) p. S1137–S1142

43 *Grunow M. Günther H.-O. Lehmann M.* Campaign planning for multistage batch processes in the chemical industry, OR Spectrum, 24 (2002) p. 281–314

44 *Hampel R.* Beyond the millenium, Chem. Ind. 10 (1997) p. 380–382

45 *Harjunkoski Jain V. Grossmann I. E.* Hybrid mixed-integer/constrained logic programming strategies for solving scheduling and combinatroial optimization problems, Comput. Chem. Eng. 24 (2000) p. 337–343

46 *Hasebe S. Hashimoto I. Ishikawa A.* General reordering algorithm for scheduling of batch processes, J. Chem. Eng. Jpn. 24 (1991) p. 483–489

47 *Hasebe S. Taniguchi S. Hashimoto I.* Automatic adjustment of crossover method in the scheduling using genetic algorithm, Kagaku Kogaku Ronbunshu, 22 (1996) p. 1039–1045

48 *Henning G. P. Cerdá J.* Knowledge based predictive and reactive scheduling in industrial environments, Comput. Chem. Eng. 24 (2000) p. 2315–2338

49 *Honkomp S. J. Mockus L. Reklaitis G. V.* Robust scheduling with processing time uncertainty, Comput. Chem. Eng. S21 (1997) p. S1055–S1060

50 *Honkomp S. J. Lombardo S. Rosen O. Pekny J. F.* The curse of reality – why process scheduling optimisation problems are difficult in practice, Comput. Chem. Eng. 24 (2000) p. 323–328

51 *Huang W. Chung P. W. H.* Scheduling of pipeless batch plants using constraint satisfaction techniques, Comput. Chem. Eng. 24 (2000) p. 377–383

52 *Ierapetritou M. G. Floudas C. A.* Short-term scheduling: new mathematical models vs algorithmic improvements, Comput. Chem. Eng. 22 (1998a) p. S419–S426

53 *Ierapetritou M. G. Floudas C. A.* Effective continuous-time formulation for short-term scheduling: I. Multipurpose batch processes, Ind. Eng. Chem. Res. 37 (1998b) p. 4341–4359

54 *Ierapetritou M. G. Floudas C. A.* Effective continuous-time formulation for short-term scheduling: II. Multipurpose/multiproduct continuous processes, Ind. Eng. Chem. Res. 37 (1998c) p. 4360–4374

55 *Ierapetritou M. G. Hene T. S. Floudas C. A.* Effective continuous-time formulation for short-term scheduling: III. Multi intermediate due dates, Ind. Eng. Chem. Res. 38 (1999) p. 3446–3461

56 *Ierapetritou M. G. Pistikopoulos E. N. Floudas C. A.* Operational planning under uncertainty, Comput. Chem. Eng. 20 (1996) p. 1499–1516

57 *Ivanescu C. V. Fransoo J. C. Bertrand J. W. M.* Makespan Estimation And Order Acceptance In Batch Process Industries When Processing Times Are Uncertain, OR Spectrum, 24 (2002) p. 467–495

58 *Iyer R. R. Grossmann I. E.* Bilevel Decomposition A lgorithm For Long-Range Planning Of Process Networks, Ind. Eng. Chem. Res. 37 (1998) p. 474–481

59 *Jain V. Grossmann I. E.* Cyclic scheudling of continuous paralle-process units with decaying performance, AIChE J. 44 (1998) p. 1623–1636

60 *Jain V. Grossmann I. E.* Algorithms for hybrid MILP/CP models for a class of optimization problems, Informs J. Comput. 13 (2001) p. 258–276

61 *Janak S. L. Lin X. Floudas C. A.* Enhanced continuous-time unit-specific event-based formulation for short-term scheduling of multipurpose batch processes: Resource constraints and mixed storage policies, Ind. Eng. Chem. Res. 43 (2004) p. 2516–2533

62 *Joly M. Moro L. F. L. Pinto J. M.* Planning and scheduling for petroleum refinerines using mathematical programming, Braz. J. Chem. Eng. 19(2) (2002) p. 207–228

63 *Joly M. Pinto J. M.* Mixed-integer programming techniques for the scheduling of fuel oil and asphalt production, Chem. Eng. Res. Des. 81 (2003) p. 427–447

64 *Kallrath J.* Planning and scheduling in the process industry, OR Spectrum, 24 (2002a) 219–250

65 *Kallrath J.* Combined strategic and operational planning – an MILP success story in chemical industry, OR Spectrum, 24 (2002b) p. 315–341

66 *Kanakamedala K. B. Reklaitis G. V.* Venkata-subramanian V. Reactive schedule modification in multipurpose batch chemical plants, Ind. Eng. Chem. Res. 33 (1994) p. 77–90

67 *Karimi I. A. McDonald C. M.* Planning and scheduling of parallel semicontinuous processes. 2. Short–term Scheduling, Ind. Eng. Chem. Res. 36 (1997) p. 2701–2714

68 *Keskinocak P. Wu F. Goodwin R. Murthy S. Akkiraju R. Kumaran S. Derebail A.* Scheduling solutions for the paper industry, Oper. Res. 50 (2002) p. 249–259

69 *Kim M. Jung J. H. Lee I.-B.* Optimal scheduling of multiproduct batch processes for various intermediate storage policies, Ind. Eng. Chem. Res. 35 (1996) p. 4048–4066

70 *Kim Y.-D. Kim J.-G. Choi B. Kim H.-U.* Production scheduling in a semiconductor wafer fabrication facility producing multiple product types with distinct due dates, IEEE Trans. Robot. Autom. 17 (2001) p. 589–598

71 *Kondili E. Pantelides C. C. Sargent R. W. H.* A general algorithm for scheduling of batch operations, Proceedings of the 3rd International Symposium on Process Systems Engineering, Sydney, Australia, (1988) pp. 62–75

72 *Kondili E. Pantelides C. C. Sargent R. W. H.* A general algorithm for short-term scheduling of batch operations – 1. Mixed integer linear programming formulation, Comput. Chem. Eng. 17 (1993) p. 211–227

73 *Krarup J. Bilde O.* Plant location, set covering and economic lot size: an O(mn) algorithm for structured problems, Int Ser. Num. Math. 36 (1977) p. 155–180

74 *Ku H. Karimi I. A.* An evaluation of simulated annealing for batch process scheduling, Ind. Eng. Chem. Res. 30 (1991) p. 163–169

75 *Kudva G. Elkamel A. Pekny J. F. Reklaitis G. V.* Heuristic algorithm for scheduling batch and semicontinuous plants with production deadlines, intermediate storage limitations and equipment changeover costs, Comput. Chem. Eng. 18 (1994) p. 859–875

76 *Kuriyan K. Reklaitis G. V.* Approximate scheduling algorithms for network flowshops, Chem. Eng. Symp. Ser. 92 (1985) p. 79–90

77 *Kuriyan K. Reklaitis G. V.* Scheduling network flowshops so as to minimise makespan, Comput. Chem. Eng. 13 (1989) p. 187–200

78 *Kuroda M. Shin H. Zinnohara A.* Robust scheduling in an advanced planning and scheduling, Int. J. Prod. Res. 40 (2002) p. 3655–3668

79 *Lee Y. G Malone M. F.* Batch processes planning for waste minimization, Ind. Eng. Chem. Res. 39 (2000) p. 2035–2044

80 *Lee Y. G Malone M. F.* Flexible batch process planning, Ind. Eng. Chem. Res. 39 (2000b) p. 2045–2055

81 *Lee Y. G Malone M. F.* Batch process schedule optimization under parameter volatility, Int. J. Prod. Res. 39 (2001a) p. 603–623

82 *Lee Y. G. Malone M. F.* A general treatment of uncertainties in batch process planning, Ind. Eng. Chem. Res. 40 (2001b) p. 1507–1515

83 *Lee K.-H. Park H. I. Lee I. B.* A novel non-uniform discrete time formulation for short-term scheduling of batch and continuous processes, Ind. Eng. Chem. Res. 40 (2001a) p. 4902–4911

84 *Lin X. Floudas C. A.* Design, synthesis and scheduling of multipurpose batch plants via en effective continuous-time formulation, Comput. Chem. Eng. 25 (2001b) p. 665–674

85 *Liu M. L. Sahinidis N. V.* Long-range planning in the process industries – a projection approach, Comput. Oper. Res. 3 (1996a) p. 237–253

86 *Liu M. L. Sahinidis N. V.* Optimization in process planning under uncertainty, Ind. Eng. Chem. Res. 35 (1996b) p. 4154–4165

87 *Majozi T. Zhu X. X.* A novel continuous-time MILP formulation for multipurpose batch plants. 1. Short-term scheduling, Ind. Eng. Chem. Res. 40 (2001) p. 5935–5949

88 *Maravelias C. T. Grossmann I. E.* New general continuous-time statetask network formulation for short-term scheduling of multipurpose batch plants, Ind. Eng. Chem. Res. 42 (2003) p. 3056–3074

89 *Maravelias C. T. Grossmann I. E.* A hybrid MILP/CP decomposition approach for the continuous time scheduling of multipurpose batch plants, Comput. Chem. Eng. 28 (2004) p. 1921–1949

90 *Mauderli A. M. Rippin D. W. T.* production planning and scheduling for multi-purpose batch chemical plants, Comput. Chem. Eng. 3 (1979) p. 199–206

91 *McDonald C. M. Karimi I. A.* Planning and scheduling of parallel semicontinuous processes. 1. Production planning, Ind. Eng. Chem. Res. 36 (1997) p. 2691–2700

92 *Méndez C. A. Cerdá J.* Optimal scheduling of a resource-constrained multiproduct batch plant supplying intermediates to nearby end-product facilities, Comput. Chem. Eng. 24 (2000) p. 369–376

93 *Méndez C. A. Henning G. P. Cerdá J.* An MILP continuous-time approach to short-term scheduling of resource-constrained multistage flowshop batch facilities, Comput. Chem. Eng. 25 (2001) p. 701–711

94 *Mignon D. J. Honkomp S. J. Reklaitis G. V.* A framework for investigating schedule robustness under uncertainty, Comput. Chem. Eng. S19 (1995) p. S615–S620

95 *Mockus L. Reklaitis G. V.* Continuous-Time Representation In Batch/Semicontinuous Process Scheduling – Randomized Heuristics Approach, Comput. Chem. Eng. S20 (1996) p. S1173–S1178

96 *Mockus L. Reklaitis G. V.* Continuous time representation approach to batch and continuous process scheduling. 1. MINLP formulation, Ind. Eng. Chem. Res. 38 (1999a) p. 197–203

97 *Mockus L. Reklaitis G. V.* Continuous time representation approach to batch and continuous process scheduling. 2. Computational issues, Ind. Eng. Chem. Res. 38 (1999b) p. 204–210

98 *Mockus L. Vinson J. M. Luo K.* The integration of production plan and operating schedule in a pharmaceutical pilot plant, Comput. Chem. Eng. 26 (2002) p. 697–702

99 *Moon S. Park S. Lee W. K.* New MILP models for scheduling of multiproduct batch plants under zero-wait policy, Ind. Eng. Chem. Res. 35 (1996) p. 3458–3469

100 *Moro L. F. L.* Process technology in the petroleum refining industry – current situation and future trends, Comput. Chem. Eng. 27 (2003) p. 1303–1305

101 *Murakami Y. Uchiyama H. Hasebe S. Hashimoto I.* Application of repetitive SA method to scheduling problems of chemical processes, Comput. Chem. Eng. S21 (1997) p. S1087–S1092

102 *Neiro S. M. S. Pinto J. M.* Supply chain optimisation of petroleum refinery complexes, Proceedings of the 4th International Conference on Foundations of Computer-Aided Process Operations, Florida, USA, Jan (2003), Cache Corp, pp. 59–72

103 *Neumann K. Schwindt C. Trautmann N.* Advanced production scheduling for batch

104 *Orcun S. Altinel I. K. Hortaçsu Ö.* General continuous time models for production planning and scheduling of batch processing plant, mixed integer linear program formulations and computational issues, Comput. Chem. Eng. 25 (2001) p. 371–38

plants in process industries, OR Spectrum, 24 (2002) p. 251–279

105 *Pantelides C. C.* Unified frameworks for optimal process planning and scheduling, Proceedings of the 2nd Conference on Foundations of Computer-Aided Process Operations, Snowmass, Colorado, USA, July 10–15 (1994) Cache. Corp., pp 253–274

106 *Pantelides C. C. Realff M. J. Shah N.* Short-term scheduling of pipeless batch plants Trans..IChemE A, 73 (1995) p. 431–444

107 *Papageorgiou L. G. Pantelides C. C.* Optimal campaign planning/scheduling of multipurpose batch/semicontinuous plants. 1. Mathematical formulation, Ind. Eng. Chem. Res. 35 (1996a) p. 488–509

108 *Papageorgiou L. G. Pantelides C. C.* Optimal campaign planning/scheduling of multipurpose batch/semicontinuous plants. 2. A mathematical decomposition approach, Ind. Eng. Chem. Res. 35 (1996b) p. 510–529

109 *Pearn W. L. Chung S. H. Chen A. Y. Yang M. H.* A case study on the multistage IC final testing scheduling problem with reentry, Int. J. Prod. Econ. 88 (2004) p. 257–267

110 *Pekny J. F. Miller D. L. McCrae G. J.* Application of a parallel travelling salesman problem to no-wait flowshop scheduling, paper presented at AIChE Annual Meeting, November 27 – December 2, Washington D.C. (1988)

111 *Pekny J. F. Miller D. L. McCrae G. J.* An exact parallel algorithm for scheduling when production costs depend on consecutive system states, Comput. Chem. Eng. 14 (1990) p. 1009–1023

112 *Petkov S. B. Maranas C. D.* Multiperiod planning and scheduling of multiproduct batch plants under demand uncertainty, Ind. Eng. Chem. Res. 36 (1997) p. 4864–4881

113 *Pinedo M.* Scheduling. theory, algorithms and systems. Prentice Hall, New York (1995)

114 *Pinto J. M. Grossmann I. E.* Optimal cyclic scheduling of multistage continuous multiproduct plants, Comput. Chem. Eng. 18 (1994) p. 797–816

115 *Pinto J. M. Grossmann I. E.* A continuous time MILP model for short-term scheduling of multistage batch plants, Ind. Eng. Chem. Res. 34 (1995) p. 3037–3051

116 *Pinto J. M. Grossmann I. E.* A logic-based approach to scheduling problems with resource constraints, Comput. Chem. Eng. 21 (1997) p. 801–818

117 *Pinto J. M. Grossmann I. E.* Assignment and sequencing models for the scheduling of chemical processes, Ann OR, 81 (1998) p. 433–466

118 *Pinto J. M. Joly M. Moro L. F. L.* Planning and scheduling models for refinery operations, Comput. Chem. Eng. 24 (2000) p. 2259–2276

119 *Reklaitis G. V.* Perspectives on scheduling and planning of process operations, Proceedings of the 4th International Symposium on Process Systems Engineering. Montebello, Canada, August 5–9 (1991)

120 *Reklaitis G. V. Mockus L.* Mathematical programming formulation for scheduling of batch operations based on non-uniform time discretization, Acta Chim. Slov. 42 (1995) p. 81–86

121 *Rippin D. W. T.* Batch process systems engineering: a retrospective and prospective review, Comput. Chem. Eng. S17 (1993) p. S1–S13

122 *Rodrigues M. T. M. Gimeno L. Passos C. A. S. Campos M. D.* Reactive scheduling approach for multipurpose batch chemical plants, Comput. Chem. Eng. S20 (1996) p. S1215–S1226

123 *Roe B. Papageorgiou L. G. Shah N.* A hybrid CLP and MILP approach to batch process scheduling, Proc. of 8th International Symposium on Process Systems Engineering, Kunming, China (2003) p. 582–587

124 *Roe B. Papageorgiou L. G. Shah N.* A hybrid MILP/CLP algorithm for multipurpose batch process scheduling, Comput. Chem. Eng. 29 (2005) p. 1277-129

125 *Romero J. Puigjaner L.* Joint financial and operating scheduling/planning in industry, Proceedings of 14th European Symposium on Computer-Aided Process Engineering, May (2004) Elsevier, p. 883–888

126 *Romero J. Puigjaner L. Holczinger T. Friedler F.* Scheduling intermediate storage multipurpose batch plants using the S-Graph, AIChE J. 50 (2004) p. 403–417

127 *Roslöf J. Harjunkoski I. Björkqvist J. Karlsson S. Westerlund T.* An MILP-based reordering algorithm for complex industrial scheduling and rescheduling, Comput. Chem. Eng. 25 (2001) p. 821–828

128 *Rotstein G. E. Lavie R. Lewin D. R.* Synthesis of Flexible and Reliable Short-Term Batch Production Plans, Comput. Chem. Eng. 20 (1994) p. 201–215

129 *Sahinidis N. V. Grossmann I. E.* Reformulation of multiperiod MILP models for planning and scheduling of chemical processes, Comput. Chem. Eng. 15 (1991) p. 255–272

130 *Sahinidis N. V. Grossmann I. E. Fornari R. E. Chathrathi M.* Optimisation model for long-range planning in the chemical industry, Comput. Chem. Eng. 15 (1991a) 255–272

131 *Sahinidis N. V. Grossmann I. E.* MINLP model for cyclic multiproduct scheduling on continuous parallel lines, Comput. Chem. Eng. 15 (1991b) 85–103

132 *Sand G. Engell S.* Modelling and solving real-time scheduling problems by stochastic integer programming, Comput. Chem. Eng. 28 (2004) p. 1087–1103

133 *Sanmarti E. Espuña A. Puigjaner L.* Effects of equipment failure uncertainty in batch production scheduling, Comput. Chem. Eng. S19 (1995) p. S565–S570

134 *Schilling G. Pantelides C. C.* A simple continuous time process scheduling formulation and a novel solution algorithm, Comput. Chem. Eng. S20 (1996) p. S1221–S1226

135 *Schilling G. H.* Algorithms for short-term and periodic process scheduling and rescheduling, PhD Thesis, University of London (1997)

136 *Schilling G. Pantelides C. C.* Optimal periodic scheduling of multipurpose plants, Comput. Chem. Eng. 23 (1999) p. 635–655

137 *Schnelle K. D.* Preliminary design and scheduling of a batch agrochemical plant, Comput. Chem. Eng. 24 (2000) p. 1535–1541

138 *Schwindt C. Trautmann N.* Batch scheduling in process industries: an application of resource-constrained project scheduling, OR Spektrum, 22 (2000) p. 501–524

139 *Shah N.* Single- and multi-site planning and scheduling: Current status and future challenges, AIChE Ser. 94 (1998) p. 75–90

140 *Shah N. Pantelides C. C.* Optimal long-term campaign planning and design of batch plants, Ind. Eng. Chem. Res. 30 (1991) p. 2308–2321

141 *Shah N. Pantelides C. C. Sargent R. W. H.* A general algorithm for short-term scheduling of batch operations – 2. Computational issues, Comput. Chem. Eng. 17 (1993a) p. 229–244

142 *Shah N. Pantelides C. C. Sargent R. W. H.* Optimal periodic scheduling of multipurpose batch plants, Ann. Oper. Res. 42 (1993b) p. 193–228

143 *Shobrys D. E. White D. C.* Planning, scheduling and control systems: why cannot they work together, Comput. Chem. Eng. 26 (2002) p. 149–160

144 *Sunol, A. K. Kapanoglu M. Mogili P.* Selected topics in artificial intelligence for planning and scheduling problems, knowledge acquisition and machine learning, Proc NATO ASI on Batch Processing Systems Engineering, Series F, Vol, 143 (1992) pp 595–630

145 *Tahmassebi T.* Industrial experience with a mathematical programming based system for factory systems planning/scheduling, Comput. Chem. Eng. S20 (1996) p. S1565–S1570

146 *van den Heever S. A. Grossmann I. E.* Comput. Chem. Eng. 27 (2003) p. 1813–1839

147 *van Hentenryck P.* Constraint satisfaction in Logic Programming, MIT Press (1989)

148 *Wallace M. G. Novello S. Schimpf J.* ECLiPSe: A platform for constraint logic programming, ICL Systems Journal 12 (1997) p. 137–158

149 *Wang K. F. Lohl T. Stobbe M. Engell S.* A genetic algorithm for online-scheduling of a multiproduct polymer batch plant, Comput. Chem. Eng. (2000) p. 393–400

150 *Wellons M. C. Reklaitis G. V.* Scheduling of multipurpose batch plants. 1. Formation of single-product campaigns, Ind. Eng. Chem. Res. 30 (1991a) p. 671–688

151 *Wellons M. C. Reklaitis G. V.* Scheduling of multipurpose batch plants. 2. Multiple-product campaign formation and production planning, Ind. Eng. Chem. Res. 30 (1991b) p. 688–705

152 *Wilkinson S. J. Shah N. Pantelides C. C.* Aggregate modelling of multipurpose plant operation, Comput. Chem. Eng. 19 (1995) p. S583–S588

153 *Wilkinson S. J. Cortier A. Shah N. Pantelides C. C.* Integrated production and distribution scheduling on a Europe-wide basis, Comput. Chem. Eng. S20 (1996) p. S1275–S1280

154 *Wu D. Ierapetritou M. G.* Decomposition approaches for the efficient solution of short-term scheduling problems, Comput. Chem. Eng. 27 (2003) p. 1261–1276

155 *Xia Q. Macchietto S.* Routing, scheduling and product mix optimization by minimax algebra, Chem. Eng. Res. Des. 72 (1994) p. 408–414

156 *Yee K. L.* Efficient algorithms for multipurpose plant scheduling, PhD Thesis, University of London (1998)

157 *Yee K. L. Shah N.* Scheduling of fast-moving consumer goods plants, J. Oper. Res. Soc. 48 (1997) p. 1201–1214

158 *Yee K. L. Shah N.* Improving the efficiency of discrete-time scheduling formulations, Comput. Chem. Eng. S22 (1998) p. S403–S410

159 *Zentner M. G. Reklaitis G. V.* An interval-based mathematical model for the scheduling of resource-constrained batch chemical processes, Proc. NATO ASI on Batch Processing Systems Engineering, Series F, Vol, 143 (1992) p. 779–807

160 *Zhang X, Sargent R. W. H.* The optimal operation of mixed production facilities-a general formulation and some approaches for the solution, Proceedings of the 5th International Symposium on Process Systems Engineering, Kyongju, Korea, June (1994) p. 171–178

161 *Zhang X Sargent R. W. H.* The optimal operation of mixed production facilities-extensions and improvements, Comput. Chem. Eng. S20 (1996) p. S1287–S1292

3
Process Monitoring and Data Reconciliation

Georges Heyen and Boris Kalitventzeff

3.1
Introduction

Measurements are needed to monitor process efficiency and equipment condition, but also to take care that operating conditions remain within an acceptable range to ensure good product quality and to avoid equipment failure and any hazardous conditions. Recent progress in automatic data collection and archiving has solved part of the problem, at least for modern, well-instrumented plants. Operators are now faced with a lot of data, but they have little means to extract and fully exploit the relevant information it contains.

Furthermore, plant operators recognize that measurements and laboratory analysis are never error-free. Using these measurements without any correction yields inconsistencies when generating plant balances or estimating performance indicators. Even careful installation and maintenance of the hardware can not completely eliminate this problem.

Model-based statistical methods, such as data reconciliation, have been developed to analyze and validate plant measurements. The objective of these techniques is to remove errors from available measurements and to yield complete estimates of all the process state variables as well as of unmeasured process parameters.

This chapter constitutes a tutorial on process monitoring and data reconciliation. First, the key concepts and issues underlying a plant data validation, sources of error and redundancy considerations are introduced. Then, the data reconciliation problem is formulated for simple stready-state linear systems and extended further to consider nonlinear cases. The role of sensibility analysis is also introduced. Dynamic data reconciliation, which is still a subject of major research interest, is treated next. The chapter concludes with a section devoted to the optimal design of the measurement system. Detailed algorithms and supporting software are presented along with the solution of some motivating examples.

Computer Aided Process and Product Engineering. Edited by Luis Puigjaner and Georges Heyen
Copyright © 2006 WILEY-VCH Verlag GmbH & Co. KGaA, Weinheim
ISBN: 3-527-30804-0

3.2
Introductory Concepts for Validation of Plant Data

Data validation makes use of a plant model in order to identify measurement errors and to reduce their average magnitude. It provides estimates of all process state variables, whether directly measured or not, with the lowest possible uncertainty. It allows one to assess the value of key performance indicators, which are target values for process operation, or is used as a soft sensor to provide estimates of some unmeasured variables, as in inferential control applications.

Especially in a framework of real-time optimal control, where model fidelity is of paramount importance, data validation is a recommended step before fine-tuning model parameters: there is no incentive in seeking to optimize a model when it does not match the actual behavior of the real plant.

Data validation can also help in gross error detection, meaning either process faults (such as leaks) or instrument faults (such as identification of instrument bias and automatic instrument recalibration).

Long an academic research topic, data validation is currently attracting more and more interest, since the amount of measured data collected by Digital Control Systems (DCS) and archived in process information management systems,exceeds what can be handled by operators and plant managers. Real-time applications, such as optimal control, also require frequent parameter updates, in order to ensure fidelity of the plant model. The economic value of extracting consistent information from raw data is recognized. Data validation thus plays a key role in providing coherent and error-free information to decision makers.

3.2.1
Sources of Error

Some sources of errors in the balances depend on the sensors themselves:

- Intrinsic sensor precision is limited, especially for online equipment, where robustness is usually considered more important than accuracy.
- Sensor calibration is seldom performed as often as desired, since this is a costly and time-consuming procedure requiring competent manpower.
- Signal converters and transmission add noise to the original measurement.
- Synchronization of measurements may also pose a problem, especially for chemical analysis, where a significant delay exists between sampling and result availability.

Other errors arise from the sensor location or the influence of external effects. For instance, the measurement of gas temperature at the exit of a furnace can be influenced by radiation from the hot wall in the furnace. Inhomogeneous flow can also cause sampling problems. A local measurement is not representative of an average bulk property.

A second source of error when calculating plant balances is the small instabilities of the plant operation and the fact that samples and measurements are not taken at exactly the same time. Using time averages for plant data partly reduces this problem.

3.2.2
Redundancy

Besides safety considerations, the ultimate goal in performing measurements is to assess the plant performance and to take actions in order to optimize the operating conditions. However, most performance indicators can not be directly measured and must be inferred from some measurements using a model. For instance, the extent of a reaction in a continuous reactor can be calculated from a flow rate and two composition measurements. In general terms, model equations that relate unmeasured variables to a sufficient number of available measurements are used.

However, in some cases, more measurements are available than are strictly needed, and the same performance indicator can be calculated in several ways using different subsets of measurements. For instance, the conversion in an adiabatic reactor where a single reaction takes place is directly related to the temperature variation. Thus the extent of the reaction can be inferred from a flow rate and two temperature measurements using the energy balance equation. In practice, all estimates of performance indicators will be different, which makes life difficult and can lead to endless discussions about "best practice."

Measurement redundancy should not be viewed as a source of trouble, but as an opportunity to perform extensive checking. When redundant measurements are available, they allow one not only to detect and quantify errors, but also to reduce the uncertainty using procedures known as data validation.

3.2.3
Data Validation

The data validation procedure comprises several steps. The first is the measurement collection. Nowadays, in well-instrumented plants, this is performed routinely by automated equipment.

The second step is conditioning and filtering: not all measurements are available simultaneously, and synchronization might be required. Some data are acquired at higher frequency and filtering or averaging can be justified.

The third step is to verify the process condition and the adequacy of the model. For instance, if a steady-state model is to be used for data reconciliation, the time series of raw measurements should be analyzed to detect any significant transient behavior.

The fourth step is gross error detection: the data reconciliation procedure to be applied later is meant to correct small random errors. Thus, large systematic errors that could result from complete sensor failure should be detected first. This is usually done by verifying that all raw data remain within the upper and lower bounds.

More advanced statistical techniques, such as principal component analysis (PCA), can also be applied at this stage. *Ad hoc* procedures are applied in case some measured value is found inadequate or missing: it can be replaced by a default value or by the previous one available.

The fifth step checks the feasibility of data reconciliation. The model equations are analyzed and the variables are sorted. Measured variables are redundant (and can thus be validated) or just determined; unmeasured variables are determinable or not. When all variables are either measured or observable, the data reconciliation problem can be solved to provide an estimate for all state variables.

The sixth step is the solution of the data reconciliation problem. The mathematical formulation of this problem will be presented in more detail later.

Each measurement is corrected as slightly as possible in such a way that the corrected measurements match all the constraints (or balances) of the process model. Unmeasured variables can be calculated from reconciled values using some model equations.

In the seventh step the systems perform a result analysis. The magnitude of the correction for each measurement is compared to its standard deviation. Large corrections are flagged as suspected gross errors.

In the final step, results are edited and may be archived in the plant information management system. Customized reports can be edited and forwarded to various users (e.g., list of suspect sensors sent to maintenance, performance indicators sent to the operators, daily balance and validated environmental figures to site management).

3.3
Formulation

Data reconciliation is based on measurement redundancy. This concept is not limited to the case where the same variable is measured simultaneously by several sensors. It is generalized with the concept of spatial redundancy, where a single variable can be estimated in several independent ways from separate sets of measurements. For instance, the outlet of a mixer can be directly measured or estimated by summing the measurements of all inlet flow rates. For dynamic systems, temporal redundancy is also available, by which repeated observations of the same variables are obtained. More generally, plant structure is additional information that can be exploited to correct measurements.

Variables describing the state of a process are related by some constraints. The basic laws of nature must be verified: mass balance, energy balance, some equilibrium constraints. Data reconciliation uses information redundancy and conservation laws to correct measurements and convert them into accurate and reliable knowledge.

Kuehn and Davidson (1961) were the first to explore the problem of data reconciliation in the process industry. Vaclavek (1968, 1969) also addressed the problem of variable classification, and the formulation of the reconciliation model. Mah et al.

(1976) proposed a variable classification procedure based on graph theory, while Crowe (1989) based an analysis on a projection matrix approach to obtain a reduced system. Joris and Kalitventzeff (1987) proposed a classification algorithm for general nonlinear equation systems, comprising mass and energy balances, phase equilibrium and nonlinear link equations. A thorough review of classification methods is available in Veverka and Madron (1996) and in Romagnoli and Sanchez (2000). A historical perspective of the main contributions on data reconciliation can also be found in Narasimhan and Jordache (2000).

3.3.1
Steady-State Linear System

The simplest data reconciliation problem deals with steady state mass balances, assuming all variables are measured, and results in a linear problem. In this case $\mathbf{x}$ is the vector of n state variables, while $\mathbf{y}$ is the vector of measurements. We assume that random errors $\mathbf{e}=\mathbf{y}-\mathbf{x}$ follow a multivariate normal distribution with zero mean.

The state variables are linked by a set of m linear constraints:

$$\mathbf{A}\mathbf{x} - \mathbf{d} = 0 \tag{1}$$

The data reconciliation problem consists of identifying the state variables $\mathbf{x}$ that verify the set of constraints and are close to the measured values in the least square sense, which results in the following objective function:

$$\min_{x}(\mathbf{y} - \mathbf{x})^{\mathrm{T}}\mathbf{W}(\mathbf{y} - \mathbf{x}) \tag{2}$$

where $\mathbf{W}$ is a weight matrix.

The method of Lagrange multipliers allows one to obtain an analytical solution:

$$\hat{\mathbf{x}} = \mathbf{y} - \mathbf{W}^{-1}\mathbf{A}^{\mathrm{T}}\left(\mathbf{A}\mathbf{W}^{-1}\mathbf{A}^{\mathrm{T}}\right)^{-1}(\mathbf{A}\mathbf{y} - \mathbf{d}) \tag{3a}$$

It is assumed that there are no linearly dependent constraints.

In order to solve practical problems and obtain physically meaningful solutions, it may be necessary to take into account inequality constraints on some variables (e.g., flow rate should be positive). However, this makes the solution more complex, and the constrained problem can not be solved analytically.

It can be shown that $\hat{\mathbf{x}}$ is the maximum likelihood estimate of the state variables if the measurement errors are normally distributed with zero mean, and if the weight matrix $\mathbf{W}$ corresponds to the inverse of the error covariance matrix $\mathbf{C}$. Equation (3) then becomes:

$$\hat{\mathbf{x}} = \mathbf{y} - \mathbf{C}\mathbf{A}^{\mathrm{T}}\left(\mathbf{A}\mathbf{C}\mathbf{A}^{\mathrm{T}}\right)^{-1}(\mathbf{A}\mathbf{y} - \mathbf{d}) = \left[\mathbf{I} - \mathbf{C}\mathbf{A}^{\mathrm{T}}\left(\mathbf{A}\mathbf{C}\mathbf{A}^{\mathrm{T}}\right)^{-1}\mathbf{A}\right]\mathbf{y} + \mathbf{C}\mathbf{A}^{\mathrm{T}}\left(\mathbf{A}\mathbf{C}\mathbf{A}^{\mathrm{T}}\right)^{-1}\mathbf{d}$$
$$\hat{\mathbf{x}} = \mathbf{M}\mathbf{y} + \mathbf{e} \tag{3b}$$

The estimates are thus related to the measured values by a linear transformation. They are therefore normally distributed with the average value and covariance matrix obtained by calculating the expected values:

$$E(\hat{x}) = M E(y) = x \tag{4}$$

$$\text{Cov}(\hat{x}) = E\left[(My)(My)^T\right] = MCM^T$$

This shows that the estimated state variables are unbiased. Furthermore, the accuracy of the estimates can easily be obtained from the measurement accuracy (covariance matrix **C**) and from the model equations (matrix **A**).

3.3.2
Steady-State Nonlinear System

The data reconciliation problem can be extended to nonlinear steady-state models and to cases where some variables **z** are not measured. This is expressed by:

$$\min_{x,z}(y - x)^T W(y - x) \tag{5}$$

$$\text{s.t. } f(x, z) = 0$$

where the model equations are mass and component balance equations, energy balance, equilibrium conditions, and link equations relating measured values to state variables (e.g., conversion from mass fractions to partial molar flow rates).

Usually the use of performance equations is not recommended, unless the performance parameters (such as compressor efficiency and overall heat transfer coefficients or fouling factors for heat exchangers) remain unmeasured and will thus be estimated by solving the data reconciliation problem. It would be difficult to justify correcting measurements using an empirical correlation, e.g., by correcting the outlet temperatures of a compressor by enforcing the value of the isentropic efficiency. The main purpose of data reconciliation is to allow monitoring of those efficiency parameters and to detect their degradation.

Equation (5) takes the form of a nonlinear constrained minimization problem. It can be transformed into an unconstrained problem using Lagrange multipliers **Λ** and the augmented objective function *L* has to be minimized:

$$L(x, z, \Lambda) = \left\{\frac{1}{2}(x-y)^T C^{-1}(x-y) + \Lambda^T \cdot f(x,z)\right\} \tag{6}$$

$$\min_{x,y,\Lambda} L(x, y, \Lambda)$$

The solution must verify the necessary optimality conditions i.e., the first derivatives of the objective function with respect to all independent variables must vanish. Thus one has to solve the system of normal equations:

$$\frac{\partial L}{\partial x} = C^{-1}(x-y) + A^T \cdot \Lambda = 0$$

$$\frac{\partial L}{\partial z} = B^T \cdot \Lambda = 0 \tag{6}$$

$$\frac{\partial L}{\partial \Lambda} = f(x,z) = 0$$

This last equation can be linearized as:

$$\frac{\partial L}{\partial \Lambda} = A \cdot x + B \cdot z + d = 0 \tag{7}$$

where A and B are partial Jacobian matrices of the model equation system:

$$A = \frac{\partial f}{\partial x}$$
$$B = \frac{\partial f}{\partial z} \tag{8}$$

The system of normal equations in Eq. (6) is nonlinear and has to be solved iteratively. Initial guesses for measured values are straightforward to obtain. Process knowledge usually estimates good initial values for unmeasured variables. No obvious initial values exist for Lagrange multipliers, but solution algorithms are not too demanding in that respect (Kalitventzeff et al., 1978). The Newton-Raphson method is suitable for small problems and requires a solution of successive linearizations of the original problem Eq. (6):

$$\begin{vmatrix} X \\ Z \\ \Lambda \end{vmatrix} = J^{-1} \begin{vmatrix} C^{-1}Y \\ 0 \\ -d \end{vmatrix} \tag{9}$$

where the Jacobian matrix J of the equation system has the following structure:

$$J = \begin{vmatrix} C^{-1} & 0 & A^T \\ 0 & 0 & B^T \\ A & B & 0 \end{vmatrix} \tag{10}$$

Numerical algorithms embedding a step size control, such as Powell's dogleg algorithm (Chen and Stadtherr 1981) are quite successful for larger problems.

When solving very large problems, it is necessary to exploit the sparsity of the Jacobian matrix and use appropriate solution algorithms, such as those described by Chen and Stadtherr (1984a). It is common to assume that measurements are independent, which reduces the weight matrix C^{-1} to a diagonal. Ideally, the elements of matrices A and B should be evaluated analytically. This is straightforward for the elements corresponding to mass balance equations, which are linear, but can be difficult when the equations involve physical properties obtained from an independent physical property package.

The solution strategy exposed above does not allow one to handle inequality constraints. This justifies the use of alternative algorithms to solve directly the nonlinear programming (NLP) problem defined by Eq. (6). Sequential quadratic programming (SQP) is the method of choice (Chen and Stadtherr 1984a; Kyriakopoulou and Kalitventzeff 1996, 1997). At each iteration, an approximation of the original problem is solved: the original objective function being quadratic is retained and the model constraints are linearized around the current estimate of the solution.

Before solving the NLP problem, some variable classification and preanalysis is needed to identify unobservable variables, parameters, and nonredundant measurements. Measured variables can be classified as *redundant* (if the measurement is absent or detected as a gross error, the variable can still be estimated from the model) or *nonredundant*. Likewise, unmeasured variables are classified as *observable* (estimated uniquely from the model) or *unobservable*. The reconciliation algorithm will correct only redundant variables. If some variables are not observable, the program will either request additional measurements (and possibly suggest a feasible set) or solve a smaller subproblem involving only observable variables. The preliminary analysis should also detect *overspecified variables* (particularly those set to constants) and *trivial redundancy*, where the measured variable does not depend at all upon its measured value but is inferred directly from the model. Finally, it should also identify model equations that do not influence the reconciliation, but are merely used to calculate some unmeasured variables. Such preliminary tests are extremely important, especially when the data reconciliation runs as an automated process. In particular, if some measurements are eliminated as gross errors due to sensor failure, nonredundant measurements can lead to unobservable values and nonunique solutions, rendering the estimates and fitted values useless. As a result, these cases need to be detected in advance through variable classification. Moreover, under these conditions, the NLP may be harder to converge.

3.3.3
Sensitivity Analysis

Solving the data reconciliation problem provides more than validated measurements. A sensitivity analysis can also be carried out. It is based on the linearization of equation system in Eq. (9), possibly augmented to take into account active inequality constraints.

Equation (9) shows that reconciled values of process variables x and z, and of Lagrange multipliers Λ are linear combinations of the measurements. Thus their covariance matrix is directly derived from the measurements covariance matrix (Heyen et al. 1996).

Knowing the variance of validated variables allows one to detect the respective importance of all measurements in the state identification problem. In particular, some measurements might appear to have little effect on the result and might thus be discarded from analysis. Some measurements may appear to have a very high impact on key validated variables and on their variance: these measurements should be carried out with special caution, and it may prove wise to duplicate the corresponding sensors.

The standard deviation of validated values can be compared to the standard deviation of the raw measurement. Their ratio measures the improvement in confidence brought by the validation procedure. A nonredundant measurement will not be improved by validation. The reliability of the estimates for unmeasured observable variables is also quantified.

The sensitivity analysis also allows one to identify all state variables dependent on a given measurement, as well as the contribution of the measurement variance to the variance of the reconciled value. This information helps locate critical sensors, whose failure may lead to troubles in monitoring the process.

A similar analysis can be carried out for all state variables, whether measured or not. For each variable, a list of all measurements used to estimate its reconciled value is obtained. The standard deviation of the reconciled variable is calculated, but also its sensitivity with respect to the measurement's standard deviation. This allows one to locate sensors whose accuracy should be improved in order to reduce the uncertainty affecting the major process performance indicators.

3.3.4
Dynamic Data Reconciliation

The algorithm described above is suitable for analyzing steady-state processes. In practice it is also used to handle measurements obtained from processes operated close to steady state, with small disturbances. Measurements are collected over a period of time and average values are treated with the steady state algorithm. This approach is acceptable when the goal is to monitor some performance parameters that vary slowly with time, such as the fouling coefficient in heat exchangers. It is also useful when validated data are needed, to fine tune a steady-state simulation model, e.g., before optimizing set point values that are updated once every few hours.

However, a different approach is required when the transient behavior needs to be monitored accurately. This is the case for regulatory control applications, where data validation has to treat data obtained with a much shorter sampling interval. Dynamic material and energy balance relationships must then be considered as a constraint.

The earliest algorithm was proposed by Kalman (1960) for the linear time-invariant system model.

The general nonlinear process model describes the evolution of the state variables **x** by a set of ordinary differential equations (ODE):

$$\dot{\mathbf{x}} = f(t, \mathbf{x}, \mathbf{u}) + w(t) \tag{11}$$

where **x** are state variables, **u** are process inputs, and $w(t)$ is white noise with zero mean and covariance matrix $\mathbf{R}(t)$.

To model the measurement process, one usually considers sampling at discrete times $t = k\mathrm{T}$, and measurements related to state variables by:

$$\mathbf{y}_k = h(\mathbf{x}_k) + v_k \tag{12}$$

where measurement errors are normally distributed random variables with zero mean and covariance matrix $\mathbf{Q}_k$. One usually considers that process noise w and measurement noise v are not correlated.

By linearizing Eqs. 11 and 12 at each time step around the current state estimates, an *extended Kalman filter* can be built (see, for instance, Narasimhan and Jordache

2000). It allows one to propagate an initial estimate of the states and the associated error covariance, and to update them at discrete time intervals using the measurement innovation (the difference between the measured values and the predictions obtained by integrating the process model from the previous time step).

An alternative approach relies on NLP techniques. As proposed by Liebman et al. (1992), the problem can be formulated as

$$\min_{\mathbf{x}} \frac{1}{2} \sum_{j=t_0}^{t_N} \left(\mathbf{y}_j - \mathbf{x}(t_j)\right)^{\mathrm{T}} \mathbf{Q}_j^{-1} \left(\mathbf{y}_j - \mathbf{x}(t_j)\right) \tag{13}$$

subject to

$$\mathbf{f}\left(\frac{d\mathbf{x}(t)}{dt}, \mathbf{x}(t)\right) = 0; \quad \mathbf{x}(t_0) = \hat{\mathbf{x}}_0 \tag{14}$$

$$h\left(\mathbf{x}(t)\right) = 0 \tag{15}$$

$$g\left(\mathbf{x}(t)\right) \leq 0 \tag{16}$$

In this formulation, we expect that all state variables can be measured. When some measurements are not available, this can be handled by introducing null elements in the weight matrix $\mathbf{Q}$. Besides enforcing process specific constraints, the equalities in Eq. (15) can also be used to define nonlinear relationships between state variables and some measurements.

All measurements pertaining to a given time horizon $[t_0...t_N]$ are reconciled simultaneously. Obviously, the calculation effort increases with the length of the time horizon, and thus with the number of measurements. A tradeoff exists between calculation effort and data consistency. If measurements are repeated N times in the horizon interval, each measured value will be reconciled N times with different neighboring measurements, as long as it is part of the moving horizon. Which set of reconciled values is the "best" and should be considered for archiving is an open question. The value corresponding to the latest time t_N will probably be selected for online control application, while a value taken in the middle of the time window might be preferred for archiving or offline calculations.

Two solution strategies can be considered. The sequential solution and optimization combines an optimization algorithm such as SQP with an ODE solver. Optimization variables are the initial conditions for the ODE system. Each time the optimizer sets a new value for the optimization variables, the differential equations are solved numerically and the objective function Eq. (13) is evaluated. This method is straightforward, but not very efficient: accurate solutions of the ODE system are required repeatedly and handling the constraints Eqs. (15) and (16) might require a lot of trial and error. An implementation of this approach in a MATLAB environment is described by Romagnoli and Sanchez (2000).

Simultaneous solution and optimization is considered more efficient. The differential constraints are approximated by a set of algebraic equations using a weighted residuals method, such as orthogonal collocation. Predicted values of the state variables are thus obtained by solving the resulting set of algebraic equations, supplemented by the algebraic constraints of Eqs. (15) and (16). With this transformation, the distinction between dynamic data reconciliation and steady state data reconciliation vanishes. However this formulation requires solving a large NLP problem. This approach was first proposed by Liebman et al. (1992).

3.4
Software Solution

Data reconciliation is a functionality that is now embedded in many process analysis and simulation packages or is proposed as a standalone software solution. Bagajewicz and Rollins (2002) present a review of eight commercial and one academic data reconciliation packages. Most of them are limited to material and component balances.

More advanced features are only available in a few packages: direct connectivity to DCS systems for online applications, access to an extensive physical property library, handling pseudocomponents (petroleum fractions), simultaneous data validation, and identification of process performance indicators, sensitivity analysis, automatic gross error detection and correction, a model library for major process unit modules, handling of rigorous energy balances and phase equilibrium constraints, evaluation of confidence limits for all estimates. The packages offering the larger sets of features are Datacon (Invensys) (2004) and Vali (Belsim) (2004).

Dynamic data reconciliation is still an active research topic (Binder et al., 1998). It is used in combination with some real-time optimization applications, usually in the form of custom-developed extended Kalman Filters (see, for instance, Musch et al. (2004)), but dedicated commercial packages have yet to reach the market.

3.5
Integration in the Process Decision Chain

Data reconciliation is just one step – although an important step – in the data processing chain. Several operations, collectively known as data validation, are executed sequentially:
- Raw measurements are filtered to eliminate some random noise. When data is collected at high frequency, a moving average might be calculated to reduce the signal variance.
- If steady state data reconciliation is foreseen, the steady state has to be detected.
- Measurements are screened in order to detect outliers, or truly abnormal values (out of feasible range, e.g., negative flow rate).
- The state of the process might be identified when the plant can operate in different regimes or with a different set of operating units. Principal Component Analysis (PCA) analysis is typically used for that purpose, and allows one to select a reference case and to assign the right model structure to the available data set. This

step also allows some gross error detection (measurement set deviates significantly from all characterized normal sets).
- Variable classification takes place in order to verify that redundancy is present in the data set and that all state variables can be observed.
- The data reconciliation problem is solved.
- A global Chi-square test can detect the presence of gross errors.
- *A posteriori* uncertainty is calculated for all variables, and corrections are compared to the measurement standard deviation. In an attempt to identify gross errors, sequential elimination of suspect measurements (those with large corrections) can possibly identify suspect sensors. Alternatively, looking at subsystems of equations linking variables with large corrections allows one to pinpoint suspect units or operations in the plant.
- Key performance indicators and their confidence limits are evaluated and made available for reporting.
- Model parameters are tuned based on reconciled measurements and made available to process optimizers.

3.6
Optimal Design of Measurement System

The quality of validated data obviously depends on the quality of the measurement. Recent studies have paid more attention to this topic. The goal is to design measurement systems allowing one to achieve a prescribed accuracy in the estimates of some key process parameters, and to secure enough redundancy to make the monitoring process resilient with respect to sensor failures. Some preliminary results have been published, but no general solution can be found addressing large-scale nonlinear systems or dynamics.

Madron (1972) solved the linear mass balance case using a graph-oriented method. Meyer et al. (1994) proposed an alternative minimum-cost design method based on a similar approach. Bagajewicz (1997) analyzed the problem for mass balance networks, where all constraint equations are linear. Bagajewicz and Sanchez (1999) also analyze reallocation of existing sensors. The design and retrofit of a sensor network was also analyzed by Benqlilou et al. (2004) who discussed both the strategy and tools structure.

3.6.1
Sensor Placement based on Genetic Algorithm

A model-based sensor location tool, making use of a genetic algorithm to minimize the investment cost of the measurement system has been proposed by Heyen et al. (2002) and further developed by Gerkens and Heyen (2004).

They propose a general mathematical formulation of the sensor selection and location problem in order to reduce the cost of the measurement system while providing estimates of all specified key process parameters within a prescribed accuracy. The goal is to extend the capability of previously published algorithms and to address a broader problem, not being restricted to flow measurements and linear constraints.

The set of constraint equations is obtained by linearizing the process model at the nominal operating conditions, assuming steady state. The process model is complemented with link equations that relate the state variables to any accepted measurements, or to key process parameters whose values should be estimated from the set of measurements. In our case, the set of state variables for process streams comprises all stream temperatures, pressures and partial molar flow rates. In order to handle total flow rate measurements, the link equation describing the mass flow rate as the sum of all partial molar flow rates weighted by the component's molar mass has to be defined. Similarly, link equations relating the molar or mass fractions to the partial molar flow rates have also to be added for any stream where an analytical sensor can be located.

Link equations also have also to be added to express key process parameters, such as heat transfer coefficients, reaction extents or compressor efficiencies.

In the optimization problem formulation, the major contribution to the objective function is the annualized operating cost of the measurement system. In the proposed approach, we will assume that all variables are measured; those that are actually unmeasured will be handled as measured variables with a large standard deviation. Data reconciliation requires a solution of the optimization problem described by Eq. (5). The weight matrix $\mathbf{W} = \mathbf{C}^{-1}$ is limited to diagonal terms, which are the inverse of the measurement variance. The constrained problem is transformed into an unconstrained one using the Lagrange formulation as previously shown.

Assuming all state variables are measured, the solution takes the following form:

$$\begin{bmatrix} \mathbf{X} \\ \mathbf{\Lambda} \end{bmatrix} = \begin{bmatrix} \mathbf{W} & \mathbf{A}^{\mathrm{T}} \\ \mathbf{A} & 0 \end{bmatrix}^{-1} \begin{bmatrix} \mathbf{WY} \\ -\mathbf{d} \end{bmatrix} = \mathbf{M}^{-1} \begin{bmatrix} \mathbf{C}^{-1}\mathbf{Y} \\ -\mathbf{d} \end{bmatrix} \tag{17}$$

The linear approximation of the constraints is easily obtained from the solution of the nonlinear model, since $\mathbf{A}$ is the Jacobian matrix of the nonlinear model evaluated at the solution.

Thus matrix $\mathbf{M}$ can be easily built, knowing the variance of measured variables appearing in submatrix $\mathbf{W}$ and the model Jacobian matrix $\mathbf{A}$ (which is constant). This matrix will be modified when assigning sensors to variables. Any diagonal element of matrix $\mathbf{W}$ will remain zero (corresponding to infinite variance) as long as a sensor is not assigned to the corresponding process variable; it will be computed from the sensor precision and the variable value when a sensor is assigned in Section 3.6.2.3. Equation (17) need not be solved, since measured values Y are not known. However the variances of the reconciled values X depend only on the variance of measurements as shown in Heyen et al. (1996):

$$\mathrm{var}\left(X_i\right) = \sum_{j=1}^{m} \frac{\left([\mathbf{M}^{-1}]_{ij}\right)^2}{\mathrm{var}\left(Y_j\right)} \tag{18}$$

The elements of $\mathbf{M}^{-1}$ are obtained by calculating a lower and upper triangular (LU) factors of matrix $\mathbf{M}$. In the case when matrix $\mathbf{M}$ is singular, we can conclude that the measurement set has to be rejected, since it does not allow observation of all variables. Row i of $\mathbf{M}^{-1}$ is obtained by back substitution using the LU factors, using a right-hand-side vector whose components are δ_{ij} (Kronecker factor: $\delta_{ij} = 1$ when $i = j$, $\delta_{ij} = 0$ otherwise).

In the summation of Eq. (18), only the variables Y_j that have been assigned a sensor are considered, since the variance of unmeasured variables has been set to infinity.

3.6.2
Detailed Implementation of the Algorithm

Solution of the sensor network problem is carried out in seven steps:

1. process model formulation and definition of link equations;
2. model solution for the nominal operating conditions and model linearization;
3. specification of the sensor database and related costs;
4. specification of the precision requirements for observed variables;
5. verification of problem feasibility;
6. optimization of the sensor network;
7. report generation.

Each of the steps is described in detail before presenting a test case.

3.6.2.1
Process Model Formulation and Definition of Link Equations

In the current implementation, the process model is generated using the model editor of the Vali 3 data validation software, which is used as the basis for this work (Belsim 2004). The model is formulated by drawing a flow sheet using icons representing the common unit operations, and linking them with material and energy streams. Physical and thermodynamic properties are selected from a range of physical property models. Any acceptable measurement of a quantity that is not a state variable (T, P, partial molar flow rate) requires the definition of an extra variable and the associated link equation, which is done automatically for standard measurement types (e.g., mass or volume flow rate, density, dew point, molar or mass fractions, etc.). Similarly, extra variables and link equations must be defined for any process parameter to be assessed from the plant measurements. A proper choice of extra variables is important, since we may note that many state variables can not be measured in practice (e.g., no device exists to directly measure a partial molar flow rate or an enthalpy flow).

In order to allow the model solution, enough variables need to be set by assigning them values corresponding to the nominal operating conditions. The set of specified variables must at least match the degrees of freedom of the model, but overspecifications are allowed, since a least square solution will be obtained by the data reconciliation algorithm.

3.6.2.2

Model Solution for the Nominal Operating Conditions and Model Linearization

The data reconciliation problem is solved either using a large-scale SQP solver, or the Lagrange multiplier approach. When the solution is found, the value of all state variables and extra variables is available, and the sensitivity analysis is carried out (Heyen et al. 1996). A dump file is generated, containing all variable values, and the nonzero coefficients of the Jacobian matrix of the model and link equations. All variables are identified by a unique tag name indicating its type (e.g., S32.T is the temperature of stream S32, E102.K is the overall heat transfer coefficient of heat exchanger E102, and S32.MFH2O is the molar fraction of component H2O in stream S32).

3.6.2.3

Specification of the Sensor Database and Related Costs

A data file must be prepared that defines for each acceptable sensor type the following parameters:

- the sensor name;
- the annualized cost of operating such a sensor;
- parameters a_i and b_i of the equation allowing to estimate the sensor accuracy from the measured value y_i, according to the relation: $\sigma_i = a_i + b_i y_i$;
- a character string pattern to match the name of any process variable that can be measured by the given sensor (e.g., a chromatograph will match any mole fraction, and will thus have the pattern MF*, while an oxygen analyzer will be characterized by the pattern MFO2).

3.6.2.4

Specification of the Precision Requirements for Observed Variables

A data file must be prepared that defines the precision requirements for the sensor network after processing the information using the validation procedure. The following information is to be provided for all specified key performance indicators or for any process variable to be assessed:

- the composite variable name (stream or unit name + parameter name);
- the required standard deviation σ_i^t, either as an absolute value, or as a percentage of the measured value.

3.6.2.5

Verification of Problem Feasibility

Before attempting to optimize the sensor network, the program first checks for the existence of a solution. It solves the linearized data reconciliation problem assuming all possible sensors have been implemented. In the case where several sensors are available for a given variable, the most precise one is adopted. This also provides an upper limit C_{max} for the cost of the sensor network.

A feasible solution is found when two conditions are met:

- the problem matrix **M** is not singular.
- the standard deviation σ_i of all selected reconciled variables is lower than the specified value σ_i^t.

When the second condition is not met, several options can be examined. One can extend the choice of sensors available in the sensor definition file by adding more precise instruments. One can also extend the choice of sensors by allowing measurement of other variable types. Finally, one can modify the process definition by adding extra variables and link equations, allowing more variables besides state variables to be measured.

3.6.2.6
Optimization of the Sensor Network

Knowing that a feasible solution exists, one can start a search for a lower cost configuration. The optimization problem as posed involves a large number of binary variables (in the order of number of streams × number of sensor types). The objective function is multimodal for most problems. However, identifying sets of suboptimal solutions is of interest, since criteria besides cost might influence the selection process. Since the problem is highly combinatorial and not differentiable, we attempted to solve it using a genetic algorithm (Goldberg 1989). The implementation we adopted is based on the freeware code developed by Carroll (1998). The selection scheme used involves tournament selection with a shuffling technique for choosing random pairs for mating. The evolution algorithm includes jump mutation, creep mutation, and the option for single-point or uniform crossover.

The sensor selection is represented by a long string (gene) of binary decision variables (chromosomes). In the problem analysis phase, all possible sensor allocations are identified by finding matches between variable names (see Section 3.6.2.2) and sensor definition strings (see Section 3.6.2.3). A decision variable is added each time a match is found. Multiple sensors with different performance and cost can be assigned to the same process variable.

The initial gene population is generated randomly. Since we know from the number of variables and the number of constraint equations the number of degrees of freedom of the problem, we can bias the initial sensor population by fixing a rather high probability of selection (typically 80 %) for each sensor. We found however that this parameter is not critical. The initial population count does not appear to be critical either. Problems with a few hundred binary variables were solved by following the evolution of populations of 10–40 genes, 20 being our most frequent choice.

Each time a population is generated, the fitness of its members must be evaluated. For each gene representing a sensor assignment, we can estimate the cost C of the network, by summing the individual costs of all selected sensors. We also have to build the corresponding matrix **M** (Eq. (3b)) and factorize it, which is done using a code exploiting the sparsity of the matrix.

The standard deviation σ_i of all process variables is then estimated using Eq. (18).

This allows calculating a penalty function P that takes into account the uncertainty affecting all observed variables. This penalty function sums penalty terms for all m target variables.

$$P = \sum_{i=1}^{m} P_i \qquad (19)$$

where $P_i = \dfrac{\sigma_i}{\sigma_i^t}$ when $\sigma_i \leq \sigma_i^t$

and $P_i = 0.01 \, \min\left(10, \dfrac{\sigma_i}{\sigma_i^t}\right)^2$ when $\sigma_i > \sigma_i^t$

The fitness function F of the population is then evaluated as follows:

- if matrix **M** is singular, return $F = -C_{max}$
- otherwise return $F = -(C + P)$.

Penalty function Eq. (5) (slightly) increases the merit of a sensor network that performs better than specified. Penalty function Eq. (6) penalizes genes that do not meet the specified accuracy, but it does not reject them totally, since some of their chromosomes might code interesting sensor subnetworks.

The population is submitted to evolution according to the mating, crossover, and mutation strategy. Care is taken that the current best gene is always kept in the population, and is duplicated in case it should be submitted to mutation. After a specified number of generations, the value of the best member of the population is monitored. When no improvement is detected for a number of generations, the current best gene is accepted as a solution. There is no guarantee that this solution is an optimal one, but it is feasible and (much) better than the initial one.

3.6.2.7
Report Generation

The program reports the best obtained configurations as a list of sensors assigned to process variables to be measured. The predicted standard deviation for all process variables is also reported, as well as a comparison between the achieved and target accuracies for all key process parameters.

3.6.3
Perspectives

The software prototype described here has been further improved by allowing more flexibility in the sensor definition (e.g., defining acceptable application ranges for each sensor type) and by addressing retrofit problems by specifying an initial instrument layout. The capability of optimizing a network for several operating conditions has also been implemented. The solution time grows significantly with the number of potential sensors. In order to address this issue, the algorithm has been parallelized (Gerkens and Heyen 2004) and the efficiency of parallel processing remains good as long as the number of processors is a divisor of the chromosome population size. Full optimization of very complex processes remains a challenge, but suboptimal feasible solutions can be obtained by requiring observability for smaller subflowsheets.

The proposed method can be easily adapted to different objective functions besides cost to account for different design objectives. Possible objectives could address the resiliency of the sensor network to equipment failures, or the capability to detect gross errors, in the line proposed by Bagajewicz (2001).

There is no guarantee that this solution found with the proposed method is an optimal one, but it is feasible and (much) better than the initial one.

3.7
An Example

A simplified ammonia synthesis loop illustrates the use of data validation, including sensitivity analysis and the design of sensor networks.

The process model for this plant is shown in Figure 3.1. The process involves a five-component mixture (N_2, H_2, NH_3, CH_4, Ar), 10 units, 14 process streams, and 4 utility streams (ammonia refrigerant, boiler feed water, and steam).

Feed stream f0 is compressed before entering the synthesis loop, where it is mixed with the reactor product f14. The mixture enters the recycle compressor C-2 and is chilled in exchanger E-1 by vaporizing ammonia. Separator F-1 allows one to recover liquid ammonia in f5, separated from the uncondensed stream f6. A purge f7 leaves

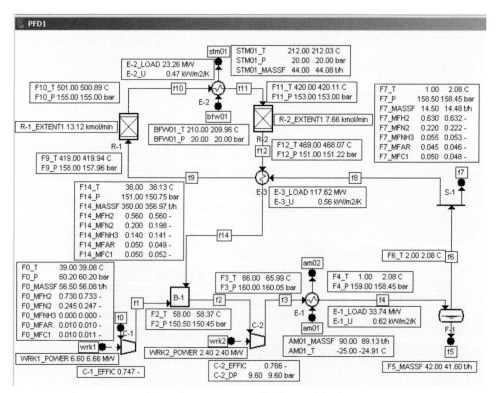

Figure 3.1 Data validation, base case. Measured and reconciled values are shown in result boxes as well as key performance indicators

the synthesis loop, while f8 enters the effluent to feed preheater E-3. The reaction takes place in two adiabatic reactors R-1 and R-2, with intermediate cooling in E-2, where steam is generated.

Energy balances and countercurrent heat transfer are considered in heat exchangers E-1, E-2, and E-3. Reactors R-1 and R-2 consider atomic balances and energy conservation. Compressors C-1 and C-2 take into account an isentropic efficiency factor (to be identified). Vapor-liquid equilibrium is verified in heat exchanger E-1 and in separator F-1.

The model comprises 160 variables, 89 being unmeasured. Overall, 118 equations have been written: 70 are balance equations and 48 are link equations relating the state variables (pressure, enthalpy and partial molar flow rates) either to variables that can be measured (temperature, molar fraction, and mass flow rate) or to performance indicators to be identified.

A set of measurements has been selected using engineering judgment. Values taken as measurements were obtained from a simulation model and disturbed by random errors.

The standard deviation assigned to the measurements was:

- $1\,°C$ for temperatures below $100\,°C$, $2\,°C$ for higher temperatures
- 1% of measured value for pressures
- 2% of measured values for flow rates
- 0.001 for molar fractions below 0.1, 1% of measured value for higher compositions
- 3% of the measured value for mechanical power.

Measured values are displayed in Figure 3.1, as are the validated results. The identified values of performance indicators are also displayed. These are the extent of the synthesis reaction in catalytic bed R-1 and R-2, the heat load and transfer coefficients in exchangers E-1, E-2 and E-3, and the isentropic efficiency of compressors C-1 and C-2.

Result analysis shows that all process variables can be observed. All measurement corrections are below 2σ, except for methane in stream f7.

The value of objective function Eq. (5) is 19.83, compared to a χ^2 threshold equal to 42.56. Thus, no gross error is suspected from the global test.

Sensitivity analysis reveals how the accuracy of some estimates could be improved. For instance, Table 3.1 shows the sensitivity analysis results for the heat transfer coefficient in unit E-1. The first line in the table reports the value, absolute accuracy and relative accuracy of this variable. The next rows in the table identify the measurements that have a significant influence on the validated value of the E-1 heat transfer coefficient. For instance, 77.57% of the uncertainty on U comes from the uncertainty of variable AM01-T (temperature of stream am01). The derivative of U with respect to AM01-T is equal to 0.12784. Thus one can conclude that the uncertainty on the heat transfer coefficient could be reduced significantly if a more accurate measurement of a single temperature is available.

Table 3.2 shows that the reaction extent in reactor R-2 can be evaluated without resorting to precise chemical analysis. The uncertainty for this variable is 4.35% of

Table 3.1 Sensitivity analysis for heat transfer coefficient in exchanger E-1

Variable		Tag Name	Value	Abs.Acc.	Rel.Acc.	Penal.	P.U.
K	U E-1	Computed	3.5950	0.14515	4.04 %		–

Measurement		Tag Name	Contrib.	Der.Val.	Rel.Gain	Penal.	P.U.
T	S AM01	AM01_T	77.57 %	0.12784	1.21 %	0.01	C
T	S AM02	AM02_T	5.75 %	−0.34800E-01	0.21 %	0.00	C
MFNH3	R F6	F7_MFNH3	4.33 %	−30.216	34.29 %	3.67	–
MASSF	R AM01	AM01_MASSF	4.05 %	0.16227E-01	46.50 %	0.23	t/h
MASSF	R F12	F14_MASSF	1.75 %	−0.27455E-02	33.79 %	0.99	t/h
T	S F7	F7_T	1.50 %	−0.17794E-01	62.36 %	1.16	C
T	S F6	F6_T	1.50 %	−0.17794E-01	62.36 %	0.01	C
T	S F4	F4_T	1.50 %	−0.17794E-01	62.36 %	1.16	C

Table 3.2 Sensitivity analysis for reaction extent in reactor R-2

Variable		Tag Name	Value	Abs.Acc.	Rel.Acc.	Penal.	P.U.
EXTENT1	U R-2	Computed	7.6642	0.33372	4.35 %		kmol min^{-1}

Measurement		Tag Name	Contrib.	Der.Val.	Rel.Gain	Penal.	P.U.
T	S F11	F11_T	26.82 %	−0.86410E-01	21.85 %	0.00	C
T	S F12	F12_T	25.13 %	0.83640E-01	26.78 %	0.22	C
T	S F9	F9_T	21.52 %	0.77397E-01	27.69 %	0.22	C
T	S F10	F10_T	19.95 %	−0.74532E-01	22.02 %	0.00	C
MASSF	R F5	F5_MASSF	1.56 %	0.49680E-01	49.64 %	0.23	t/h
MASSF	R BFW01	STM01_MASSF	1.51 %	0.46591E-01	35.39 %	0.01	t/h
MASSF	R AM01	AM01_MASSF	0.81 %	0.16647E-01	46.50 %	0.23	t/h
MASSF	R F0	F0_MASSF	0.77 %	0.25907E-01	58.25 %	0.14	t/h
MFNH3	R F12	F14_MFNH3	0.58 %	18.215	29.41 %	0.15	–

the estimated value and results mainly from the uncertainty in four temperature measurements. Better temperature sensors for streams f9, f10, f11 and f12 would allow one to better estimate the reaction extent.

This sensor network provides acceptable estimates for all process variables.

However the application of the sensor placement optimization using a genetic algorithm can identify a cheaper alternative.

Table 3.3 Cost, accuracy, and range for available sensors

Measured Variable	Relative cost	Standard deviation σ	Acceptable range
T	1	1 °C	$T < 150\,°C$
T	1	2 °C	$T > 150\,°C$
P	1	1 %	1–300 bar
Flow rate	5	2 %	$1-100 \text{ kg s}^{-1}$
Power	1	3 %	1–10,000 kW
Molar composition (all components in stream)	20	0.001 1 %	$x_i < 0.1$ $x_i > 0.1$

A simplified sensor data base has been used for the example. Only six sensor types were defined, with accuracies and cost as defined in Table 3.3.

Accuracy targets are specified for seven variables:

- two compressor efficiencies, target σ = 4 % of estimated value
- three heat transfer coefficients, target σ = 5 % of estimated value
- two reaction extents, target σ = 5 % of estimated value.

The program detects that up to 59 sensors could be installed. When all of them are selected, the cost is 196 units, compared to 42 sensors and 123 cost units for our initial guess shown in Figure 3.1. Thus the solution space involves $2^{59} = 5.76 \times 10^{17}$ solutions (most of them being unfeasible).

We let the search algorithm operate with a population of 20 chromosomes, and iterate until no improvement is noticed for 200 consecutive generations. This requires a total of 507 generations and 10,161 evaluations of the fitness function, which runs in 90 s on a laptop PC (1 GHz Intel Pentium III processor, program compiled with Compaq FORTRAN compiler, local optimization only). Figure 3.2 shows

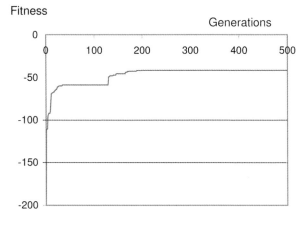

Figure 3.2 Evolution of fitness function with number of generations

that the fitness function value varies sharply in the first generations and later improves only marginally. A solution with a cost similar to the final one is obtained after 40 % of the calculation time.

The proposed solution involves only 26 sensors, for a total cost reduced to 53 cost units. The number of sensors is reduced from 16 to 11 for T, from 15 to 12 for P, from 6 to 2 for flow, and from 3 to 1 for composition. Thus the algorithm has been able to identify a solution satisfying all requirements with a considerable cost reduction.

3.8
Conclusions

Efficient and safe plant operation can only be achieved if the operators are able to monitor key process variables. These are the variables that either contribute to the process economy (e.g., yield of an operation) or are linked to the equipment quality (fouling in a heat exchanger, activity of a catalyst), to safety limits (departure from detonation limit), or to environmental considerations (amount of pollutant rejected).

Most performance parameters are not directly measured and are evaluated by a calculation based on several experimental data. Random errors that always affect any measurement also propagate in the estimation of performance parameters. When redundant measurements are available, they allow one to estimate the performance parameters based on several data sets, leading to different estimates, which may lead to confusion.

Data reconciliation allows one to address the state estimation and measurement correction problems in a global way by exploiting the measurement redundancy. Redundancy is no longer a problem, but an asset. The reconciled values exhibit a lower variance compared to original raw measurements; this allows process operation closer to limits (when this results in improved economy).

Benefits from data reconciliation are numerous and include:
- improvement of measurement layout;
- decrease of number of routine analyses;
- reduced frequency of sensor calibration: only faulty sensors need to be calibrated;
- removal of systematic measurement errors;
- systematic improvement of process data;
- clear picture of plant operating condition and reduced measurement noise in trends of key variables;
- early detection of sensor deviation and of equipment performance degradation;
- actual plant balances for accounting and performance follow-up;
- safe operation closer to the limits;
- quality at process level.

Current developments aim at combining online data acquisition with data reconciliation. Reconciled data are displayed in control rooms in parallel with raw measurements. Departures between reconciled and measured data can trigger alarms. Analysis of time variation of those corrections can draw attention to drifting sensors that need recalibration.

Data reconciliation can also be viewed as a virtual instrument; this approach is particularly developed in biochemical processes, where direct measurement of the key process variables (population of microorganisms and yield in valuable by-products) is estimated from variables that are directly measured online, such as effluent gas composition.

Current research aims at easing the development of data reconciliation models by employing libraries of predefined unit operations, automatic equation generation for typical measurement types, analyses of redundancy and observability, analyses of error distribution of reconciled values, interfaces to online data collection systems and archival data bases, and developing specific graphical user interfaces.

References

1 *Albuquerque J. S. Biegler L. T.* Data reconciliation and gross-error detection for dynamic systems. AIChE J. 42 (1996) p. 2841–2856

2 *Bagajewicz M. J.* Process Plant Instrumentation: Design and Upgrade. Chap. 6, Technomic Publishing Company, Lancaster PA (USA) (1997)

3 *Bagajewicz M. J.* Design and retrofit of sensor networks in process plants. AIChE J. 43(9) (2001) p. 2300–2306

4 *Bagajewicz M. J. Rollins D. K.* Data reconciliation. In: B.G. Liptak (ed.) Instrument Engineers' Handbook (3rd edn.), Vol. 3: Process Software and Digital Networks. CRC, Taylor and Francis, Boca Raton, FL (USA) (2002).

5 *Bagajewicz M. J. Sanchez M. C.* Design and upgrade of nonredundant and redundant linear sensor networks. AIChE J. 45(9) (1999) p. 1927–1938

6 *Belsim*, VALI 4 User's Guide. Belsim Saint-Georges-sur-Meuse, Belgium (2004).

7 *Benqlilou C., Graells M. Puigjaner L.* Decision-making strategy and tools for sensor networks design and retrofit. Ind. Eng. Chem. Res. 43 (2004) p. 1711–1722

8 *Binder T. Blank L. Dahmen W. Marquardt W.* Towards multiscale dynamic data reconciliation. In: R. Berber (ed.) Nonlinear Model-Based Process Control. NATO ASI series, Kluwer, Dordrecht (1998)

9 *Carroll D. L.* FORTRAN Genetic Algorithm Driver, version 1.7, http://www.staff.uiuc.edu/carroll/ga.html (1998) accessed July 2001.

10 *Chen H. S. Stadherr M. A.* A modification of Powell's dogleg algorithm for solving

systems of non-linear equations. Comput. Chem. Eng. 5(3) (1981) p. 143–150

11 *Chen H. S. Stadherr M. A.* On solving large sparse nonlinear equation systems. Comput. Chem. Eng. 8(1) (1984a) p. 1–6

12 *Chen H. S. Stadherr M. A.* Enhancements of Han-Powell method for successive quadratic programming. Comput. Chem. Eng. 8(3/4) (1984b) p. 299–234 (1984b)

13 *Crowe C. M.* Observability and redundancy of process data for steady state reconciliation. Chem. Eng. Sci. 44 (1989) p. 2909–2917

14 *Crowe C. M.* Data reconciliation – progress and challenges. J. Process Control. (6) (1996) p. 89–98

15 *Datacon Invensys, http://www.simsci.com/products/datacon.stm, cited 3 May (2004)*

16 *Gerkens C. Heyen G.* Use of Parallel Computers in Rational Design of Redundant Sensor Networks. 14th European Symposium on Computer Aided Process Engineering, Lisbon (2004)

17 *Goldberg D. E.* Genetic Algorithms in Search, Optimization and Machine Learning. Addison-Wesley, Reading, MA (USA) (1989)

18 *Heyen G. Dumont M. N. Kalitventzeff B.* Computer aided design of redundant sensor networks. In: Grievnik J. Dumont M. N. Kalitventzeff B. 12th European symposium on Computer Aided Process Engineering The Hague, Elsevier Science, Amsterdam (2002)

19 *Heyen G. Maréchal E. Kalitventzeff B.* Sensitivity calculations and variance analysis in plant measurement reconciliation. Comput. Chem. Eng. 20S (1996) p. 539–544

20 *Joris P. Kalitventzeff B.* Process measurements analysis and validation. In: Proceedings Chemical Engineering Fundamentals Conference (CEF'87): Use of Computers in Chemical Engineering, Italy, pp. 41–46 (1987)

21 *Kalitventzeff B., Laval P. Gosset R. Heyen G.* The validation of industrial measurements, a necessary step before the parameter identification of the simulation model for large chemical engineering systems. In: Proceedings of International Congress "Contribution des calculateurs électroniques au développement du génie chimique," Sociètè de Chimie Industrielle, Paris (1978)

22 *Kalman R. E.* A new approach to linear filtering and prediction problems. Trans. ASME J. Basic Eng. 82D (1960) p. 35–45

23 *Kuehn D. R. Davidson H.* Computer control: mathematics of control. Chem. Eng. Prog. 57 (1961) p. 44–47

24 *Kyriakopoulou D. J. Kalitventzeff B.* Data reconciliation using an interior point sqp, ESCAPE-6, Rhodes, Greece, 26–29 May, Pergamon Press (1996)

25 *Kyriakopoulou D. J.Kalitventzeff B.* Reduced Hessian interior point SQP for large-scale process optimization. First European Congress on Chemical Engineering, Florence, Italy, 4–7 May AIDIC Milano (1997)

26 *Liebman M. J. Edgar T. F. Lasdon L. S.* Efficient data reconciliation and estimation for dynamic processes using nonlinear programming techniques. Comput. Chem. Eng. 16 (1992) p. 963–986

27 *Madron F.* Process Plant Performance: Measurement and Data Processing for Optimization and Retrofits. Ellis Horwood, London (1992)

28 *Mah R. S. H. Stanley G. M. Downing D. W.* Reconciliation and Rectification of Process Flow and Inventory Data. Ind. and Eng. Chem. Proc. Des. Dev. 15 (1976) p. 175–183

29 *Meyer M. Le Lann J. M. L. Koehert B. Enjalbert M.* Optimal selection of sensor location on a complex plant using graph-oriented approach. In: Moser F. Schnitzer H. Bart H.-J. (eds.) European Symposium on Cumputer-Aided Process Engineering-3, Graz, Austria (1993). Suppl. to Computers and Chemical Engineering, Pergamon Press, Oxford (1993)

30 *Musch H. List T. Dempf D. Heyen G.* On-line Estimation of Reactor Key Performance Indicators: an Industrial Case Study. (2004)

29 *Narasimhan S. Jordache C.*, Data Reconciliation and Gross Error Detection, an Intelligent use of Process Data. Gulf, Publishing Company, Houston, TX (USA) (2000)

30 *Romagnoli J. A. Sanchez M. C.* Data Processing and Reconciliation for Chemical Process Operations. Academic Press, San Diego, CA (USA) (2000)

31 *Vaclavek V.* Studies on system engineering: on the application of the calculus of observations in calculations of chemical engineering balances. Coll. Czech. Chem. Commun. 34 (1968) p. 3653–3660

32 *Vaclavek V.* Studies on System Engineering: Optimal Choice of the Balance Measurements in Complicated Chemical Engineering Systems. Chem. Eng. Sci. 24 (1969) p. 947–955

33 *Vali Belsim* http://www.belsim.com/Products-main.htm, cited 3 May 2004

34 *Veverka V. V. Madron F.* Material and energy balancing in the process industries. From microscopic balances to large plants, Computer-Aided Chemical Engineering. Elsevier Science, Amsterdam (1996)

4
Model-based Control

Sebastian Engell, Gregor Fernholz, Weihua Gao, and Abdelaziz Toumi

4.1
Introduction

As explained in several chapters of this volume, rigorous process models can be used to optimize the design and the operating parameters of chemical processing plants. However, optimal settings of the parameters do not guarantee optimal operation of the real plant. The reasons for this are the inevitable plant-model mismatches, the effects of disturbances, changes in the plant behavior over time, etc. Usually not even the constraints on process or product parameters are met at the real plant if operating parameters that were obtained from offline optimization are applied.

The only effective way to cope with the effect of plant-model mismatch, disturbances etc. is to use some sort of feedback control. Feedback control means that (some of) the degrees of freedom of the plant are modified based on the observation of measurable variables. These measurements may be performed quasicontinuously or with a certain sampling period, and accordingly the operation parameters (termed inputs in feedback control terminology) may be modified in a quasicontinuous fashion or intermittently. Often, key process parameters cannot be measured online at a reasonable cost. One important use of process models in process control is the model-based estimation of such parameters from the available measurements. This topic has been dealt with in the previous chapter. In this chapter, we focus on the use of rigorous process models for feedback control by model-based online optimization.

Feedback control can be combined with model-based optimization in several different ways. The simplest, and most often used, approach is to perform an offline optimization and to divide the degrees of freedom into two groups. The variables in the first group are applied to the real process as they were computed by the offline optimization. The variables in the second group are used to control some other variables to the values which resulted from the offline optimization, e.g., requirements on purities are met by controlling the product concentration by manipulating the feed rate to a reactor or the reflux in a distillation column. In the design of these feedback controllers, dynamic plant models are used, in most cases obtained from a line-

Computer Aided Process and Product Engineering. Edited by Luis Puigjaner and Georges Heyen
Copyright © 2006 WILEY-VCH Verlag GmbH & Co. KGaA, Weinheim
ISBN: 3-527-30804-0

arization of the rigorous model around the optimal operating regime or process trajectory. If nonlinear process models are available from the design stage, these models can be used directly in model-based control schemes. This leads to nonlinear model-predictive control (NMPC) where the future values of the controlled variables are predicted over a finite horizon (the prediction horizon) using the model, and the future inputs are optimized over a certain horizon (the control horizon). The first inputs' values are applied to the plant. Thereafter, the procedure is repeated, taking new measurements into account. A major advantage of this approach is the ability to include process constraints in the optimization, thus exploiting the full potential of the plant and the available actuators (pumps, valves) and respecting operating limits of the equipment. In Section 4.2, NMPC around a precomputed trajectory of the process is presented in more detail and its application to a reactive semibatch distillation process is discussed.

When closed-loop control is used to track a precomputed trajectory and the controllers perform satisfactorily, the process is kept near the operating point that was computed as the optimal one offline. Those variables which are under feedback control track their precomputed set-point even in the presence of disturbances and plant-model mismatch. However, the overall operation will in general no longer be optimal, because the precomputed operating regime is optimal for the nominal plant model, but not for the real plant.

As an extension of this concept, feedback control can be combined with model adaptation and reoptimization. At a lower sampling rate than the one used for control, some model parameters are adapted based upon the available measurements. After the model has been updated, it is used for a reoptimization of the operating regime. The new settings can be implemented directly or be realized by feedback. In Section 4.3, such a control scheme is presented for the example of batch chromatographic separations, including experimental results.

A serious problem in practice is structural plant-model mismatch. This means that an adaptation of the model parameters, even for an infinite number of noise-free measurements, will not give a model that accurately represents the real process. Therefore if the structurally incorrect model is used in optimization, the resulting operating parameters will not be optimal; often, not even the constraints will be met by the real process unless the constrained variables are under feedback control with some safety margin that reflects the attainable control performance, which again causes a suboptimal operation.

A solution to the problem of plant-model mismatch is the use of optimization strategies that incorporate feedback directly, i.e., use the information gained by online measurements not only to update the model but also to modify the optimization problem. In Section 4.4, this idea is presented in detail and the application to batch chromatography is used to demonstrate its potential.

NMPC involves online optimization on a finite horizon based upon a nonlinear plant model. This approach can be employed not only to keep some process variables at their precomputed values or make them track certain trajectories, but also to perform online predictive optimization of the plant performance. Bounds, e.g., on product specifications, can be included in the formulation as constraints rather than setting up a separate feedback control layer to meet the specifications at the real plant. In this spirit, the problem of controlling quasicontinuous (simulated moving bed) chromatographic separations is formulated in Section 4.5 as an online optimization problem, where the measured outputs have to meet the constraints on the product purities but the optimization goal is not tracking of a precomputed trajectory, but optimal process operation.

4.2
NMPC Applied to a Semibatch Reactive Distillation Process

4.2.1
Formulation of the Control Problem

In NMPC, a process model is used to predict the future process outputs $\hat{\mathbf{y}}$ over a fixed prediction time horizon H_p for given sequences of H_r changes of the manipulated variables $\mathbf{u}$. The aim of the controller is to minimize a quadratic function of the deviation between the process outputs $\hat{\mathbf{y}}$ and their desired trajectories $\mathbf{y}^{\text{ref}}$ as well as of the changes of the manipulated variables. The control move at the sampling point $k+1$ is given by the optimization problem Eq. (1). The parameters γ_{ij} and λ_{ij} allow scaling the controlled and manipulated variables and shifting the weight either on good setpoint tracking or on smooth controller actions. Bounds on the manipulated variables can be enforced by using sufficiently large penalties λ_{ij} or by adding inequality constraints (Eq. (2)) to the optimization problem Eq. (1):

$$\min_{\mathbf{u}_{k+1},\ldots,\mathbf{u}_{k+H_r}} \left\{ \sum_{i=1}^{H_p} \sum_{j=1}^{m} \gamma_{ij} \left(y_{j,k+i}^{\text{ref}} - \hat{y}_{j,k+i} \right)^2 + \sum_{i=1}^{H_r} \sum_{j=1}^{r} \lambda_{ij} \left(u_{j,k+i}^{\text{ref}} - u_{j,k+j} \right)^2 \right\} \tag{1}$$

$$u_j^{\min} \leq u_{k+j} \leq u_j^{\max} \quad \forall j = 1, \ldots, H_r. \tag{2}$$

If the control scheme is applied to a real plant, plant-model mismatch or disturbances will lead to differences between the predicted and the real process outputs. Therefore a time-varying disturbance model, as proposed by Draeger et al. (1995), is included in the process model. The formal representation of the complete model that is used by the model predictive controller is

$$\hat{y}_{(k+i)} = y_{(k+i)}^{\text{model}} + \mathbf{d}_{k,i} \quad \forall i = 1, \ldots, H_p \tag{3}$$

where $\mathbf{d}_{k,i}$ denotes the estimated disturbances, y_k^{model} denotes the model outputs given by the physical process model, and $\hat{y}_k$ the model prediction of the controller used in the optimization problem (1). The disturbances $\mathbf{d}_{k,i}$ are recalculated at every time k for each time step i. The process model is simulated from time k−i until time k taking into account the actual control actions giving the model outputs $y^{\text{model}}(k|k\text{-}i)$. The errors $\mathbf{e}_{k,i}$ computed as the differences between the measurements y_k^{meas} and the model outputs $y^{\text{model}}(k|k\text{-}i)$:

$$e_{k,i} = y_k^{meas} - y^{model}(k \mid k - i).$$

(4)

The new estimates of the disturbances are calculated by a first order filter:

$$d_{k,i} = \alpha e_{k,i} + (1 - \alpha)d_{k-1,i}.$$

(5)

4.2.2
The Methyl Acetate Process

Methyl acetate is produced from acetic acid and methanol in an esterification reaction. The conventional process consists of a reactor and a complex distillation column configuration, while using reactive distillation, high purity methyl acetate can be produced in a single column (Agreda et al. 1990). The reaction can either by catalyzed homogeneously by sulfonic acid or heterogeneously using a solid catalyst. The latter avoids material problems caused by the sulfonic acid as well as the removal of the catalyst at the end of the batch. This process is investigated here. A scheme of the process is shown in Figure 4.1.

The column consists of three parts. Two structured catalytic packings of 1 m height are located in the lower part of the column while the upper part contains a noncatalytic packing. Methanol is filled into the reboiler before the beginning of the batch and heated until the column is filled with methanol. Acetic acid is fed to the column above the reactive section. Since acetic acid is the highest boiling component, it is necessary to feed it above the catalytic packing in order to ensure that both raw materials are present in the catalytic area in sufficient concentrations. The upper section purifies the methyl acetate. The azeotropes of the mixture are overcome because water and acetic acid are present in the stream that enters the separation stages. The plant considered here is a pilot plant in the Department of Biochemical and Chemical Engineering at Universität Dortmund. A batch run takes approximately 17 h.

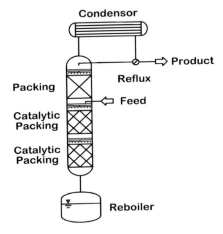

Figure 4.1 Scheme of the semibatch column

A more detailed description of the process and a rate-based model and its validation are presented in Kreul et al. (1998) and Noeres (2003). The latter pointed out that
for this process the accuracy of a rate-based model is not significantly higher than
that of an equilibrium stage model, and thus the equilibrium stage model was used
to determine the optimal operation of the process (Fernholz et al. 2000) and as a
basis for controller design. Mass and energy balances for all parts of the plant result
in a differential-algebraic equation system consisting of more than 2000 equations.
The main assumptions in the model are:

- The structured packings can be treated as a number of theoretical plates using the
  HETP-value (height equivalent to theoretical plate).
- The vapor and the liquid phase are in thermodynamic equilibrium.
- All chemical properties depend on the temperature and the composition.
- The phase equilibrium is calculated using the Wilson equations. The dimerization
  of acetic acid in the vapor phase is taken into consideration.
- The reaction kinetics are formulated by a quasihomogeneous correlation.
- The pressure drop of the packing is calculated by the equation of Máckowiak
  (1991).
- The hold-up of the packing is determined by an experimentally verified correla-
  tion.
- Negligible vapor hold-up.
- Ideal vapor behavior.
- Constant molar hold-up in the condenser.
- The dynamics of the tray hydraulics and the liquid enthalpy are taken into consid-
  eration.

The aim of the controller is to ensure the tracking of the optimal trajectory in the
presence of model inaccuracies and disturbances acting on the process.

4.2.3
Simplified Solution of the Model Equations

Generally, any process model can be used to predict the future process outputs $\hat{y}_{k+1,\,n}$
as long as the model is sufficiently accurate. A straightforward approach would be to
use the same model that was used to calculate the optimal operation. Unfortunately
the integration of this differential algebraic model is too time-consuming to solve the
optimization problem given by Eqs. (1) and (2) within one sampling interval. Thus,
a different model had to be developed to make sure that the solution of Eqs. (1) and
(2) is found between two sampling points.

The physical process model is based on heat and mass balances resulting in a set
of differential equations. A large number of algebraic equations is needed to calculate the physical properties, the phase equilibrium, the reaction kinetics and the tray
hold-ups, as well as the connections between the different submodels. Various
numerical packages are now available to solve large differential-algebraic equation
(DAE) systems like gPROMS (1997) or the Aspen Custom Modeler (ACM). Even

though they are designed to solve large and sparse DAEs in an efficient way, general-purpose solvers do not take advantage of the mathematical structure of a special problem. Our aim was to find a way to reduce the numerical effort required to calculate the solution of the DAE system which describes the reactive distillation process.

The main idea is to split up the equation system into a small section that is treated by the solver in the usual manner and a large subsystem containing mainly the algebraic equations. An independent solver that communicates with the DAE solver calculates the solution of this subsystem. Generally, this sequential approach may not be advantageous since the solution of the algebraic part must be provided in each step of the iteration of the DAE solver. It will only be superior if the solution of the second part is calculated in a highly efficient way. Therefore an analysis of the system equations for one separation tray is given in the sequel. Similar considerations can be easily made for the reactive trays as well as the other submodels of the process.

The core of the model for each separation tray consists of the mass balances of the components (Eq. (6)) the heat balance (Eq. (7)), and the constitutive equation for the liquid mole fractions (Eq. (8)):

$$\frac{d}{dt}(x_i N) = L^{in} x_i^{in} - L^{out} x_i + V^{in} y_i^{in} - V^{out} y_i, \tag{6}$$

$$\frac{d}{dt}(h_{lig} N) = L^{in} h_{liq}^{in} - L^{out} h_{liq} + V^{in} h_{vap}^{in} - V^{out} h_{vap}, \tag{7}$$

$$1 = \sum x_i. \tag{8}$$

In addition to the core Eqs. (6)–(8), empirical correlations are used to calculate the molar hold-ups of the trays (Eq. (9)), the liquid enthalpy (Eq. (10)), the vapor enthalpy (Eq. (11)), and the density (Eq. (12)):

$$N = c \rho^{2/3} L^{1/3}, \tag{9}$$

$$h_{liq} = f_1(x_i, T), \tag{10}$$

$$h_{vap} = f_2(y_i, T), \tag{11}$$

$$\rho = f_3(x_i, T). \tag{12}$$

Finally the phase equilibrium is calculated by using a four-parameter Wilson activity coefficient model for the liquid phase and a vapor-phase model which takes into consideration the dimerization of the acetic acid in the vapor phase (Noeres 2003). This phase equilibrium model (Eq. (13)) is an implicit set of equations in contrast to Eqs. (9)–(12) which are explicit functions of the composition and the temperature.

$$y_i = \frac{x_i \cdot \gamma_i \cdot p_i^0}{\varphi_i \cdot p}, \tag{13a}$$

$$1 = \sum y_i. \tag{13b}$$

Even though the formal description of Eqs. (9)–(13) feigns that its size is similar to the core model Eqs. (6)–(8), the opposite is true. Owing to the necessity of introducing a lot of auxiliary variables, especially for the phase equilibrium, Eqs. (9)–(13) make up the largest part of the overall system. Thus the idea is to move as many algebraic equations as possible, especially the parts containing the auxiliary variables, from the part which is handled by the DAE solver to an additional solver that exploits the mostly explicit structure of the equations. The DAE solver used in this work is DASOLVE, a standard solver in gPROMS for stiff DAEs (gPROMS, 1997). The proposed architecture of the algorithm is shown in Figure 4.2.

The main task of the external software is to solve the implicit phase equilibrium Eq. (13a,b) in an efficient manner. Solving Eq. (13a,b) for given pressure and liquid composition means to find the temperature T such that condition Eq. (13b) is fulfilled for the values calculated by Eq. (13a). Thus, the phase equilibrium calculation can be treated as solving a nonlinear equation with one unknown variable. Once Eq. (13a) and Eq. (13b) are solved, the remaining variables can be calculated straightforwardly by the explicit Eqs. (9)–(12). All values are passed back from the DAE solver via the foreign object interface.

In order to minimize the number of equations handled by the DAE-solver, the dynamics of the tray hold-up N and the liquid enthalpy h_{liq} were neglected. This causes deviations between the original model and the model with neglected dynamics. Several case studies were performed to check the differences between the original model and the model with neglected dynamics. In many cases the predictions of both models can hardly be distinguished. In some cases, however, noticeable differences in the dynamic behaviors result. These inaccuracies have to be handled by the disturbance estimation of Eqs. (3)–(5). By applying this scheme to the complete column model, the time required to calculate the solution for typical model predictive control scenarios could be reduced by a factor of 6–10.

The use of a simplified model and the special solution algorithm enable the online solution of the optimal control problem of Eqs. (1)–(3). The optimization algorithm L-BFGS-B of Byrd et al. (1994) is used to solve the optimization problem. This code solves nonlinear optimization problems with simple bounds on the decision variables and ensures a decrease of the goal function in each iteration step. The user of this code has to supply the values of the goal function as well as its derivatives with respect to the decision variables. The value of the cost function is calculated by integrating the model, the derivatives are obtained by perturbation. Using perturbations offers the opportunity to parallelize the calculation. Within a sampling period of

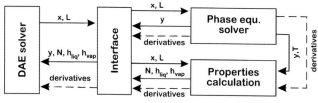

Figure 4.2 Scheme of the algorithm

6 min, about 100 function and gradient evaluations can be performed. The maximal number of function and gradient evaluations which were needed for the cases investigated was 33. Thus, the algorithm is able to find the optimal solution within the sampling time.

4.2.4
Controller Performance

The analysis of Fernholz et al. (1999a) showed that a suitable control structure for this process is to control the concentrations of methyl acetate and water in the product stream by the reflux ratio and the heat duty of the reboiler. At our pilot plant, NIR (near infrared spectroscopy) measurements of the product concentrations are available. The nonlinear model predictive controller was tested in several simulation cases. Here the original model is used as the simulated process, whereas the simplified model is used in the controller. In order to explore the benefits of the nonlinear controller, a linear controller was designed as well (Engell and Fernholz 2003). The linear controller was chosen based on an averaged linear model calculated from several linear models which were obtained by linearization of the nonlinear model at several points on the optimal trajectory. The controller design was done using the frequency response approximation technique (Engell and Müller 1993). The details of the linear controller design are beyond the scope of this book, they can be found in Fernholz et al. (1999b).

The parameters of the cost function in Eq. (1) were chosen such that deviations of both controlled variables give the same contribution to the objective functions. Additional bounds on the manipulated variables were added. The reflux ratio is physically bounded by the values 0 and 1, while the heat duty is bounded to a lower value of 1 kW and an upper one of 8 kW to ensure proper operation of the column. In order to avoid undesired abrupt changes of the manipulated variables, small penalties on these changes were added. The values of the penalty parameters λ_{ij} were selected in a way that large changes are possible for large deviations of the controlled variables but are unfavorable if they are close to their set-points[1]. Preliminary work on the model predictive control of this process had shown that the choice of a control horizon of $H_r = 2$ and of a prediction horizon of $H_p = 5$ gave good results. The closed loop responses for a set-point change of the methyl acetate concentration from a mole fraction of 0.8 down to 0.6 and back to 0.8 are shown in Figure 4.3. The use of the nonlinear controller reduces the time required to decrease the methyl acetate concentration drastically. The price that has to be paid for this reduction is a larger deviation of the water concentration. For the set-point change back to the original value, the differences between the two controllers are small.

Next, the performance of both controllers was checked for set-points of methyl acetate and water which force the process into a region where a sign change of the static gain occurs. If the set-points of the mole fractions of methyl acetate and water are

---

[1] The value of all λ_{ij} is 0.01, while γ is set to one. The physical units of the controlled variables are mole mole^{-1}, the reflux ratio is dimensionless and the heat duty is given in kilowatts.

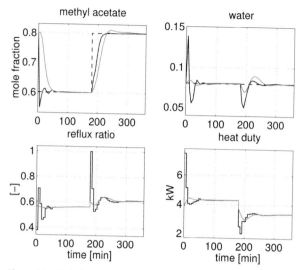

Figure 4.3 Methyl acetate set-point tracking. *Black:* nonlinear controller; *grey:* linear controller

changed simultaneously to 0.97 and 0.02 respectively, both controllers drive the process in the correct direction (Figure 4.4), but only the nonlinear controller is able to track both concentrations accurately. If the set-points are set back to their original values, the linear controller becomes unstable while the nonlinear controller works properly.

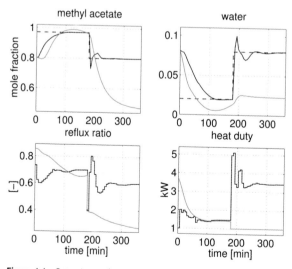

Figure 4.4 Set-point tracking in a region of a sign change of the static gain. *Black:* nonlinear controller; *grey:* linear controller

4.2.4.1

Disturbance Rejection

The main goal of the controller is to track the optimal trajectory of the process in the case of disturbances and plant-model mismatch. In the case of an accurate model and the absence of disturbances, no feedback controller would be necessary. Thus, two disturbances are imposed on the process during the simulation to test the disturbance rejection capabilities of the controllers.

First the influence of disturbances of the heat supply is considered. After 200 min the heat supply is decreased by 0.7 kW (which is about 20 % of the nominal value), set back to its nominal value at $t = 300$ min and increased by 0.7 kW at $t = 550$ min until it is again reset to the nominal value at $t = 700$ min. The simulation results for both controllers are depicted in Figure 4.5. The nonlinear controller rejects the disturbance much faster than the linear controller, especially for the product methyl acetate.

The second disturbance investigated is a failure of the heating system of the column. In order to minimize heat losses across the column surface, the plant is equipped with a supplementary heating system. A malfunction of this system will change the heat loss across the surface. The heat loss is increased by 50 W per stage, set back to 0 and decreased by 50 W per stage at the same times at which the disturbances of the heat duty were imposed before. The simulation results in Figure 4.6 show that the nonlinear controller rejects this disturbance more efficiently than the linear controller. Thus, the disturbance rejection can be significantly improved by using the nonlinear predictive controller.

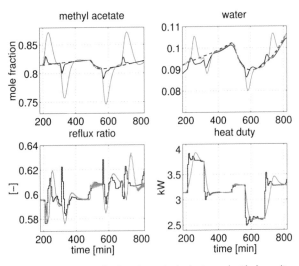

Figure 4.5 Rejection of a disturbance in the heat supply. *Black:* nonlinear controller; *grey:* linear controller; *dashed:* optimal trajectory

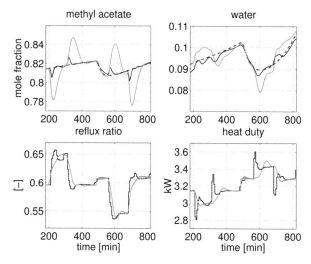

Figure 4.6 Rejection of a disturbance in the heat loss. *Black*: nonlinear controller; *grey*: linear controller; *dashed*: optimal trajectory

4.2.5
Summary
In this section, we presented the principle of NMPC to track a precomputed trajectory of a complex process and discussed the application to a semibatch reactive distillation column. Neglecting the dynamics of the molar hold-ups and of the enthalpies enabled splitting up the original DAE system into a small DAE part, which is treated by the numerical simulator, and an algebraic part, which is solved by an external algorithm. This approach reduced the time needed to solve the model equations by a factor of 6–10. These reductions made the use of a process model that is based on heat and mass balances possible for a model predictive controller.

The resulting nonlinear controller showed superior set-point tracking properties compared to a carefully designed linear controller. The nonlinear controller is able to track set-points that lie in regions where the process shows sign changes in the static gains and any linear controller becomes unstable. The nonlinear controller rejects disturbances faster than the linear controller. Moreover, since the nonlinear controller makes use of a model the range of validity of which is not restricted to a fixed operating region in contrast to the linear one, the nonlinear controller might be used for different trajectories giving more overall flexibility. The superior performance of the controller is due to the fact that a nonlinear process model is used. On the other hand, its stability and performance depend on the accuracy of the rigorous process model. If, e.g., the change of the gain of the process (which is caused by the fact that the product purity is maximized for certain values of reflux and heat duty) occurs for different values of the reflux and the heat duty than predicted by the model, the controller may fail to stabilize the process.

4.3
Control of Batch Chromatography Using Online Model-based Optimization

4.3.1
Principle and Optimal Operation of Batch Chromatography

The chromatographic separation is based on the different adsorptivities of the components to a specific adsorbent which is fixed in a chromatographic column. The most widespread process, batch chromatography, involves a single column which is charged periodically with pulses of the feed solution. These feed injections are carried through the column by pure desorbent. Owing to different adsorption affinities, the components in the mixture migrate at different velocities and therefore they are gradually separated. At the outlet of the column, the purified components are collected between cutting points, the locations of which are decided by the purity requirements on the products (Figure 4.7).

For a chromatographic batch process with given design parameters (combination of packing and desorbent, column dimensions, maximum pump pressure), the determination of the optimal operating regime can be posed as follows: a given amount (or flow) of raw material has to be separated into the desired components at minimal cost while respecting constraints on the purities of the products. The operation cost may involve the investment into the plant and the packing, labor and solvent cost, the value of lost material (valuable product in the nonproduct fractions) and the cost of the further processing, e.g., removal of the solvent. The free operating parameters are:

- the throughput of solvent and feed material, represented by the flow rate Q or the interstitial velocity u, constrained to the maximum allowed throughput which in turn is limited by the efficiency of the adsorbent or the pressure drop;
- the injection period t_{inj}, representing the duration of the feed injection as a measure of the size of the feed charge;
- the cycle period t_{cyc}, representing the duration from the beginning of one feed injection to the beginning of the next;
- the fractionating times.

The mathematical modeling of single chromatographic columns has been extensively described in the literature by several authors, and is in most cases based on differential mass balances (Guiochon 2002). The modeling approaches can be classified by the physical phenomena they include and thus by their level of complexity. Details

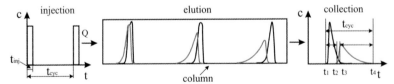

Figure 4.7 Principle of batch chromatography

on models and solution approaches can be found e.g., in (Dünnebier and Klatt 2000). The most general one-dimensional model (ignoring radial inhomogeneities) is the general rate model (GRM)

$$\frac{\partial c_{b,i}}{\partial t} + \frac{(1-\varepsilon_b)3k_{1,i}}{\varepsilon_b R_p}\left(c_{b,i} - c_{p,i|r=R_p}\right) + r_{kin,i}^{liq} = D_{ax}\frac{\partial^2 c_{b,i}}{\partial x} + u\frac{\partial c_{b,i}}{\partial x},\tag{14}$$

$$(1-\varepsilon_p)\frac{\partial q_i}{\partial t} + \varepsilon_p\frac{\partial c_{p,i}}{\partial t} - \varepsilon_p D_{p,i}\left[\frac{1}{r^2}\frac{\partial}{\partial r}\left(r^2\frac{\partial c_{p,i}}{\partial r}\right)\right] - r_{kin,i}^{sol} = 0,\tag{15}$$

where also reaction terms in the liquid and in the solid phase were included.

These two partial differential equations describe the concentrations in the mobile phase ($c_{b,i}$) and in the stationary phase (q_i and $c_{p,i}$). The adsorption isotherms relate the concentrations q_i (substance i adsorbed by the solid) and $c_{p,i}$ (substance i in the stationary liquid phase). A commonly utilized isotherm functional form is bi-Langmuir isotherm:

$$q_i = \frac{a_1 c_{p,i}}{1 + \sum_j b_{1j}c_{p,i}} + \frac{a_2 c_{p,i}}{1 + \sum_j b_{2j}c_{p,j}}.\tag{16}$$

An efficient numerical solution for the GRM incorporating arbitrary nonlinear isotherms was proposed by Gu (1995). The mobile phase and the stationary phase are discretized using the finite element and the orthogonal collocation method. The resulting ordinary differential equation (ODE) system is solved using an ODE solver which is based on the Gear's method for stiff ODEs. The numerical solution yields the concentrations of the components in the column at different locations and times. The concentration information at the outlet of the column is used to generate the chromatogram from which the production rate and the recovery yield can be computed.

The requirements on the products can usually be formulated in terms of minimum purities, minimum recoveries or maximum losses. In the case of a binary separation without intermediate cuts, these constraints can be transformed into each other, so either the recovery yield or the product purity may be constrained. The production cost is determined by many factors, in particular the throughput, the solvent consumption and the cost of downstream processing. A simple objective function is the productivity Pr, i.e., the amount of product produced per amount of adsorbent. This formulation results in the following nonlinear dynamic optimization problem:

$$\max \qquad \mathrm{Pr}\left(u, t_{cyc}, t_{inj}\right) = \frac{\dot{m}_{Product}}{m_{Adsorbent}}$$

$$\text{such that} \quad \mathrm{Rec}_i \geq \mathrm{Rec}_{min,i}, \quad i = 1,\ldots,n_{sp}\tag{17}$$
$$0 \leq u \leq u_{max},$$
$$0 \leq t_{inj}, t_{cyc}$$

where Rec_i denotes the revovery yield of product i.

This type of problem can be solved by standard optimization algorithms. In order to reduce the computation times to enable online optimization, Dünnebier et al. (2001) simplified the optimization problem and decomposed it in order to enable a more efficient solution. They exploited the fact that the recovery constraints are always active at the optimal solution and consider them as equalities. The resulting solution algorithm consists of two stages, the iterative solution of the recovery equality constraints, and the solution of the remaining unconstrained static nonlinear problem.

4.3.2
Model-based Control with Model Adaptation

In industrial practice, chromatographic separations are usually controlled manually. However, automatic feedback control leads to a uniform process operation closer to the economic optimum, and it can include online reoptimization. Dünnebier et al. (2001) proposed the model-based online optimization strategy shown in Figure 4.8.

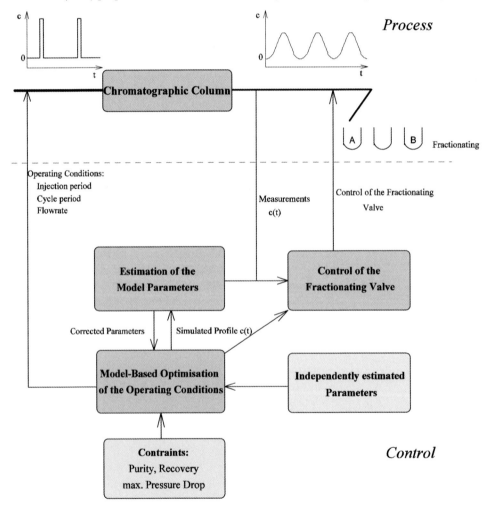

Figure 4.8 Control scheme for chromatographic batch separations

Essentially, this scheme performs the above optimization of the operating parameters online. To improve the model accuracy and to track changes in the plant, an online parameter estimation is performed. A similar run-to-run technique has been proposed by Nagrath et al. (2003).

Note that this scheme contains feedback only in the parameter estimation path. Therefore it will lead to good results only if the model is structurally correct so that the parameter estimation leads to a highly accurate model.

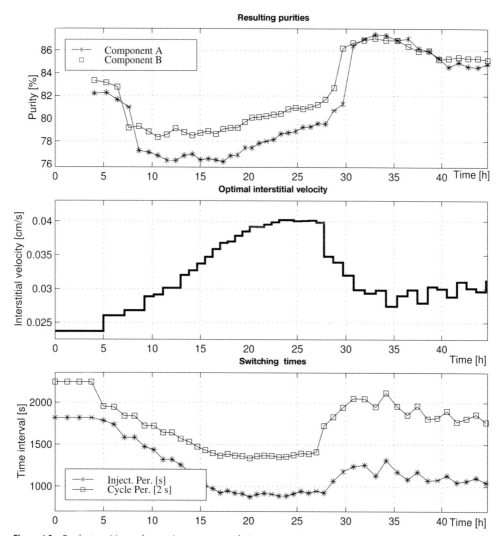

Figure 4.9 Product purities and operating parameters during an experimental run (set-point change for required purity at approximately 28 h from 80 to 85 %) (from Dünnebier et al., 2001)

The scheme was tested successfully at the pilot scale for a sugar separation with linear adsorption isotherm by Dünnebier et al. (2001). The product concentrations were measured using a two-detector concept as first proposed by Altenhöhner et al. (1997). A densimeter was used for the measurement of the total concentration of fructose and glucose and a polarimetric detector for the determination of the total rotation angle. Both devices were installed in series at the plant outlet.

Figure 4.9 shows an experimental result. First the operating parameters are modified in order to meet the product purity and recovery of 80 % each. After about 28 h, the controlled operating parameters reach a stable steady state. At this point a set-point change takes place in the product specifications: purity and recovery are now required to be 86 %. The control scheme reacts immediately, reducing the interstitial velocity and increasing the injection and cycle intervals. This leads to a better separation of the two peaks and to an increase in purity as desired. The controlled system quickly converges to a new steady state.

4.3.3
Summary

The key idea of the approach described in this section is to use model-based set-point optimization for model-based closed-loop control. Plant-model mismatch is tackled by adapting key model parameters to the available measurements so that the concentration profiles at the output which are predicted by the model match the observed ones. Experimental results showed that this approach works very well in the case of sugar separations where the model is structurally correct. The optimization algorithm was tailored to the structure of the problem so that convergence problems were avoided. Owing to the use of a tailored algorithm and the fact that the process is quite slow, computation times were not a problem.

4.4
Control by Measurement-based Online Optimization

In the two-step approach described in the previous section, the model parameters are updated by a parameter estimation procedure so that the model represents the plant at current operating conditions as accurately as possible. The updated model is used in the optimization procedure to generate a new set-point. This method works well for parametric mismatch between the model and the real plant. However, it does not guarantee an improvement of the set-point when structural errors in the model are present. In chromatographic separations, structural errors result e.g., from the approximation of the real isotherm by the Bi-Langmuir function. One important cause of plant-model mismatch can be the presence of small additional impurities in the mixture which may lead to considerable deviations of the observed concentration profiles at the output. The model-based optimization then generates a suboptimal operating point which in general does not satisfy the constraints on purity or recovery. The conventional solution is to introduce an additional control loop that regulates the product purities, as proposed and tested by Hanisch (2002). However, the changes of the operating parameters caused by this control loop may conflict with the goal of optimizing performance.

4.4.1
The Principle of Iterative Optimization

To cope with structural plant-model mismatch, the available measurements can be used not only to update the model but also to modify the optimization problem in such a manner that the gradient of the (unknown) real process mapping is driven to zero, in contrast to satisfying the optimality conditions for the theoretical model. Such an iterative two-step method was proposed by Roberts (1979), termed integrated system optimization and parameter estimation (ISOPE). A gradient-modification term is added to the objective function of the optimization problem. ISOPE generates set-points which converge towards the true optimum despite parametric and structural model mismatch. Theoretical optimality and convergence of the method were proven by Brdyś et al. (1987). From a practical point of view, the key element of ISOPE is the estimation of the gradient of the plant outputs with respect to the optimization variables.

The general model-based set-point optimization problem can be stated as

$$\min_{\mathbf{u}} \quad J(\mathbf{u}, \mathbf{j})$$
$$\text{such that} \quad \mathbf{g}(\mathbf{u}) \leq 0 \tag{18}$$
$$\mathbf{u}_{\min} \leq \mathbf{u} \leq \mathbf{u}_{\max}$$

where $J(\mathbf{u},\mathbf{y})$ is a scalar objective function, $\mathbf{u}$ is a vector of optimization variables (set-points), $\mathbf{y}$ is a vector of output variables, and $\mathbf{g}(\mathbf{u})$ is a vector of constraint functions. The relationship between $\mathbf{u}$ and $\mathbf{y}$ is represented by a model

$$\mathbf{y} = \mathbf{f}(\mathbf{u}, \alpha) \tag{19}$$

where α is a vector of model parameters. ISOPE is an iterative algorithm, where at each step of the iteration measurement information (i.e., the plant output $\mathbf{y}^*$, which was measured after the last set-point was applied) is used to update the model and to modify the optimization problem. The updating of the model can be realized as a parameter estimation procedure. A vector of gradient modifiers is computed using the gradient of the updated model and of the plant at set-point $\mathbf{u}^{(k)}$:

$$\lambda^{(k)} = \left((\mathbf{y}^*)'_{\mathbf{u}} - \mathbf{y}'_{\mathbf{u}} \right) J'_{\mathbf{y}}(\mathbf{u}, \mathbf{y}) \Big|_{\mathbf{u}^{(k)}}. \tag{20}$$

The optimization problem of Eq. (18) is modified by adding a gradient-modification term to the objective function:

$$\min_{\mathbf{u}} \quad J(\mathbf{u}, \mathbf{y}) + \lambda^{(k)^{\mathrm{T}}} \mathbf{u}$$
$$\text{such that} \quad \mathbf{g}(\mathbf{u}) \leq 0 \tag{21}$$
$$\mathbf{u}_{\min} \leq \mathbf{u} \leq \mathbf{u}_{\max}.$$

Assuming that the constraint function $\mathbf{g}(\mathbf{u})$ is known, the optimization problem can be solved by any nonlinear optimization algorithm. Let $\hat{\mathbf{u}}^{(k)}$ denote the solution to Eq. (21), then the next set-point is chosen as:

$$\mathbf{u}^{(k+1)} = \mathbf{u}^{(k)} + \mathbf{K}\left[\hat{\mathbf{u}}^{(k)} - \mathbf{u}^{(k)}\right] \tag{22}$$

where $\mathbf{K}$ is a diagonal gain matrix, the diagonal elements are in the interval [0,1], i.e., $\mathbf{K}$ is a damping term. Starting from an initial set-point, ISOPE will generate a sequence of set-points which, for an appropriate gain matrix, will converge to a set-point which satisfies the necessary optimality conditions of the actual plant. It can be proven that the modification term leads to the satisfaction of the optimality conditions at the true plant optimum.

Tatjewski (2002) redesigned the ISOPE method resulting in a new algorithm that does not require the parameter estimation procedure. The key idea is to introduce a model shift term in the modified objective function:

$$\min_{\mathbf{u}} = J\left(\mathbf{u}, \mathbf{y} + \mathbf{a}^{(k)}\right) + \lambda^{(k)^{\mathrm{T}}} \mathbf{u} \tag{23}$$

with the following definition of the modifier

$$\lambda^{(k)} = \left(\left(\mathbf{y}^*\right)'_{\mathbf{u}} - \mathbf{y}'_{\mathbf{u}}\right) J'_{\mathbf{y}}\left(\mathbf{u}, \mathbf{y} + \mathbf{a}^{(k)}\right)\Big|_{\mathbf{u}^{(k)}} \tag{24}$$

Here

$$\mathbf{a}^{(k)} = \mathbf{y}^{*(k)} - \mathbf{y}^{(k)} . \tag{25}$$

Although the parameter α is not updated, it is can be proven that the optimality conditions are satisfied. Parameter adaptation thus is no longer necessary, although it may be beneficial to the convergence of the procedure. As the optimality of the result is solely due to the gradient-modification in the optimization problem, the redesigned algorithm could be termed *iterative gradient-modification optimization*.

4.4.2
Handling of Constraints

If constraint functions depend on the behavior of the real plant, they cannot be assumed to be precisely known, and using a model for the computation of the constraint functions will not assure that the constraints are actually satisfied. In the original derivation of the ISOPE method, constraints were assumed to be process-independent. An extension of the ISOPE strategy which considers process-dependent constraints can be found in Brdyś et al. (1986). In this formulation, a recursive Lagrange multiplier is used. Tatjewski et al. (2001) also proposed using a follow-up constraint controller that is responsible for satisfying the output constraints.

A different method to handle the process-dependent constraints was proposed in Gao and Engell 2005. It is based on the idea of using plant information acquired at the last set-point $g^*(u^{(k)})$ to modify the model-based constraint functions $g(u)$ at the current iteration. The modified constraint functions approximate the true constraint functions of the plant in the vicinity of the last set-point. The modified constraint function is formulated as:

$$\hat{g}^{(k)}(u) = g(u) + g^*(u^{(k)}) - g(u^{(k)}) + \left((g^*)'_u(u^{(k)}) - g'_u(u^{(k)}) \right)(u - u^{(k)}). \tag{26}$$

The modified constraint function has the following properties at $u^{(k)}$:

- The modified constraint has the same value as the real constraint function, $\hat{g}^{(k)}$ $(u^{(k)}) = g^*(u^{(k)})$.
- The modified constraint has the same first order derivative as the real constraint function, $(\hat{g}^{(k)})'_u(u^{(k)}) = (g^*)'_u(u^{(k)})$.

As the modified constraint is only valid in the vicinity of $u^{(k)}$, a bound $u^{(k)} - \Delta u \leq u \leq u^{(k)} + \Delta u$ is added to the optimization problem to limit the search range in the next iteration. This guarantees that the constraints are not violated greatly.

4.4.3
Estimation of the Gradient of the Plant Mapping

A key element of the iterative gradient-modification optimization method is to estimate the gradient of the plant mapping. Several methods for this have been proposed during the last 20 years. These methods can be grouped into two categories according to whether set-point perturbations are used or not. Early versions of the ISOPE technique used finite difference techniques to obtain the plant gradient by applying perturbations to the current set-point. Later versions used dynamic perturbations and linear system identification methods to estimate the gradient (Lin et al. 1989, Zhang and Roberts 1990). Both methods have the disadvantage of requiring additional perturbations. In Roberts (2000), Broyden's formula was used to estimate the required gradient from current and past measurement information. The Broyden estimate is updated at each iteration using a formula of the form:

$$D^{(k)} = D^{(k-1)} + \frac{\left[\Delta Y^{(k)} - D^{(k-1)} \Delta X^{(k)}\right]\left(\Delta X^{(k)}\right)^{\mathrm{T}}}{\left(\Delta X^{(k)}\right)^{\mathrm{T}} \Delta X^{(k)}} \tag{27}$$

with

$$\Delta Y^{(k)} = Y^{(k)} - Y^{(k-1)}$$
$$\Delta X^{(k)} = X^{(k)} - X^{(k-1)}$$

where D is the estimate of $\partial Y(X)/\partial X$, and the superscript k refers to the iteration index. Although no additional perturbation is needed, care must be taken to avoid ill-conditioning as $\Delta X^{(k)} \to 0$. It should also be noted that the updating formula requires to be initialized with $D^{(0)}$. Brdyś and Tajewski (1994) proposed a different way of implementing a finite difference approximation of the gradient without additional set-point perturbations. This method uses set-points in past iterations instead of additional set-point perturbations. The gradient at set-point $\mathbf{u}^{(k)}$ is approximated as:

$$\left. (\mathbf{y}^*)'_{\mathbf{u}} \right|_{\mathbf{u}^{(k)}} \approx \left(\mathbf{S}^{(k)} \right)^{-1} \cdot \begin{bmatrix} \mathbf{y}^{*(k)} - \mathbf{y}^{*(k-1)} \\ \vdots \\ \mathbf{y}^{*(k)} - \mathbf{y}^{*(k-m)} \end{bmatrix} \tag{28}$$

with

$$\mathbf{S}^{(k)} = \left[\mathbf{u}^{(k)} - \mathbf{u}^{(k-1)}, \ldots, \mathbf{u}^{(k)} - \mathbf{u}^{(k-m)} \right]^{\mathrm{T}}$$

where m is the dimension of the vector $\mathbf{u}$. Theoretically, the smaller the difference between the set-points, the more accurate will the approximation of the gradient be. On the other hand, because the measurements of the plant outputs $\mathbf{y}^{*(k-i)}$, $i = 0,1,\ldots,$ m are usually corrupted by errors, the matrix $\mathbf{S}^{(k)}$ should be sufficiently well-conditioned to obtain a good approximation of the gradient. Let

$$d^{(k)} = \frac{\sigma_{\min}\left(\mathbf{S}^{(k)} \right)}{\sigma_{\max}\left(\mathbf{S}^{(k)} \right)} \tag{29}$$

denote the conditioning of $\mathbf{S}^{(k)}$ in terms of its singular values. If $d^{(k)}$ is too small, the errors in the measurements will be amplified considerably and the gradient estimation will be corrupted by noise. In Brdyś and Tajewski (1994) the optimization problem is reformulated to take into account future requirements of the gradient estimation. An inequality constraint

$$d^{(k)} \geq \delta \tag{30}$$

(where $0 < \delta < 1$) is added to the optimization problem at the $(k-1)^{\text{th}}$ iteration so that the set-point $\mathbf{u}^{(k)}$ will give a good approximation of the gradient. The advantage of this method is that no additional set-point perturbations are needed, but a loss of optimality will be observed at the current iteration because the inequality constraint reduces the feasible set of set-points. Therefore more iterations are required to attain the optimum, especially for a bigger values of δ.

A novel method was proposed for the gradient estimation in Gao and Engell (2005). It follows the same idea as Brdyś's method, i.e., using the past set-points in the finite difference approximation of the gradient. But the conditioning of $\mathbf{S}^{(k)}$ is included not as a constraint in the optimization problem, but as an indicator to decide whether an additional set-point perturbation should be added. At the $(k-1)^{\text{th}}$ iteration, after a new set-point $\mathbf{u}^{(k)}$ is acquired, $d^{(k)}$ is computed using $\{ \mathbf{u}^{(k)}, \mathbf{u}^{(k-1)}, \ldots,$ $\mathbf{u}^{(k-m)} \}$. If it is less than the given constant δ, an additional set-point $\mathbf{u}_a^{(k)}$ will be added to formulate a new set-point set $\{ \mathbf{u}^{k)}, \mathbf{u}_a^{(k)}, \mathbf{u}^{(k-1)}, \ldots, \mathbf{u}^{(k-m-1)} \}$ for the gradient approximation. The gradient at $\mathbf{u}^{(k)}$ is approximated by:

$$\left. (\mathbf{y}^*)'_{\mathbf{u}} \right|_{\mathbf{u}^{(k)}} \approx \left(\mathbf{S}_a^{(k)} \right)^{-1} \cdot \begin{bmatrix} \mathbf{y}^{*(k)} - \mathbf{y}_a^{*(k)} \\ \mathbf{y}^{*(k)} - \mathbf{y}^{*(k-1)} \\ \vdots \\ \mathbf{y}^{*(k)} - \mathbf{y}^{*(k-m-1)} \end{bmatrix} \tag{31}$$

with

$$S_a^{(k)} = \left[\mathbf{u}^{(k)} - \mathbf{u}_a^{(k)}, \; \mathbf{u}^{(k)} - \mathbf{u}^{(k-1)}, \ldots, \mathbf{u}^{(k)} - \mathbf{u}^{(k-m-1)} \right]^{\mathrm{T}}.$$

The additional set-point provides an additional perturbation around the current set-point. Its location is optimized by solving

$$\max_{\mathbf{u}_a^{(k)}} \quad d_a^{(k)} = \frac{\sigma_{\min}\left(\mathbf{S}_a^{(k)} \right)}{\sigma_{\max}\left(\mathbf{S}_a^{(k)} \right)}$$

$$\text{such that} \quad \hat{\mathbf{g}}^{(k-1)}\left(\mathbf{u}_a^{(k)} \right) \leq 0 \tag{32}$$

$$\mathbf{u}^{(k-1)} - \Delta\mathbf{u} \leq \mathbf{u}_a^{(k)} \leq \mathbf{u}^{(k-1)} + \Delta\mathbf{u}$$

$$\mathbf{u}_{\min} \leq \mathbf{u}_a^{(k)} \leq \mathbf{u}_{\max}.$$

Therefore, by introducing the additional set-point, $\mathbf{S}_a^{(k)}$ is kept well-conditioned and the optimal set-point $\mathbf{u}^{(k)}$ can be used in the gradient approximation. This method does not compromise optimality, and it is not as expensive as finite difference techniques with set-point perturbations in each iteration, because an additional set-point perturbation is added only when $d^{(k)} < d$.

The procedure can be summarized as follows:
1. Select starting set-points which include the initial set-point and m other set-points for the gradient estimation at the initial set-point. Initialize the parameters of the algorithm, i.e., K, δ and $\Delta\mathbf{u}$.
2. At the kth iteration, apply set-point $\mathbf{u}^{(k)}$ (and $\mathbf{u}_a^{(k)}$ if needed) to the plant. Measure the steady-state outputs.
3. Approximate the gradient using the proposed method. Modify the objective function and the constraint functions in the optimization problem and add the additional bound.
4. Solve the modified optimization problem Eqs. (23)–(26) using any nonlinear optimization algorithm and generate the next set-point.
5. Check the termination criterion $\| \mathbf{u}^{(k+1)} - \mathbf{u}^{(k)} \| < \varepsilon$ and decide whether to continue or to stop the optimization procedure.
6. If the termination criterion is not satisfied, check the conditioning of $\mathbf{S}^{(k)}$ in terms of its singular values

$$d^{(k)} = \frac{\sigma_{\min}\left(\mathbf{S}^{(k)} \right)}{\sigma_{\max}\left(\mathbf{S}^{(k)} \right)},$$

and if $d^{(k)} \geq \delta$ return to step 2, otherwise go to step 7.
7. Add an additional set-point by solving the optimization problem Eq. (32), then return to step 2.

4.4.4

Application to a Batch Chromatographic Separation with Nonlinear Isotherm

The iterative gradient-modification optimization method was tested in a simulation study of a batch chromatographic separation of enantiomers with highly nonlinear adsorption isotherms that had been used as a test case in laboratory experiments before (Hanisch 2002). A model with a bi-Langmuir isotherm that was fitted to measurement data is considered as the "real plant" in the simulation study. A model with isotherms of a different form is used in the set-point optimization.

The flow rate Q and the injection period t_{inj} are considered as the manipulated variables here. The cycle period t_{cyc} is fixed to the duration of the chromatogram. The performance criterion is the production rate Pr:

$$Pr = m_{product}/t_{cyc} \qquad (33)$$

The recovery yield Rec is constrained to a minimal value. This results in the optimization problem

$$\begin{aligned}
&\max_{Q, t_{inj}} && Pr(Q, t_{inj}) \\
&\text{such that} && Rec(Q, t_{inj}) \geq Rec_{min} \qquad (34) \\
& && 0 \leq Q \leq Q_{max} \\
& && t_{inj} \geq 0
\end{aligned}$$

Figure 4.10 shows the chromatograms of the "real" and the perturbed model for the same set-point. Note that such differences can be generated by rather small errors in the adsorption isotherms.

The second component is considered to be the valuable product. The purity requirement is 98 %. The recovery yield should be greater than 80 %. There is an upper limit of the flow rate of 2.06 cm³ s⁻¹. The flow rate and the injection period are normalized to the interval [0, 1] in the optimization. The gain coefficients in **K** are set to 1. The bound $\Delta\mathbf{u}$ is $[.06 \ .06]^T$. The recovery constraint was handled by the method

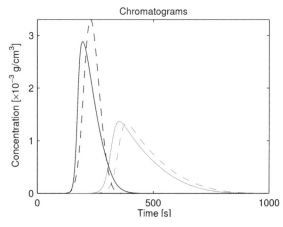

Figure 4.10 Illustration of the influence of model mismatch on the chromatogram. *Solid line:* "real" model; *dashed line:* nominal optimization model

proposed above. The iterations were stopped when the calculated set-point change was less than a predefined tolerance value ($\varepsilon = 0.006$) or the optimization algorithm did not terminate successfully.

Different gradient estimation methods were used in the iterative optimization procedure:

- the finite difference method, i.e., applying perturbations to each set-point (FDP);
- Brdyś's method, where an additional constraint is added to the optimization problem so that the next set-point can be used in the estimation of the gradient, no perturbations;
- finite difference method with additional set-point perturbations when necessary (FDPN).

Several runs of the set-point optimization were simulated, first without measurement errors and then with errors. In the case without errors, the optimization procedures with FDP and FDPN terminated successfully, while the optimization procedure with Brdyś's method stopped early because the optimization algorithm could not find a feasible point in the given number of iterations. The optimization procedure with FDPN used one iteration more than the optimization procedure with FDP, but it used only six additional set-points ($\delta = 0.2$). The optimization procedure with FDP perturbed the set-point eight times at each iteration to estimate the gradient so that it generated 80 additional set-points overall. The trajectories of the production rate Pr and of the recovery yield Rec are depicted in Figure 4.11. The recovery constraint was met by all three optimization procedures. Figure 4.12 shows the set-point trajectories and the production rate and recovery contours of the real model and the optimization model. Although a considerable mismatch exists between the real model and the optimization model, the iterative gradient-modification optimization method generates set-points which converge to the real optimum.

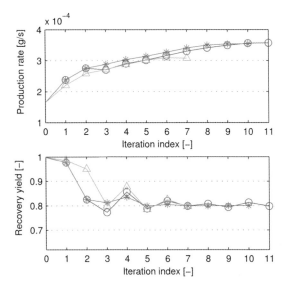

Figure 4.11 Trajectories of production rate and recovery yield, simulations without errors. * Set-points using the finite difference method (FDP), Δ set-points using Brdyś's method, o set-points using the finite difference method with additional set-point perturbations when necessary (FDPN). Recovery limit: 80%, $\delta = 0.2$

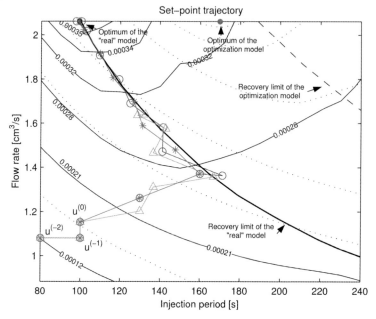

Figure 4.12 Illustration of set-point trajectories. *Solid lines*: contours of the "real" model; *dotted lines*: contours of the nominal optimization model, * set-points using FDP method, Δ set-points using Brdyś's method, o set-points using FDPN method, $u^{(0)}$ initial set-point, $u^{(1)}$ and $u^{(2)}$ additional initial set-points for the gradient estimation at $u^{(0)}$

Table 4.1 shows the results of simulations with measurement errors. Different values of δ were tried and all simulations were stopped at the optimum of the "real" model. With increasing δ, more additional set-points were used, which improved the accuracy of the gradient estimations. Therefore, fewer iterations were needed to arrive at the optimum. Considering the total number of set-points used, $\delta = 0.1$ gives a good result.

Table 4.1 Optimization results of the simulations with errors

δ	Number of iterations	Additional set-points	Final set-point	Optimum of the "real" model
0.2	13	7	(2.06, 99.73).	
0.1	13	6	(2.06, 99.79).	
0.05	15	5	(2.06, 99.53).	(2.06, 99.35).
0.01	22	4	(2.06, 99.10).	

4.4.5
Summary
The identification of an accurate model requires considerable efforts, especially for chemical and biochemical processes. In practice, inaccurate models must be used for online control and optimization. A purely model-based optimization will generate a

suboptimal or even infeasible set-point. We described a modified iterative gradient-modification optimization strategy that converges to the real optimum in a few steps while respecting the constraints. A few additional set-points are introduced to reduce the effect of measurement errors on the gradient approximation. The example of a batch chromatographic separation with highly nonlinear isotherms demonstrated the impressive improvements that can be obtained by this approach.

4.5
Nonlinear Model-based Control of a Reactive Simulated Moving Bed (SMB) Process

4.5.1
Principle and Optimization of Chromatographic SMB Separations

Batch chromatography has the usual drawbacks of a batch operation, and leads to highly diluted products. On the other hand, it is extremely flexible, several components may be recovered from a mixture during one operation and varying compositions of the desorbent can be used to enhance separation efficiency. The idea of a continuous operation with countercurrent movement of the solid led to the development of the simulated moving bed (SMB) process (Broughton 1966). It is gaining increasing attention due to its advantages in terms of productivity and eluent consumption (Guest 1997, Juza et al. 2000). A simplified description of the process is given in Figure 4.13. It consists of several chromatographic columns connected in series which constitute a closed loop. A countercurrent motion of the solid phase relative to the liquid phase is simulated by periodically and simultaneously moving the inlet and outlet lines by one column in the direction of the liquid flow.

After a start-up phase, SMB processes reach a cyclic steady state (CSS). Figure 4.13 shows the CSS of a binary separation along the columns plotted for a fixed time instant within a switching period. At every axial position, the concentrations vary as

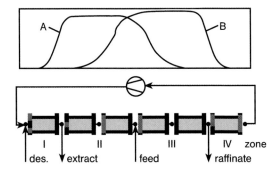

Figure 4.13 Simulated Moving-Bed Process. At the top, the concentration profiles at the cyclic steady state are shown. Pure a is withdrawn at the extract port and pure B is withdrawn at the raffinate port

a function of time, and the values reached at the end of each switching period are equal to those before the switching, relative to the port positions.

In order to exploit the full potential of SMB processes, recent research has focused on the design of the process, in particular the choice of the operation parameters for a given selection of adsorbent, solvent and column dimensions, using mathematical optimization. As the optimum should be determined precisely while meeting all constraints, rigorous models which include the discrete dynamics are used (Klatt et al. 2000, Zhang et al. 2003). In addition to a higher reliability compared to shortcut methods, this approach is applicable to a broad variety of SMB-like operating regimes. The optimization problem can be stated as (Toumi et al. 2004c):

$$
\begin{aligned}
\min_{Q_j, N_i, \tau} \quad & \text{Cost}_{\text{spec}} \\
\text{such that} \quad & \Gamma\big(\mathbf{c}_{\text{ax}}(0)\big) - \mathbf{c}_{\text{ax}}(\tau) \leq \varepsilon, \\
& \text{Pur}_{\text{Ex}} \geq \text{Pur}_{\text{Ex,min}}, \\
& \text{Pur}_{\text{Raf}} \geq \text{Pur}_{\text{Raf,min}}, \\
& 0 \leq Q_j \leq Q_{\text{max}},
\end{aligned}
\tag{35}
$$

where Pur_{EX} and Pur_{Raf} denote the purities at the extract and the raffinate ports and Γ summarizes of dynamics of the process from one switching period to the next, including the shifting of the ports and c_{ax} denotes the axial concentration profile along the columns.

The goal is to operate the process at the optimal CSS with minimal separation costs $\text{Cost}_{\text{spec}}$ while the purity requirements at both product outlets are fulfilled. Equation (34) constitutes a complex dynamic optimization problem the solution of which critically depends on an efficient and reliable computation of the CSS defined by

$$
\Gamma\big(\mathbf{c}_{\text{ax}}(0)\big) - \mathbf{c}_{\text{ax}}(\tau) \leq \varepsilon,
\tag{36}
$$

The free optimization variables are the flow rates in the sections Q_j and the switching period τ. They are transformed to the so-called β-factors, which represent the ratio between the flow rates Q_j and the hypothetical solid flow rate. This nonlinear transformation leads to a better conditioned optimization problem (Dünnebier et al. 2001). An additional constraint takes the maximum pressure drop into account. The main difficulty of the optimization problem results from the large dimension of the CSS equations when a first-principle plant model is used. A simple and robust optimization approach consists of integration of the model equations starting from initial values until the CSS is reached (sequential approach). At the CSS, the objective function as well as the constraints are evaluated and returned to an optimizer. This yields a small number of free parameters and hence a relatively simple optimization problem. The number of cycles required to reach a CSS usually is not too large (about 100) in contrast to other periodic processes like pressure swing adsorption where 1000 or more periods have to be simulated. The computational effort is therefore reasonable.

4.5.2
Model-based Control

Klatt et al. (2002) proposed a two-layer control architecture similar to the one used for batch chromatography, where the optimal operating trajectory is calculated at a low sampling rate by dynamic optimization based on a rigorous process model. The model parameters are adapted based on online measurements. The low-level control task is to keep the process on the optimal trajectory despite disturbances and plant/model mismatch. The controller is based on identified models gained from simulation data of the rigorous process model along the optimal trajectory. For the linear adsorption isotherm case, linear models are sufficient (Klatt et al. 2002), whereas in the nonlinear case neural networks (NN) were applied successfully (Wang et al. 2003). A disadvantage of this two-layer concept is that the stabilized front positions do not guarantee the product purities if plant-model mismatch occurs. Thus an additional purity controller is required.

Toumi and Engell (2004a) recently presented a nonlinear model-predictive control scheme and applied it to a three-zones reactive SMB (RSMB) process for glucose isomerization (Toumi and Engell 2004b, 2005). The key feature of this approach is that the production cost is minimized online, while the product purities are considered as constraints, thus real online optimization is performed, not trajectory tracking.

The following optimal control problem is formulated over the finite control horizon H_r:

$$\min_{[\boldsymbol{\beta}_k,\dots,\boldsymbol{\beta}_{k+H_r}]} \quad \Omega = \sum_{j=k}^{k+H_p} \left(\text{Cost}(j) + \Delta\boldsymbol{\beta}_j^{\mathsf{T}} \mathbf{R}_j \Delta\boldsymbol{\beta}_j \right)$$

such that
$$\begin{cases} \dot{\mathbf{x}}_j = f(\mathbf{x}_j, \boldsymbol{\beta}_j) \\ \mathbf{x}_{j+1,0} = \mathbf{P}\mathbf{x}_j(\tau(j)) \end{cases},$$

$$\text{Pur}_{\text{Ex},H_r} + \Delta\text{Pur}_{\text{Ex}} \geq \text{Pur}_{\text{Ex,min}},$$
$$\text{Pur}_{\text{Ex},H_p} + \Delta\text{Pur}_{\text{Ex}} \geq \text{Pur}_{\text{Ex,min}},$$
$$g(\boldsymbol{\beta}_j) \geq 0,$$
$$0 \leq Q_{I,j} \leq Q_{\max}, \quad j = k, \dots, k + H_p.$$

(37)

The prediction horizon is discretized in cycles, where a cycle is a switching time $\tau(k)$ multiplied by the total number of columns. Equation (37) constitutes a dynamic optimization problem with the transient behavior of the process as a constraint. The objective function Ω is the sum of costs incurred for each cycle (e.g., desorbent consumption) and a regularizing term added in order to smooth the input sequence in order to avoid high fluctuations in the input sequence from cycle to cycle. The first equality constraint represents the plant model evaluated over the finite prediction horizon H_p. The switching dynamics are introduced via the permutation matrix $\mathbf{P}$. Since the maximal attainable pressure drop by the pumps must not be exceeded, constraints are imposed on the flow rates in zone I. Further inequality constraints $g(\boldsymbol{\beta}_j)$ are added in order to avoid negative flow rates during the optimization.

The control objective is reflected by the purity constraint over the control horizon H_r which is corrected by a bias term $\Delta \mathrm{Pur}_{\mathrm{Ex}}$ resulting from the difference between the last simulated and the last measured process output to compensate unmodeled effects:

$$\Delta \mathrm{Pur}_{\mathrm{Ex},k} = \mathrm{Pur}_{\mathrm{Ex},(k-1)} - \mathrm{Pur}_{\mathrm{Ex},k,\mathrm{meas}}. \tag{38}$$

A second purity constraint over the whole prediction horizon acts similar to a terminal constraint forcing the process to converge towards the optimal CSS. It should be pointed out that the control goal (i.e., to fulfil the extract purity) is introduced as a constraint. A feasible path SQP algorithm is used for the optimization (Zhou et al. 1997) which generates a feasible point before it starts to minimize the objective function.

4.5.3
Online Parameter Adaptation

The concentration profiles in the recycling line are measured and collected during a cycle. Since this measurement point is fixed in the closed-loop arrangement, the sampled signal includes information of all zones. During the start-up phase, an online estimation of the actual model parameters is started in every cycle. The quadratic cost functional $J_{\mathrm{est}}(\mathbf{p})$:

$$J_{\mathrm{est}} = \sum_{i=1}^{n_{\mathrm{sp}}} \left(\int_0^{N_{\mathrm{col}}} \left(c_{i,\mathrm{meas}}(t) - c_{i,\mathrm{Re}}(t) \right)^2 \right) \tag{39}$$

is minimized with respect to the parameters p. For this purpose, the least squares solver E04UNF from the NAG-library is used. A by-product of the parameter estimation is the actual value $\mathbf{x}_0(k)$ of the state vector which is given back to the NMPC controller.

4.5.4
Simulation Study

Figure 4.14 shows a simulation scenario where the desired extract purity was set to 70 % at the beginning of the experiment. The desired extract purity was then changed to 60 % at cycle 60. At cycle 120, the desired extract purity was increased to 65 %. The enzyme activity and nonce the reaction rate is assumed to decay exponentially during the experiment. A fast response of the controller in both directions can be observed. Compared to the uncontrolled case, the controller can control the product purity and compensate the drift in the enzyme activity. The evolution of the optimizer iterations is plotted as a dashed line and shows that a feasible solution is found rapidly and that the concept can be realized in real time. In this example, the control horizon was set to two cycles and the prediction horizon to ten cycles. A diagonal matrix $\mathbf{R}_j = 0.02\, \mathbf{I}_{(3,3)}$ was chosen for regularization.

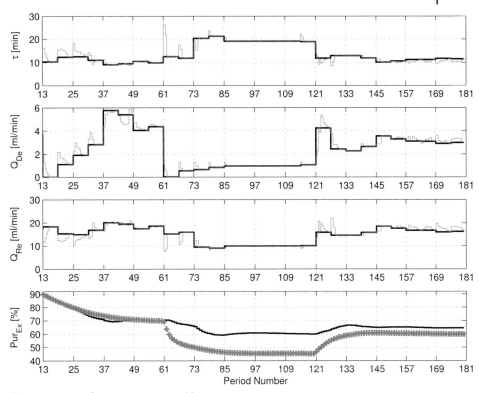

Figure 4.14 Control scenario $H_r = 2$, $H_p = 10$

Figure 4.15 shows the result of the parameter estimation. A good fit was achieved and the estimated parameter follows the drift of the reaction rate adequately.

4.5.5
Experimental Study

A sensitivity analysis showed that the process is highly sensitive to the values of the Henry coefficients, the mass transfer resistances and the reaction rate. These are therefore key parameters of the reactive SMB process. These parameters are reestimated online at every cycle (a cycle is equal to switching time multiplied by the number of columns). In Figure 4.16, the concentration profiles collected in the recycling line are compared to the simulated ones. At the end of the experiment all system parameters have converged towards stationary values as shown by Figure 4.17. The developed mathematical model describes the behavior of the RSMB process well.

The formulation of the optimization problem (37) was slightly modified for the experimental investigation. The sampling time of the controller was reduced to one switching period instead of one cycle, so that the controller reacts faster. The switching time was still used as a controlled variable, but modified only from cycle to

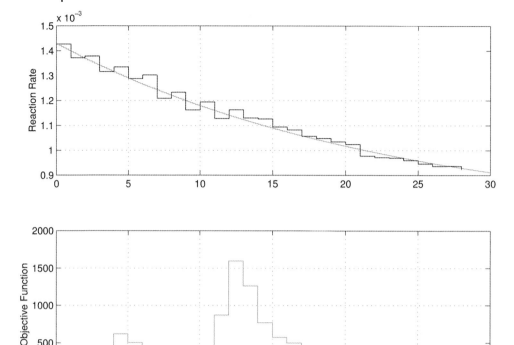
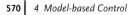

Figure 4.15 Estimation of the reaction rate

cycle. This is due to the asymmetry of the RSMB process that results from the dead volume of the recycling pump in the closed loop. It disturbs the overall performance of the process and is corrected by adding a delay for the switching of the inlet/outlet line passing the recycling pump. A detailed description of this method is provided in the patent (Hotier 1996). Therefore the shift of the valves is not synchronous to compensate for the technical imperfection of the real system and to get closer to the ideal symmetrical SMB system. In order to avoid port overlapping, the switching time must be held constant during a cycle.

In the real process, the enzyme concentration changes from column to column. The geometrical lengths of the columns also differ slightly. Moreover, the temperature is not constant over the columns due to the inevitable gradient of the closed heating-circuit. These problems cause a fluctuation of the concentration profiles at the product outlet. Even at the CSS, the product purity changes from period to period. Using the bias term given by Eq. (38) causes large variations of the controlled inputs from period to period. This effect was damped by using the minimal value over the last cycle:

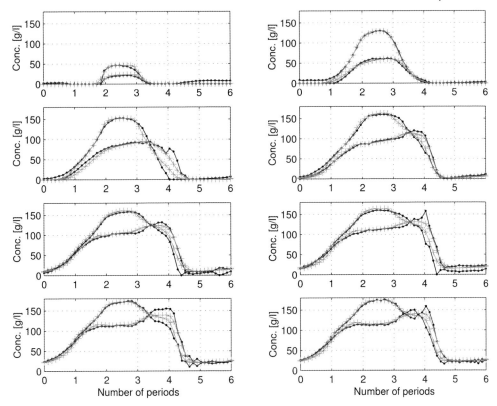

Figure 4.16 Comparison of experimental and simulated concentration profiles collected at the recycle line

$$\Delta \text{Pur}_{\text{Ex},k} = \min_{j=(k-1,\dots,k-1-N_{\text{col}})} \left(\text{Pur}_{\text{Ex},(k-1)} - \text{Pur}_{\text{Ex},k,\text{meas}}\right). \tag{40}$$

The desired). purity for the experiment reported below was 55.0 % and the controller was started at the 60th period. As in the simulation study, a diagonal matrix $\mathbf{R}_j = 0.02$ $\mathbf{I}_{(3,3)}$ was chosen for regularization. The control horizon was set to $H_r = 1$ and the prediction horizon is $H_p = 60$ periods. Figure 4.18 shows the evolution of the product purity as well as of the controlled variables. In the open-loop mode where the operating point was calculated based on the initial model, the product purity was violated at the periods numbered 48 and 54. After a cycle the controller was able to drive the purity above 55.0% and to keep it there. The controller first reduces the desorbent consumption. This action seems to be in contradiction to the intuitive idea that more desorbent injection should enhance the separation. In the presence of a reaction this is not true, as shown by this experiment. The controlled variables converge towards a steady state, but they still change from period to period, due to the nonideality of the plant.

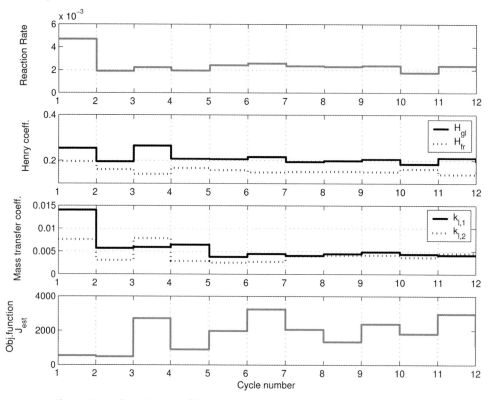

Figure 4.17 Online estimation of the model parameters

4.5.6
Summary

Closed-loop control of SMB processes is a challenging task because of the complex dynamics of the process and the large order of the discretized model. By formulating the control task as an online optimization problem on a receding horizon, the process can be at an optimal operating point while meeting constraints on the product purities. The feasibility of the approach has been demonstrated on a real pilot-scale plant using an industrial PLC-based in a well-known term.

4.6
Conclusions

In this chapter, it was demonstrated by means of several examples how rigorous, first-principles-based models can be used in process control. In the reactive distillation case study, a NMPC was presented that is based upon a slightly simplified rigorous process model. For reasons of computational efficiency, the solution of the algebraic equations was separated from the solution of the balance equations, resulting

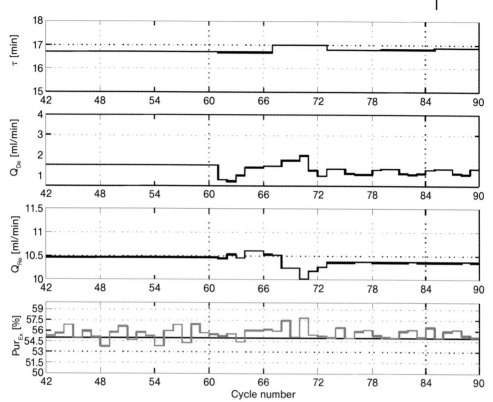

Figure 4.18 Control experiment for a target purity of 55.0%

in a performance gain by about 10. The NMPC controller not only gave a much better performance than the linear controller but it could also control the process in a region where the gains change their signs so that a linear controller inevitably fails. In related work, we used a neural net approximation of the rigorous process model in an NMPC controller, giving a slightly inferior performance with a much reduced computational effort (Engell and Fernholz 2003).

Online optimization using measurement information in many cases is an attractive alternative to the tracking of precomputed references because the process can be operated much closer to its real optimum, while still meeting hard bounds on the specifications. The measurement information can be used in the control scheme in various ways. The weakest form of feedback is to use the measurements for parameter adaptation only which requires a structurally correct model. In the control of the SMB process, this was combined with updating a disturbance model so that the desired product purities were maintained even for plant-model mismatch. Measurement information can also be used to modify the gradients in the optimization problem, ensuring convergence to the true optimum even in the case of structural model mismatch.

The biggest obstacle to the widespread use of model-based control is the effort needed to obtain faithful dynamic models of complex processes. While it has become routine to base process design on rigorous stationary process models, the effort to develop dynamic models is usually avoided. Process designers tend to neglect dynamic effects and to believe that control will somehow deal with them. As shown for the reactive distillation example, however, standard methods may fail, especially if a process is run at an optimal point, because near such an operating point, some variables will exhibit a change of the sign of the gain unless the optimum is only defined by constraints. A combination of first principles-based and black box models, the parameters of which are estimated from operational data, may be a way to obtain sufficiently accurate models without excessive effort. In combination with this approach the application of optimization techniques which take model mismatch explicitly into account, as presented in Section 4.4, is very promising.

References

1 *Agreda V. H. Partin L. R. Heise W. H.* High purity methyl acetate via reactive distillation, Chem. Eng. Prog. 86 (1990) p. 40–46

2 *Altenhöner U. Meurer M. Strube J. Schmidt-Traub H.* Parameter estimation for the simulation of liquid chromatography, J. Chromatogr A 769 (1997) p. 59–69

3 *Brdyś M. Chen S. Roberts P. D.* An extension to the modified two-step algorithm for steady-state system optimization and parameter estimation, Int. J. Syst. Sci. 17 (1986) p. 1229–1243

4 *Brdyś M. Ellis J. E. Roberts P. D.* Augmented integrated system optimization and parameter estimation technique: derivation, optimality and convergence, IEEE Proc. 134 (1987) p. 201–209

5 *Brdyś M. Tajewski P.* An algorithm for steady-state optimizing dual control of uncertain plants, Proceedings of the 1st IFAC Workshop on New Trends in Design of Control Systems, Slovakia, (1994) pp. 249–254

6 *Broughton D.* (1966) Continuous simulated counter-current sorption process employing desorbent made in said process. US Patent 3.291.726

7 *Byrd R. H. Lu P. Nocedal J. Zhu C.* (1994) A limited memory algorithm for bound constrained optimization, Technical Report NAM-08, Department of Electrical Engineering and Computer Science, Northwestern University, USA

8 *Draeger A. Ranke H Engell S.* Model predictive control using neural networks. IEEE Control Syst. Mag. 15 (5) (1995) p. 61–66

9 *Dünnebier G. Engell S. Epping A. Hanisch F. Jupke A. Klatt K.-U. Schmidt-Traub H.*

Model-based control of batch chromatography, AIChE J. 47 (2001) p. 2493–2502

10 *Dünnebier G. Klatt K.-U.* Modelling and simulation of nonlinear chromatographic separation processes: a comparison of different modelling approaches, Chem. Eng. Sci. 55 (2000) p. 373–380

11 *Engell S. Fernholz G.* Control of a Reactive Separation Process, Chem. Eng. Process. 42 (2003) p. 201–210

12 *Engell S. Müller R.* Multivariable controller design by frequency response approximation, Proceedings of the 2nd European Control Conference ECC2, Groningen, (1993) pp. 1715–1720

13 *Fernholz G. Engell S. Fougner K.* (1999a) Dynamics and Control of a Semibatch Reactive Distillation Process. Proc. 2nd European Congress on Chemical Engineering (CD-ROM), Montpellier

14 *Fernholz G. Engell S. Kreul L.-U. Górak A.* (2000) Optimal Operation of a Semibatch Reactive Distillation Column. Proc. 7th Int. Symposium on Process Systems Engineering, Keystone, Colorado. In: Computers & Chemical Engg. 24, 1569–1575

15 *Fernholz. G. Wang W. Engell S. Fougner K. Bredehöft J.-P.* Operation and control of a semi-batch reactive distillation column. Proceedings of the 1999 IEEE CCA, Kohala Coast, Hawaii, August 22–27, (1999b) pp. 397–402, IEEE Press

16 *Gao W. Engell S.* (2005) Iterative Set-Point Optimization of Batch Chromatography, *Computers and Chemical Engineering* 29, 1401–1410

17 *gPROMS User's Guide* (1997) Process System Enterprise, London, United Kingdom

18 *Gu T.* (1995) Mathematical Modelling and Scale Up of Liquid Chromatography, Springer, New York

19 *Guest D. W.* Evaluation of simulated moving bed chromatography for pharmaceutical process development, J.Chromatogr. A 760 (1997) p. 159–162

20 *Guiochon G.* Preparative liquid chromatography, J.Chromatogr. A 965 (2002) p. 129–161

21 *Hanisch F.* (2002) Prozessführung präparativer Chromatographieverfahren (Operation of Preparative Chromatographic Processes), Dr.-Ing. dissertation, University of Dortmund, and Shaker Verlag, Aachen (in German)

22 *Hotier G. Nicoud R. M.* (1996) Chromatographic simulated mobile bed separation process with dead volume correction using period desynchronization. US Patent 5 578 215

23 *Juza M. Mazzotti M. Morbidelli M.* Simulated moving-bed chromatography and its application to chirotechnology, Trends Biotechnol. 18 (2000) p. 108–118

24 *Klatt K.-U. Hanisch F. Dünnebier G.* Model-based control of a simulated moving bed chromatographic process for the separation of fructose and glucose, J. Process Control 12 (2002) p. 203–219

25 *Klatt K.-U. Dünnebier G. Hanisch F. Engell S.* (2002) Optimal Operation and Control of Simulated Moving Bed Chromatography: A Model-based Approach. *Invited Plenary Paper*, CACHE/AIChE Conference Chemical Process Control 6, 2001, Tucson. In: J. B. Rawlings, B. A. Ogunnaike, and J. W. Eaton (Eds.): *Chemical Process Control VI*, AIChE Symposium Series No. 326, Vol. 98 CACHE Publications, 2002, 239–254

26 *Kreul L. U. Górak A. Dittrich C. Barton P. I.* Dynamic catalytic distillation: advanced simulation and experimental validation, Comput. Chem. Eng. 22 (1998) p. 371–378

27 *Lin J. C. Han P. D. Roberts P. D. Wan B. W.* New approach to stochastic optimizing control of steady-state systems using dynamic information, Int. J. Control 50 (1989) p. 2205–2235

28 *Máckowiak J.* (1991) Fluiddynamik von Kolonnen mit modernen Füllkörpern und Packungen für Gas-/Flüssigsysteme, 1. Auflage, Otto-Salle-Verlag, Frankfurt (in German)

29 *Nagrath D. Bequette B. Cramer S.* Evolutionary operation and control of chromatographic processes, AIChE J. 49 (2003) p. 82–95

30 *Noeres C.* (2003) Catalytic distillation: dynamic modelling, simulation and experimental validation, Dr.-Ing. dissertation, University of Dortmund, and VDI Verlag, Düsseldorf

31 *Roberts P. D.* An algorithm for steady-state system optimization and parameter estimation, Int. J. Syst. Sci. 10 (1979) p. 719–734

32 *Roberts P. D.* Broyden derivative approximation in ISOPE optimizing and optimal control algorithms, Proceedings of the 11th IFAC Workshop on Control Applications of Optimization CAO'2000, Elsevier (2000) pp. 283–288

33 *Tatjewski P.* (2002) Iterative optimizing set-point control – the basic principle redesigned, Proceedings of the 15th Triennial IFAC World Congress, CD-ROM, Barcelona

34 *Tatjewski P. Brdyś M. A. Duda J.* Optimizing control of uncertain plants with constrained feedback controlled outputs, Int. J. Control 74 (2001) p. 1510–1526

35 *Toumi A. Engell S.* (2004c) A software package for optimal operation of continuous moving bed chromatographic processes. In: H. G. Bock, E. Kostina, H. X. Phu, and R. Rannacher (Eds.): *Modelling Simulation and Optimization of Complex Processes (Proceedings of the International Conference on High Performance Scientific Computing, Hanoi, 2003)*, Springer, 471–484

36 *Toumi A. Engell S.* Optimal operation and control of a reactive simulated moving bed process, Proceedings of the IFAC Symposium on Advanced Control of Chemical Processes, Hong Kong, Elsevier (2004a) pp. 243–248

37 *Toumi A. Engell S.* (2004b) Optimization-based Control of a Reactive Simulated Moving Bed Process for Glucose Isomerization, Chemical *Engineering Science* 59, 3777–3792

38 *Toumi A. Engell S.* (2005) Advanced control of simulated moving bed processes. In: H. Schmidt-Traub (Ed.): *Preparative Chromatography of fine chemicals and pharmaceuticals agents*, Wiley-VCH, Weinheim.

39 *Toumi A. Engell S. Ludemann-Hombourger O. Nicoud R. M. Bailly M.* Optimization of simulated moving bed and VARICOL processes, J.Chromatogr. A 1006 (2003) p. 15–31

40 *Wang C. Klatt K. Dünnebier G. Engell S. Hanisch F.* Neural network based identification of SMB chromatographic processes, Control Eng. Practice 11 (2003) p. 949–959

41 *Zhang Z. Mazzotti M. Morbidelli M.* Power-Feed operation of simulated moving bed units: changing flow-rates during the switching interval, J. Chromatogr. A 1006 (2003) p. 87–99

42 *Zhang H. Roberts P. D.* On-line steady-state optimization of nonlinear constrained processes with slow dynamics, Trans. Inst. Measure. Control. 12 (1990) p. 251–261

43 *Zhou J. L. Tits A. L. Lawrence C. T.* (1997) User's Guide for FFSQP Version 3.7: a FORTRAN code for solving constrained nonlinear (minimax). optimization problems, generating iterates satisfaying all inequality and linear constraints, University of Maryland

5
Real Time Optimization

Vivek Dua, John D. Perkins, and Efstratios N. Pistikopoulos

Abstract

This chapter considers two real time optimization (RTO) problems. The first problem is concerned with the model based control of linear discrete time systems and the second problem considers the case when logical conditions are also involved in the first problem. These RTO problems are reformulated as multiparametric programs to obtain control variables as an explicit function of the state of the system. This reduces the real time Optimization problems to simple function evaluations.

5.1
Introduction

Real Time Optimization (RTO) of a system is typically concerned with the solution of the following problem (Marlin and Hrymak, 1997; Perkins, 1998):

$$J(x) = \min_{u} f(x, u)$$
$$\text{s.t. } h(u, x) = 0$$
$$g(u, x) \leq 0 \tag{1}$$
$$x \in X$$

where x is the vector of the state of the system, u is the vector of control variables, f is a scalar objective function, such as cost, to be minimized, h is a vector representing the model of the system, g is a vector representing constraints, such as lower and upper bounds on x and u and X is a compact and convex set. Note that this problem is solved repetitively at regular time intervals.

Model Based Predictive Control (MPC) (Morari and Lee, 1999) is widely used by industry to address real time optimization problems with constraints on u and x. It is based on a receding horizon approach where a sequence of future control actions is computed based on a prediction of the future evolution of the system and applied to the system until new measurements become available. Then, a new sequence is determined which replaces the previous one – see Figure 5.1 where x^* is the desired

Computer Aided Process and Product Engineering. Edited by Luis Puigjaner and Georges Heyen
Copyright © 2006 WILEY-VCH Verlag GmbH & Co. KGaA, Weinheim
ISBN: 3-527-30804-0

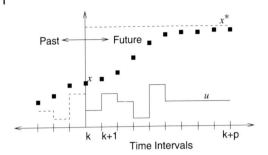

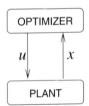

Figure 5.1   Model Based Predictive Control

state of the plant, k is the current time interval and $k + 1, \ldots, k + p$ are the future time intervals. Each sequence is evaluated by solving the optimization problem (1).

Real time optimization offers tremendous benefits but has large real time compu-tational requirements which involve a repetitive solution of problem (1) at regular time intervals (see Figure 5.2). The rest of the chapter is organised as follows. In the next section a parametric programming approach is introduced which can be used to compute u as an explicit function of x. Section 5.3 considers the case when h is given by linear discrete state space equations and the case when u also involves 0–1 binary variables is addressed in section 5.4. The solution approaches presented in sections 5.3 and 5.4 reduce RTO to simple function evaluations.

Figure 5.2   Real Time optimization

5.2
Parametric Programming

In an optimization framework, where the objective is to minimize or maximize a performance criterion subject to a given set of constraints and where some of the parameters in the optimization problem vary between specified lower and upper bounds, parametric programming is a technique for obtaining (i) the objective func-tion and the optimization variables as a function of these parameters and (ii) the regions in the space of the parameters where these functions are valid (Fiacco, 1983; Gal, 1995; Acevedo and Pistikopoulos, 1996, 1997; Pertsinidis et al., 1998; Papalex-andri and Dimkou, 1998; Acevedo and. Pistikopoulos, 1999; Dua and Pistikopoulos, 1999). Considering u as optimization variables and x as parameters in (1), parametric programming provides.

u(x)

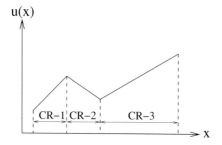

$$
u(x) = \begin{cases}
u^1(x) & \text{if } x \in CR^1 \\
u^2(x) & \text{if } x \in CR^2 \\
\vdots \\
u^i(x) & \text{if } x \in CR^i \\
\vdots \\
u^N(x) & \text{if } x \in CR^N
\end{cases}
$$

Figure 5.3 Parametric Optimization

such that $CR^i \cap CR^j = \phi$, $i \neq j$, $\forall i, j = 1, ..., N$ and $CR_i \subseteq X$, $\forall i = 1, ..., N$. A CR^i is known as a Critical Region. For the case when f, g and h are linear and separable in u and x, the CRs are polyhedra and each CR corresponds to a unique set of active constraints (Dua et al., 2002). See Figure 5.3, where u is plotted as a function of x.

The procedure for obtaining $u^i(x)$ and CR^i depends upon whether f, g and h are linear, quadratic, nonlinear, convex, differentiable, or not, and also whether u is vector of continuous or mixed – continuous and integer – variables (Dua and Pistikopoulos, 2000; Dua et al., 2002; Dua and Pistikopoulos, 1999; Dua et al., 2003; Sakizlis et al., 2002b). Recently algorithms for the case when (1) involves (i) differential and algebraic equations (Sakizlis et al., 2002a) and (ii) uncertain parameters (Sakizlis et al., 2004) have also been proposed. The engineering significance of solving parametric programming problems is highlighted in the next motivating example.

5.2.1
Example 1

Consider the refinery blending and production problem depicted in Figure 5.4 (Edgar and Himmelblau, 1989). The objective is to maximize the profit for the operating conditions given in Table 5.1, where x_1 and x_2 are the parameters representing the additional maximum allowable production of gasoline and kerosene production respectively. This results in a multi-parametric linear programming problem given in Table 5.2, where u_i and u_2 are the flowrates of the crude oils-1 and 2 respectively, in bbl/day and the units of profit are \$/day. The solution of this problem by using the algorithm of Gal and Nedoma (1972) is given in Table 5.3. The engineering significance of obtaining this solution is as follows:

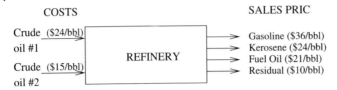

Figure 5.4   Crude Oil Refinery

(i)   A complete map of all the optimal solutions, profit and crude oil flowrates as a function of x_1 and x_2, is available.

(ii) The space of x_1 and x_2 has been divided into two regions, CR^1 and CR^2, where the profiles of profit and flowrates of crude oils remain optimal and hence (a) one does not have to exhaustively enumerate the complete space of x_1 and x_2 and (b) the optimal solution can be obtained by simply substituting the value of x_1 and x_2 into the parametric profiles without any further optimization calculations.

(iii) The sensitivity of the profit to the parameters can be identified. In CR^1 the profit is more sensitive to x_2, whereas in CR^2 it is not sensitive to x_2 at all. Thus, for any value of x that lies in CR^2, any expansion in kerosene production will not affect the profit.

This type of Information is quite useful for solving real time optimization problems. In the next section it is shown that real time model based control and optimization problems can be reformulated as multi-parametric quadratic programming problems, the solution of which is given by optimal control variables as a function of the state variables. The real time optimization problem thus reduces to simple function evaluations.

Table 5.1   Refinery Data

	Volume % Yield		Maximum allowable production (bbl/day)
	Crude # 1	Crude # 2	
Gasoline	80	44	24 000 + x_1
Kerosene	5	10	2 000 + x_2
Fuel Oil	10	36	6 000
Residual	5	10	–
Processing Cost ($/bbl)	0.50	1.00	–

Table 5.2   Refinery Model

Profit = max 8.1 u_1 + 10.8 u_2
s.t.   0.80 u_1 + 0.44 $u_2 \leq$ 24 000 + x_1
        0.05 u_1 + 0.10 $u_2 \leq$ 2 000 + x_2
        0.10 u_1 + 0.36 $u_2 \leq$ 6 000
        $u_1 \geq 0, u_2 \geq 0$
        $0 \leq x_1 \leq 6000$
        $0 \leq x_2 \leq$ 500

Table 5.3 Solution of the Refinery Example

i	CR^i	Optimal Solution
1	$-0.14\,x_1 + 4.21\,x_2 \le 896.55$ $0 \le x_1 \le 6000$ $0 \le x_2$	Profit (x) = 4.66 x_1 + 87.52 x_2 + 286758.6 $u_1 = 1.72\,x_1 - 7.59\,x_2 + 26206.90$ $u_2 = -0.86\,x_1 + 13.79\,x_2 + 6896.55$
2	$-0.14\,x_1 + 4.21\,x_2 \le 896.55$ $0 \le x_1 \le 6000$ $x_2 \le 500$	Profit (x) = 7.53 x_1 + 305409.84 $u_1 = 1.48\,x_1 + 24590.16$ $u_2 = -0.41\,x_1 + 9836.07$

5.3
Parametric Control

Consider the following state-space representation of a given process model (Pistiko-poulos et al., 2002):

$$\begin{cases} x(t+1) = Ax(t) + Bu(t) \\ \quad\;\; y(t) = Cx(t), \end{cases} \tag{2}$$

subject to the following constraints:

$$\begin{aligned} y_{min} &\le y(t) \le y_{max} \\ u_{min} &\le u(t) \le u_{max}, \end{aligned} \tag{3}$$

where $x(t) \in R^n$, $u(t) \in R^m$, and $y(t) \in R^p$ are the state, input, and output vectors respectively, subscripts *min* and *max* denote lower and upper bounds respectively and (A, B) is stabilizable. Model based control problems for regulating to the origin can then be posed as the following optimization problems:

$$\min_{U} J(U, x(t)) = x_{t+N_y|t}^T P x_{t+N_y|t} + \sum_{k=0}^{N_y-1} \left[x_{t+k|t}^T Q x_{t+k|t} + u_{t+k}^T R u_{t+k} \right]$$

$$\begin{aligned} \text{s.t. } & y_{min} \le y_{t+k|t} \le y_{max}, \; k = 1, \ldots, N_c \\ & u_{min} \le u_{t+k} \le u_{max}, \; k = 0, 1, \ldots, N_c \\ & x_{t|t} = x(t) \\ & x_{t+k+1|t} = Ax_{t+k|t} + Bu_{t+k}, \; k \ge 0 \\ & y_{t+k|t} = Cx_{t+k|t}, \; k \ge 0 \end{aligned} \tag{4}$$

where $U \triangleq \{u_t, \ldots, u_{t+N_u-1}\}$, $Q = Q^T \ge 0$, $R = R^T > 0$, $P \ge 0$, $N_y \ge N_u$ and the superscript T denotes the transpose of the corresponding vector or matrix. The problem (4) is solved repetitively at each time t for the current measurement $x(t)$ and the vector of predicted state variables, $x_{t+1|t}, \ldots, x_{t+k|t}$ at time $t + 1, \ldots, t + k$ respectively and corresponding control actions $u_t, \ldots, u_{t+k-1}$ is obtained.

In the following paragraphs, a parametric programming approach which avoids a repetitive solution of (4) is presented. First, we do some algebraic manipulations to recast (4) in a form suitable for using and developing some new parametric programming concepts. By making the following substitution in (4):

$$x_{t+k|t} = A^k x(t) + \sum_{j=0}^{k-1} A^j B u_{t+k-1-j} \tag{5}$$

the objective $J(U, x(t))$ can be formulated as the following Quadratic Programming (QP) problem:

$$\min_{U} \frac{1}{2} U^T HU + x^T(t) FU + \frac{1}{2} x^T(t) Yx(t) \tag{6}$$

s.t. $GU \le W + Ex(t)$

where $U \triangleq [u_t^T, ..., u_{t+N_u-1}^T]^T \in R^s$, $s \triangleq mN_u$, is the vector of optimization variables, $H = H^T > 0$, and H, F, Y, G, W, E are obtained from Q, R and (4)–(5). The QP problem (6) can now be formulated as the following Multi-parametric Quadratic Program (mp-QP):

$$\mu(x) = \min_{z} \frac{1}{2} z^T Hz \tag{7}$$

s.t. $Gz \le W + Sx(t)$

where $z \triangleq U + H^{-1} F^T x(t)$, $z \in R^s$, represents the vector of optimization variables, $S \triangleq E + GH^{-1}$ and x represents the vector of parameters. The main advantage of writing (4) in the form given in (7) is that z (and therefore U) can be obtained as an affine function of x for the complete feasible space of x. To derive these results, we first state the following theorem.

Theorem 1 For the problem in (7) let x_0 be a vector of parameter values and (z_0, λ_0) a KKT pair, where $\lambda_0 = \lambda(x_0)$ is a vector of nonnegative Lagrange multipliers, λ, and $z_0 = z(x_0)$ is feasible in (7). Also assume that (i) linear independence constraint qualification and (ii) strict complementary slackness conditions hold. Then,

$$\begin{bmatrix} z(x) \\ \lambda(x) \end{bmatrix} = -(M_0)^{-1} N_0 (x - x_0) + \begin{bmatrix} z_0 \\ \lambda_0 \end{bmatrix} \tag{8}$$

where,

$$M_0 = \begin{pmatrix} H & G_1^T & \cdots & G_q^T \\ -\lambda_1 G_1 & -V_1 & & \\ \vdots & & \ddots & \\ -\lambda_p G_q & & & -V_q \end{pmatrix}$$

$$N_0 = (Y, \lambda_1 S_1, ..., \lambda_p S_p)^T$$

where G_i denotes the ith row of G, S_i denotes the ith row of S, $V_i = G_i z_0 - W_i - S_i x_0$, W_i denotes the ith row of W and Y is a null matrix of dimension $(s \times n)$.

See Pistikopoulos et al. (2002) for the proof. The space of x where this solution, (8), remains optimal is defined as the Critical Region (CR^0) and can be obtained as follows. Let CR^R represent the set of inequalities obtained (i) by substituting $z(x)$ into the inequalities in (7) and (ii) from the positivity of the Lagrange multipliers, as follows:

$$CR^R = \{Gz(x) \leq W + Sx(t), \ \lambda(x) \geq 0\},$$
(9)

then CR^0 is obtained by removing the redundant constraints from CR^R as follows:

$$CR^0 = \Delta\{CR^R\}$$
(10)

where Δ is an operator which removes the redundant constraints – for a procedure to identify the redundant constraints, see Gal (1995). Since for a given space of state-variables, X, so far we have characterized only a subset of X i.e. $CR^0 \subseteq X$, in the next step the rest of the region CR^{rest}, is obtained as follows (Pistikopoulos et al., 2002):

$$CR^{rest} = X - CR^0.$$
(11)

The above steps, (8–11) are repeated and a set of $z(x)$, $\lambda(x)$ and corresponding CR^0s is obtained. The solution procedure terminates when no more regions can be obtained, i.e. when $CR^{rest} = \phi$. For the regions which have the same solution and can be unified to give a convex region, such a unification is performed and a compact representation is obtained. The continuity and convexity properties of the optimal solution are summarized in the next theorem.

Theorem 2 For the mp-QP problem, (7), the set offeasible parameters $X_f \subseteq X$ is convex, the optimal solution, $z(x) : X_f \to R^s$ is continuous and piecewise affine, and the optimal objective function $\mu(x) : X_f \to R$ is continuous, convex and piecewise quadratic.

See Pistikopoulos et al. (2002) for the proof. Based upon the above theoretical developments, an algorithm for the solution of an mp-QP of the form given in (7) to calculate U as an affine function of x and characterize X by a set of polyhedral regions, CRs, has been developed which is summarized in Table 5.4.

This approach provides a significant advancement in the solution and real time implementation of model based control problems. Since its application results in a complete set of control variables as a function of state-variables (from (8)) and the corresponding regions of validity (from (10)), which are computed off-line. Therefore during on-line optimization, no optimizer needs to be called and instead for the current state of the plant, the region, CR^0, where the value of the state variables is valid, can be identified by substituting the value of these state variables into the inequalities which define the regions. Then, the corresponding control variables can be computed by using a function evaluation of the corresponding affine function (see Figure 5.5). Figure 5.6 demonstrates how advanced controllers can be implemented on a simple hardware.

5.4
Hybrid Systems

Hybrid systems can be defined as systems comprising a number of interconnected continuous subsystems where the interconnections are determined by logical or discrete switchings. Each subsystem is governed by a unique set of differential and/or algebraic equations. In this section we focus on piecewise affine (PWA) systems (Bemporad and Morari, 1999). PWA systems are defined by partitioning the state and input space into polyhedral regions and associating with each region a different linear state update equation

$$x(t+1) = A^i x(t) + B^i u(t) + f^i \tag{12}$$

$$\text{if } \begin{bmatrix} x(t) \\ u(t) \end{bmatrix} \in p^i$$

where $i = 1, ..., s$, $x \in R^{n_c} \times \{0, 1\}^{n_l}$, $u \in R^{m_c} \times \{0, 1\}^{m_l}$, $\{P_i\}_{i=1}^{s}$ is a polyhedral partition of the set of the state and input space $P \subset R^{n+m}$, $n \triangleq n_c + n_l$, $m \triangleq m_c m_j$. P is assumed to be closed and bounded and $x_c \in R^{n_c}$ and $u_c \in R^{m_c}$ denote the continuous components of the state and input vector, respectively; $x_l \in \{0, 1\}^{n_l}$ and $u_l \in \{0, 1\}^{m_l}$ similarly denote the binary components.

Note that PWA models are not suitable for recasting analysis/synthesis problems into more compact optimization problems. For this purpose the Mixed Logical Dynamical (MLD) framework (Bemporad and Morari, 1999) is used. The general MLD form of a hybrid system is:

$$x(t+1) = Ax(t) + B_1 u(t) + B_2 \delta(t) + B_3 z(t) \tag{13}$$

$$y(t) = Cx(t) + D_1 u(t) + D_2 \delta(t) + D_3 z(t) \tag{14}$$

$$E_2 \delta(t) + E_3 z(t) \leq E_1 u(t) + E_4 x(t) + E_5 \tag{15}$$

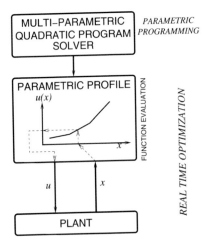

Figure 5.5 Real time optimization via parametric programming

Table 5.4 Solution Steps of the mp-QP Algorithm

Step 1	For a given space of x solve (7) by treating x as a free variable and obtain $[x_0]$.
Step 2	In (7) fix $x = x_0$ and solve (7) to obtain $[z_0, \lambda_0]$.
Step 3	Obtain $[z(x), \lambda(x)]$ from (8).
Step 4	Define CR^R as given in (9).
Step 5	From CR^R remove redundant inequalities and define the region of optimality CR^0 as given in (10).
Step 6	Define the rest of the region, CR^{rest}, as given in (11).
Step 7	If no more regions to explore, go to the next step, otherwise go to Step 1.
Step 8	Collect all the solutions and unify a convex combination of the regions having the same solution to obtain a compact representation.

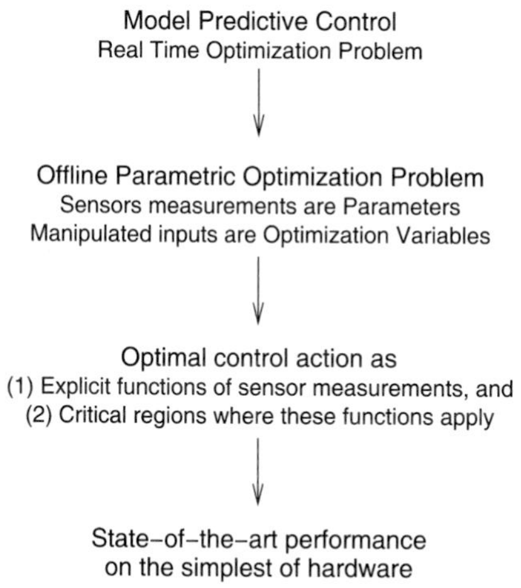

Model Predictive Control
Real Time Optimization Problem

$\downarrow$

Offline Parametric Optimization Problem
Sensors measurements are Parameters
Manipulated inputs are Optimization Variables

$\downarrow$

Optimal control action as
(1) Explicit functions of sensor measurements, and
(2) Critical regions where these functions apply

$\downarrow$

State–of–the–art performance
on the simplest of hardware

Figure 5.6 Achieving state-of-the-art control performance on simple hardware

where $x = [x_c^T \, x_l^T]^T \in R^{n_c} \times \{0, 1\}^{n_l}$ are the continuous and binary states, $u = [u_c^T \, u_l^T]^T \in R^{m_c} \times \{0, 1\}^{m_l}$ are the inputs, $y = [y_c^T \, y_l^T]^T \in R^{p_c} \times \{0, 1\}^{p_l}$ the outputs, and $\delta \in \{0, 1\}^n$, $z \in R^{r_c}$ represent auxiliary binary and continuous variables respectively. All constraints on the states, the inputs, the z and δ variables are summarized in the inequalities (15). Note that, although the description (13)–(14)–(15) seems to be linear, nonlinearity is hidden in the integrality constraints over the binary variables. MLD systems are a versatile framework to model various classes of systems. For a detailed description of such capabilities we defer the reader to Morari et al. (2003).

5.4.1
Predictive Control of MLD Systems

Let t be the current time, and $x(t)$ the current state. Consider the following optimal control problem

$$\min_{\{v_0^{T-1}\}} J\left(v_0^{T-1}, x(t)\right)$$

$$\triangleq \sum_{k=0}^{T-1} ||v(k)||_{Q_1}^2 + ||\delta(k|t)||_{Q_2}^2 + ||z(k|t)||_{Q_3}^2 + ||x(k|t)||_{Q_4}^2 + ||y(k|t)||_{Q_5}^2 \tag{16}$$

s.t. $x(k+1|t) = Ax(k|t) + B_1v(k) + B_2\delta(k|t) + B_3z(k|t)$

$y(k|t) = Cx(k|t) + D_1v(k) + D_2\delta(k|t) + D_3z(k|t)$ $\tag{17}$

$E_2\delta(k|t) + E_3z(k|t) \leq E_1v(k) + E_4x(k|t) + E_5$

where $v_0^{T-1} \triangleq [v^T(0), ..., v^T(T-1)]^T$, $Q_1 = Q_1^T > 0$, $Q_2 = Q_2^T \geq 0$, $Q_3 = Q_3^T \geq 0$, $Q_4 = Q_4^T > 0$ and $Q_5 = Q_5^T \geq 0$. $x(k|t) \triangleq x(t+k, x(t), v_0^{k-1})$ is the state predicted at time $t+k$ resulting from the input $u(t+k) = v(k)$ to (13–15) starting from $x(0|t) = x(t)$. $\delta(k|t)$, $z(k|t)$ and $y(k|t)$ are similarly defined. Assume for the moment that the optimal solution $\{v_t^*(k)\}_{k=0,...,T-1}$ exists. According to the *receding horizon* philosophy mentioned above, set

$$u(t) = v_t^*(0), \tag{18}$$

disregard the subsequent optimal inputs $v_t^*(1), ..., v_t^*(T-1)$, and repeat the whole optimization procedure at time $t+1$. Note that (16–17) is a Mixed Integer Quadratic Program (MIQP). This problem can be formulated as a Mixed Integer Linear Program (MILP) if 1 norm instead of the 2 norm is considered in the objective function. The repetitive somtion of the MIQP or MILP can be avoided by formulating (16–17) as a multiparametric program and solving it to obtain the control variables as a set of explicit functions of the current state of the system and the regions in the space of the state variables where the explicit functions remain valid (Bemporad et al., 2000; Sakizlis et al., 2002a). This is achieved by recasting (16–17) in a compact form as follows:

$$J(x(t)) = \min_{\pi_c, \pi_d} \pi_c^T Q_c \pi_c + \phi^T \pi_d$$

$$\text{s.t. } G_c\pi_c + G_d\pi_d \leq S + Fx(t) \tag{19}$$

where π_c and π_d are continuous and discrete variables of (16–17), Q_c, ϕ^T, G_c, G_d, S, F are constant matrices and vectors of appropriate dimensions and Q_c is symmetrie and positive definite. $x(t)$ is the state at the current time t. The objective is to obtain π_c and π_d as a function of $x(t)$ without exhaustively enumerating the entire space of $x(t)$. This can be achieved by using parametric programming. In the next section an algorithm for Multiparametric Mixed Integer Linear Programs (mp-MILP) is described. This reduces the real time hybrid system control problem to a function evaluation problem (Figure 5.7).

5.4.2
Multiparametric Mixed-Integer Linear Programming

Consider a multiparametric Mixed Integer Linear Programming (mp-MILP) problem of the following form:

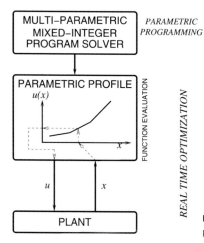

Figure 5.8   Real time optimization of hybrid systems via parametric programming

$$J(x(t)) = \min_{\pi_c, \pi_d} \phi_1^T \pi_c + \phi_2^T \pi_d$$

$$\text{s.t. } G_c \pi_c + G_d \pi_d \leq S + Fx(t)$$

(20)

where ϕ_1 and ϕ_2 are constant vectors.

5.4.2.1
Initialization

An initial feasible π_d is obtained by solving the following MILP:

$$\min_{\pi_c, \pi_d, x(t)} \phi_1^T \pi_c + \phi_2^T \pi_d$$

$$\text{s.t. } G_c \pi_c + G_d \pi_d \leq S + Fx(t)$$

(21)

where $x(t)$ is treated as a vector of free variable to find a starting feasible integer solution. Let the solution of (21) be given by $\pi_d = \overline{\pi}_d$.

5.4.2.2

Multiparametric LP Subproblem

Fix $\pi_d = \overline{\pi}_d$ (20) to obtain a multiparametric LP problem of the following form:

$$\hat{J}(x(t)) = \min_{\pi_c} \phi_1^T \pi_c + \phi_2^T \overline{\pi}_d$$

$$\text{s.t. } G_c\pi_c + G_d\overline{\pi}_d \leq S + Fx(t)$$
(22)

The solution of (22) is given by a set of linear parametric profiles, $\hat{J}(x(t))^i$, where $\hat{J}(x(t))$ is convex, and corresponding critical regions, CR^i (Gal, 1995).

The final solution of the multiparametric LP subproblem in (22) which represents a parametric upper bound on the final solution is given by (i) a set of parametric profiles, $\hat{J}(x(t))^i$, and the corresponding critical regions, CR^i, and (ii) a set of infeasible regions where $\hat{J}(x(t))^i = \infty$.

5.4.2.3

MILP Subproblem

For each critical region, CR^i, obtained from the solution of the multiparametric LP subproblem in (22), an MILP subproblem is formulated as follows:

$$\min_{\pi_c,\pi_d,x(t)} \phi_1^T \pi_c + \phi_2^T \pi_d$$

$$\text{s.t. } G_c\pi_c + G_d\pi_d \leq S + Fx(t)$$

$$\phi_1^T \pi_c + \phi_2^T \pi_d \leq \hat{J}(x(t))^i$$
(23)

$$\pi_d \neq \overline{\pi}_d$$

$$x \in CR^i$$

The integer solution, $\pi_d = \overline{\pi}_d^1$, and the corresponding CRs, obtained from the solution of (23), are then recycled back to the multiparametric LP subproblem – to obtain another set of parametric profiles. Note that the integer cut, $\pi_d \neq \overline{\pi}_d$, and the parametric cut, $\phi_1^T \pi_c + \phi_2^T \pi_d \leq \hat{J}(x(t))^i$ are accumulated at every iteration.

If there is no feasible solution to the MILP subproblem (23) in a CR^i, that region is excluded from further consideration and the current upper bound in that region represents the final solution. Note also that the integer solution obtained from the solution of (23) is guaranteed to appear in the final solution, since it represents the minimum of the objective function at the point, in $x(t)$, obtained from the solution of (23). The final solution of the MILP subproblem is given by a set of integer solutions and their corresponding CR^is.

5.4.2.4

Comparison of Parametric Solutions

The set of parametric solutions corresponding to an integer solution, $\pi_d = \overline{\pi}_d$, which represents the current upper bound are then compared to the parametric solutions corresponding to another integer solution, $\pi_d = \overline{\pi}_d^1$, in the corresponding $C Rs$ in order to obtain the lower of the two parametric solutions and update the upper bound. This is achieved by employing the procedure proposed by Acevedo and Pistikopoulos (1997b).

5.4.2.5
Multiparametric MILP Algorithm

Based upon the above theoretical developments, the steps of the algorithm can be stated as follows:

Step 0 (Initialization) Define an initial region of $x(t)$, CR, with best upper bound $\hat{J}^*(x(t)) = \infty$, and an initial integer solution $\overline{\pi}_d$.

Step 1 (Multiparametric LP Problem) For each region with a new integer solution, $\overline{\pi}_d$:

- Solve multiparametric LP subproblem (22) to obtain a set of parametric upper bounds $\hat{J}(x(t))$ and corresponding critical regions, CR.
- If $\hat{J}(x(t)) \leq \hat{J}^*(x(t))$ for some region of $x(t)$, update the best upper bound function, $\hat{J}^*(x(t))$, and the corresponding integer solutions, π_d^*,
- If an infeasibility is found in some region CR, go to Step 2.

Step 2 (Master Subproblem) For each region CR, formulate and solve the MILP master problem in (23) by (i) treating $x(t)$ as a variable bounded in the region CR, (ii) introducing an integer cut, $\pi_d \neq \overline{\pi}_d$ and (iii) introducing a parametric cut, $\phi_1^T \pi_c + \phi_2^T \pi_d \leq \hat{J}(x(t))^i$. Return to Step 1 with new integer solutions and corresponding CRs.

Step 3 (Convergence) The algorithm terminates in a region where the solution of the MILP subproblem is infeasible. The final solution is given by the current upper bounds $\hat{J}^*(x(t))$ in the corresponding CRs. The $\pi_c(x(t))$ and $\pi_d(x(t))$ corresponding to $\hat{J}^*(x(t))$ are then used to obtain $u(x(t))$.

Note that the algorithms presented in this chapter have been implemented and tested on a number of real time optimization problems (PAROS, 2004).

5.5
Concluding Remarks

In this chapter it was shown how real time optimization problems can be recast as multiparametric programs. Linear discrete time optimization problems are recast as multiparametric quadratic programs and problem involving logical decisions as multiparametric mixed integer programs. Algorithms for solving the multiparametric programs were then presented to compute the optimal control actions as an explicit function of the state of the system. This reduces real time optimization problems to simple function evaluations.

References

1 *Acevedo J. Pistikopoulos E. N.* A parametric MINLP algorithm for process synthesis problems under uncertainty. Industrial and Engineering Chemistry Research 35 (1996) p. 147–158

2 *Acevedo J. Pistikopoulos E. N.* A multiparametric programming approach for linear process engineering problems under uncertainty. Industrial and Engineering Chemistry Research 36 (1997) p. 717–728

3 *Acevedo J. Pistikopoulos E. N.* An algorithm for multiparametric mixed integer linear programming problems. Operations Research Letters 24 (1999) p. 139–148

4 *Bemporad A. Borrelli F. Morari M.* Piecewise linear optimal controllers for hybrid systems, proceedings of the American Control Conference (2000) p. 1190–1194

5 *Bemporad A. Morari M.* Control of systems integrating logic, dynamics, and constraints. Automatica 35 (1999) p. 407–427

6 *Dua V. Bozinis N. A. Pistikopoulos E. N.* A multiparametric programming approach for mixed-integer quadratic engineering problems. Computers & Chemical Engineering 26 (2002) p. 715–733

7 *Dua V. Papalexandri K. P. Pistikopoulos E. N.* Global optimization issues in multiparametric continuous and mixed-integer optimization problems, accepted for publication in the Journal of Global Optimization, 30 (2004) p. 59

8 *Dua V. Pistikopoulos E. N.* Algorithms for the solution of multiparametric mixed-integer nonlinear optimization problems. Industrial and Engineering Chemistry Research 38 (1999) p. 3976–3987

9 *Dua V. Pistikopoulos E. N.* An algorithm for the solution of multiparametric mixed integer linear programming problems. Annals of Operations Research 99 (2000) p. 123–139

10 *Edgar T. F. Himmelblau D. M.* Optimization of chemical processes. McGraw Hill Book Co, Singapore 2000

11 *Fiacco A. V.* Introduction to Sensitivity and Stability Analysis in Nonlinear Programming. Academic Press, New York 1983

12 *Gal T.* Postoptimal Analyses, Parametric Programming, and Related Topics. de Gruyter, New York 1995

13 *Gal T. Nedoma J.* Multiparametric linear programming. Management Science 18 (1972) p. 406–422

14 *Marilin T. E. Hrymak A. N.* Real-time operations optimization of continuous processes. In: Kantor J. C. Garcia C. E. Carnahan B. (Eds.), 5th Int. Conf. Chem. Proc. Control. Vol. 93 of AIChE Symposium Series 1997

15 *Morari M. Baotic M. Borrelli F.* Hybrid systems modeling and control. European Journal of Control 9 (2003) p. 177–189

16 *Morari M. Lee J.* Model predictive control: past, present and future. Computers & Chemical Engineering 23 (1999) p. 667–682

17 *Papalexandri K. P. Dimkou T. I.* A parametric mixed integer optimization algorithm for multi-objective engineering problems involving discrete decisions. Industrial and Engineering Chemistry Research 37 (5) (1998) p. 1866–1882

18 *PAROS* Parametric Optimization Solutions Ltd. http://www.parostech.com 2004

19 *Perkins J. D.* Plant-wide optimization: opportunities and challenges. In: Pekny J. Blau G. (Eds.), 3rd Int. Conf. Foundations of Computer Aided Process Operations. Vol. 94 of AIChE Symposium Series 1998

20 *Pertsinidis A. Grossmann I. E. McRae G. J.* Parametric optimization of MILP programs and a framework for the parametric optimization of MINLPs. Computers & Chemical Engineering 22 (1998) p. S205

21 *Pistikopoulos E. N. Dua V. Bozinis N. A. Bemporad A. Morari M.* On-line optimization via off-line parametric optimization tools. Computers & Chemical Engineering 26 (2002) p. 175–185

22 *Sakizlis V. Dua V. Perkins J. D. Pistikopoulos E. N.* The explicit control law for hybrid systems via parametric programming, proceedings of the 2002 American Control Conference, Anchorage 2002a

23 *Sakizlis V. Dua V. Perkins J. D. Pistikopoulos E. N.* The explicit model-based control law for continuous time systems via parametric programming, proceedings of the 2002 American Control Conference, Anchorage 2002b

24 *Sakizlis V. Kakalis N. Dua V. Perkins J. D. Pistikopoulos E. N.* Design of robust model based controllers via parametric programming. Automatica 40 (2004) p. 189–201

6
Batch and Hybrid Processes

Luis Puigjaner and Javier Romero

6.1
Introduction

Although historically, chemical engineers achieved their professional distinction with the design and operation of continuous processes [1], as we move into the new millennium, it comes somewhat as a surprise to realize that outside the petroleum and petrochemical industries, batch operation is still a common if not dominant mode of operation. Moreover, most batch processes are unlikely to be replaced by continuous processes [2, 3].

The reason is that as the production of chemicals undergoes a continuous specialization, to address the diversifying needs of the marketplace, the continuous evolution of product recipes implies a much shorter life cycle for a growing number of chemicals, than has been traditionally the case, leading to a perpetual product/process evolution [4, 5].

This situation has been matched and in part funded by a developing research interest in batch process systems engineering, batch production being the most suitable way of manufacturing the relatively large number of low-volume high-value-added products commonly found in the fine and specialty chemicals industry. Moreover, the coexistence of continuous and discrete parts in both strictly speaking batch processes and nominal continuous processes has motivated an increased research interest in further exploiting the inherent flexibility of batch procedures and the high productivity of continuous parts of the production system [6]. Thus, chemical plants constitute large hybrid systems, making it necessary to consider the continuous-discrete interactions taking place within an appropriate framework for plant and process simulation and optimization [7].

This chapter briefly discusses existing modeling frameworks for discrete/hybrid production systems embodying different approaches, before introducing a very recent framework for process recipe initialization that integrates a recipe model into the batch plant-wide model. Next, online and offline recipe adaptation from real-time plant information is presented, and finally, a model-based integrated advisory system

Computer Aided Process and Product Engineering. Edited by Luis Puigjaner and Georges Heyen
Copyright © 2006 WILEY-VCH Verlag GmbH & Co. KGaA, Weinheim
ISBN: 3-527-30804-0

is described. This system gives online advice to operators on how to react in case of process disturbances. In this way, an enhanced overall process flexibility and productivity is achieved. Application of this promising approach is illustrated through examples of increasing complexity.

6.1.1
Plant and Process Simulation

The discrete transitions occurring in chemical processing plants have only recently been addressed in a systematic manner. Barton and Pantelides [8] did pioneering work in this area. A new formal mathematical description of the combined discrete/continuous simulation problem was introduced to enhance the understanding of the fundamental discrete changes required to model processing systems. The modeling task is decomposed into two distinct activities: modeling fundamental physical behavior, and modeling the external actions imposed on this physical system resulting from interaction of the process with its environment by disturbances, operation procedures, or other control actions.

The physical behavior of the system can be described in terms of a set of integral and partial differential and algebraic equations (IPDAE). These equations may be continuous or discontinuous. In the latter case, the discontinuous equations are modeled using state-task networks (STNs) and resource-task networks (RTNs), which are based on discrete models. Otherwise, other frameworks based on a continuous representation of time have appeared more recently (event operation network among others). The detailed description of the different representation frameworks is the topic of the next section.

6.1.2
Process Representation Frameworks

The representation of a state-task network (STN) proposed by Kondili et al. [9] was originally intended to describe complex chemical processes arising in multiproduct/multipurpose batch chemical plants. The established representation is similar to the flow sheet representation of continuous plants, but is intended to describe the process itself rather than a specific plant.

The distinctive characteristic of the STN is that it has two types of nodes; mainly, the *state* nodes, representing the feeds, intermediates and final products and the *task* nodes, representing the processing operations which transform material from input states to output states (Fig. 6.1).

This representation is free from the ambiguities associated with recipe networks where only processing operations are represented. Process equipment and its connectivity are not explicitly shown. Other available resources are not represented.

The STN representation is equally suitable for networks of all types of processing tasks, continuous, semicontinuous or batch. The rules followed in its construction are:

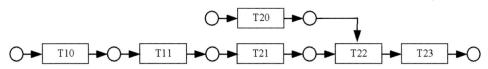

Figure 6.1 State-task network representation of chemical processes.
Circles: state nodes; *rectangles*: task nodes

- A task has as many input (output) states as different types of input (output) material.
- Two or more streams entering the same state are necessarily of the same material. If mixing of different streams is involved in the process, then this operation should form a separate task.

The STN representation assumes that an operation consumes material from input states at a fixed ratio and produces material for the output state also at a known fixed proportion. The processing time of each operation is known *a priori* and considered to be independent of the amount of material to be processed. Otherwise, the same operation may lead to different states (products) using different processing times.

States may be associated to four main types of storage policy:

- unlimited intermediate storage
- finite intermediate storage
- no intermediate storage
- zero wait (the product is unstable).

An alternative representation; the *resource-task network* (RTN) was proposed by Pantelides [10]. In contrast to the STN approach, where a task consumes and produces materials while using equipment and utilities during its execution, in this representation, a task is assumed only to consume and produce resources. Processing items are treated as though consumed at the start of a task and produced at the end. Furthermore, processing equipment in different conditions can be treated as different resources, with different activities consuming and generating them; this enables a simple representation of changeover activities. Pantelides [10] also proposed a discrete-time scheduling formulation based on the RTN, which, due to the uniform treatment of resources, only requires the description of three types of constraint, and does not distinguish between identical equipment items. He demonstrated that the integrality gap could not be worse than the most efficient form of STN formulation, but the ability to capture additional problem features in a straightforward fashion is attractive. Subsequent research has shown that these conveniences in formulation are overshadowed by the advantages offered by the STN formulation in allowing explicit exploitation of constraint structure through algorithm engineering.

The STN and RTN representations use discrete-time models. Such models suffer from a number of inherent drawbacks [11]:

- The discretization interval must be fine enough to capture all significant events, which may result in a very large model.

- It is difficult to model operations where the processing time is dependent on the batch size.
- The modeling of continuous operations must be approximated and minimum run-lengths give rise to complicated constraints.

Therefore, attempts have been made to develop frameworks based on a continuous-time representation. Reklaitis and Mockus [12] developed a continuous-time formulation based on the STN representation. A common resource grid is need, with the timing of the grid points ("event orders" in their terminology) determined by optimization. The same authors introduced an alternative solution procedure based on Bayesian heuristics in a later work [13]. Zhang and Sargent [14] describe a continuous-time representation based on RNT representation for both batch and continuous operations. The poor relaxation performance of the continuous-time models is the main obstacle to their large scale application. To avoid this deficiency, Shilling and Pantelides [11] modify the model by Zhang and Sargent (1996). A global linearization gives rise to a mixed-integer linear programming (MILP) which is solved by a hybrid branch-and-bound procedure. Recent reviews on these approaches can be found in Shah [15] and Silver et al. [16].

A realistic and flexible description of complex recipes has been recently improved using a flexible modeling environment [17] for the scheduling of batch chemical processes. The process structure (individual tasks, entire subtrains or complex structures of manufacturing activities) and related materials (raw, intermediate or final products) is characterized by means of a processing network which describes the material balance. In the most general case, the activity carried out in each process constitutes a general activity network. Manufacturing activities are considered at three different levels of abstraction: the process level, the stage level and the operation level.

This hierarchical approach permits the consideration of material states (subject to material balance and precedence constraints) and temporal states (subject to time constraints) at different levels.

At the process level, the process and materials network (PMN) provides a general description of production structures (such as synthesis and separation processes) and materials involved, including intermediates and recycled materials. An explicit material balance is specified for each of the processes in terms of a stoichiometric-like equation relating raw materials, intermediates and final products (Fig. 6.2). Each process may represent any kind of activity necessary to transform the input materials into the derived outputs.

Between the process level and the detailed description of the activities involved at the operation level, there is the stage level. At this level, the block of operations to be executed in the same equipment is described. Hence, at the stage level each process is split into a set of the blocks (Fig. 6.3). Each stage implies the following constraints:

- The sequence of operations involved requires a set of implicit constraints (links).
- Unit assignment is defined at this level. Thus, for all the operations of the same stage, the same unit assignment must be made.

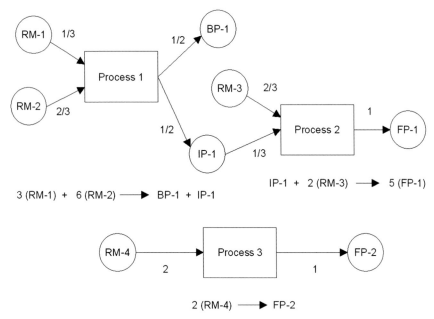

$$3 \text{ (RM-1)} + 6 \text{ (RM-2)} \longrightarrow \text{BP-1} + \text{IP-1}$$

$$\text{IP-1} + 2 \text{ (RM-3)} \longrightarrow 5 \text{ (FP-1)}$$

$$2 \text{ (RM-4)} \longrightarrow \text{FP-2}$$

Figure 6.2 A process and materials network (PMN) describing the processing of two products. RM are row materials, IP are intermediate products, BP are by-products and FP are final products

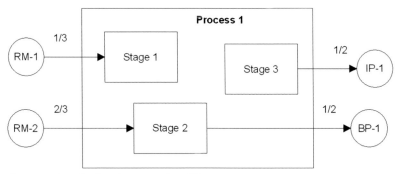

Figure 6.3 Stage level. Each stage involves different unit assignment opportunities

- A common size factor is attributed to each stage. This size factor summarizes the contribution of all the operations involved.

The operation level contains the detailed description of the activities contemplated in the network (tasks and subtasks), while implicit time constraints (links) must be also met at this level. The detailed representation of the structure of activities defining the different processes is called the event operation network (EON). It is also at this level that the general utility requirements (renewable, nonrenewable, storage) are represented.

The event operation network representation model describes the appropriate timing of process operations. A continuous-time representation of process activities is made using three basic elements: events, operations and links [18, 19].

Events designate those time instants where some change occurs. They are represented by nodes in the EON graph, and may be linked to operations or other events. Each event is associated to a time value and a lower bound.

Operations comprise those time intervals between events (Fig. 6.4). Each operation m is represented by a box linked with solid arrows to its associated nodes: initial $NI\ m$ and final $NF\ m$ nodes. Operations establish the equality links between nodes (two) in terms of the characteristic properties of each operation: the operation time, TOP and the waiting time TW. The operation time will depend on the amount of materials to be processed; the unit model and product changeover. The waiting time is the lag time between operations, which is bounded.

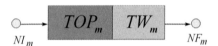

$$NI_m \quad TOP_m \quad TW_m \quad NF_m$$

Figure 6.4 The time description for operations. *TOP* Operation time, *TW* waiting time, *NI m* initial node of operation *m*, *NF m* final node of operation *m*

Finally, links are established between events by precedence constraints. A dashed arrow represents each link K from its node of origin NO_k to its destiny node ND_k and an associated offset time ΔT_K.

$$NO_k \quad \cdots\ \Delta T_k\ \cdots\ \rightarrow \quad ND_k$$

Figure 6.5 Event to event link and associated offset time representation. The *dashed arrow* represents each link K from its node of origin NO_k to its destiny node ND_k

Despite its simplicity, the EON representation is very general and flexible and it allows the handling of complex recipes (Fig. 6.6). The corresponding TOP, according to the batch size and material flow rate, also represents transfer operations between production stages. The necessary time overlapping of semicontinuous operations with batch units is also contemplated in this representation through appropriate links.

Other resources required for each operation (utilities, storage, capacity, manpower, etc.) can also be considered associated to the respective operation and timing.

Simulation of plant operation can be performed in terms of the EON representation from the following information contained in the process recipe and production structure characteristics:

- A sequence of production runs or jobs associated to a process or recipe.
- A set of assignments associated to each job and consistent with the process p.
- A batch size associated to each job and consistent with the process.
- A set of shifting times for all the operations involved.

These decisions may be generated automatically by using diverse procedures for the determination of an initial feasible solution. Hence, simulation may be executed by

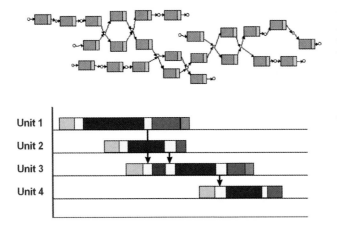

solving the corresponding EON to determine the timing of the operations and other resources requirements.

The flexibility and potential of the EON representation has been further exploited by incorporating the flexible recipe concept, which is the subject of the next section.

6.2
The Flexible Recipe Concept

The simulation environments described in the previous section assume operating at nominal conditions following fixed recipes. Moreover, these nominal conditions are determined only once and sometimes considering only one stage of the process recipe. However, batch and hybrid manufacturing systems' optimum performance require an integrated modeling environment capable of incorporating systematic information and of adapting to changing plant scenarios. Very recently, the flexible recipe concept has been introduced as an appropriate mechanism that permits the simultaneous optimization of recipe and plant operation [20, 21].

This concept arose from the fact that batch processes normally do not operate at the plant-wide optimal nominal conditions of the fixed batch recipes, but the traditional fixed recipe does not allow for adjustment to plant resource availability or to variations in both quality of raw materials and in the actual process conditions. However, the industrial process is often subject to various disturbances and to constrained plant resources availability. Therefore, the fixed recipe is in practice approximately adapted, but in a rather unsystematic way depending on the experience and intuition of operators. As an alternative, the concept of flexible recipe operation is introduced, and a general framework is presented to systematically deal with the required adaptations at a plant-wide level.

The flexible recipe concept was considered for the first time in the context of evolutionary operation [22]. The main objective of that approach was to gain statistical insight into the problem behavior in order to gradually improve process efficiency through suggestions of minor recipe modifications in each batch-run. However, it was not until the work of Rijnsdorp appeared [20] that the concept of flexible recipes was adequately introduced. Here, the term recipe is understood in a more abstract way as referring to the selected set of adjustable elements that control the process output generating the flexible recipe. According to this concept, a flexible recipe philosophy to operate batch processes was described in the work of Verwater-Lukszo [21]. This philosophy distinguishes two main levels in the flexible recipe: (a) the recipe initialization level, where different aspects of a master flexible recipe are adjusted to actual process conditions and availability of resources at the beginning of the batch, thus giving the initialized control recipe; and (b) the recipe correction level, where the initialized control recipe is adjusted to run-time process deviations, thus generating corrected control recipes. A flexible recipe improvement system tool (called COMBO) was developed by TNO TPD (Netherlands Organization for Applied Scientific Research) for application of the flexible recipes concept in industrial practice.

However, in this approach, only one critical stage of the process is considered and hence, no interaction with plant-wide optimization is, in fact, attempted. More recently, the application of the flexible recipe concept to an entire batch train was attempted for the multistage case [23]. However, standard quality models were assumed for process operations, and hence, no insight into recipe behavior was obtained. A new framework for recipe initialization that integrates a recipe model into the batch plant-wide model has been recently introduced [24]. The aim of this approach is to optimize the entire batch process, from recipe set-point adjustment to product sequencing. For this purpose, a recipe model and a plant-wide production model are required to build the flexible recipe model. Moreover, fulfillment of present standards (ISA S88) should be a requirement for implementation in industrial practice.

6.2.1
The Flexible Recipe and the Framework of ISA S88

Batch-processes flexibility may be mainly exploited at the level of the recipe formulation. Here, the set of process parameters is adjusted to warrant process outputs as a function of uncertain process inputs. Each one of such parameters, whose value may be changed for each batch, is called a recipe item. These items can be quantitative or qualitative, time-dependent or time-independent. The equipment requirement level, as defined in ISA-S88 [25], is already a flexible category in itself. In fact, ISA-S88 defines this level as an equipment choice constraint. Finally, considering flexibility in the recipe procedure would only be contemplated when some unexpected event happens, which is out of the scope of the batch process flexibility enhancement sought here.

In a company four types of recipes are typically found:

- General recipe and site recipe; which basically describe the technique and are equipment independent.
- Master recipe, a recipe which is equipment-dependent and which provides specific and unique batch-execution information describing how a product is to be produced in a given set of process equipment.

- Control recipe, which starting as a copy of the master recipe, contains detailed information for minute-to-minute process operation of a single batch.

The flexible recipe might be derived from a master recipe and subsequently used for generating and updating a control recipe. Verwater and Keesman [26] introduced the concept of different levels between these two stages defined at ISA-S88. With these new levels a better description of the different possible functionalities of the flexible recipe is obtained:

- Master control recipe, that is, a master recipe valid for a number of batches, but adjusted to the actual conditions (actual prices or quality requirements) from which the individual control recipes per batch are derived.
- Initialized control recipe, that is, the adjustment of the still-adjustable process conditions of a master control recipe to the actual process conditions at the beginning of the batch, i.e., the adjustment of variables such as temperature, pressure, catalyst addition and processing time in the face of deviations in the initial temperature of the batch, equipment fouling, available processing time and so on.
- Corrected control recipe, the result of adjusting the initialized control recipe to process deviations during the batch.
- And finally, for monitoring and archiving purposes, it is also useful to define the accomplished control recipe.

Therefore, on the basis of this basic philosophy, a novel flexible recipe approach [24] has been recently proposed that excerpts a flexible recipe model from a total master control recipe. This model describes the whole batch process train. However, it is only concerned with the critical batch process variables. Besides, it also considers the possible interactions between different batches because of scheduling purposes.

Regarding the different levels between the master recipe and the initialized control recipe described, it can be concluded that four different flexible-recipe systems may be useful:

- A system for adjusting the master recipe to the actual prices and quality requirements, defining the master control recipe.
- A system for defining the initialized control recipe from the master control recipe as a function of the actual process conditions, availability of resources at the beginning of the batch and of the availability of the plant equipment.
- A model to generate the corrected control recipe in the face of deviations during each batch.
- A system for updating and improving the master control recipe as the database of accomplished control recipes increases. This model will also improve the preceding models.

The interaction of these systems in a real-plant environment is described in Fig. 6.7.

These systems will have to be developed in laboratory experiments, pilot plant operation, during normal production by a systematic introduction of acceptable small changes in certain inputs and parameters, or by adjusting white models and simulating them under different operating conditions.

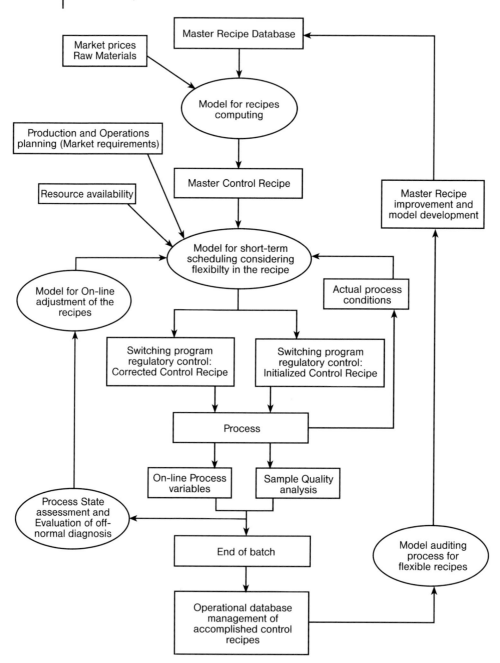

Figure 6.7 Optimal flexible recipe environment information flow proposed

6.3
The Flexible Recipe Model

The flexible recipe model is the tool that permits us to integrate a recipe optimization procedure with a batch plant optimization level. It represents the relationship that correlates a batch process output as a function of the selected input items of the recipes for different batch plant production scenarios. Therefore, it is a recipe description model that incorporates plant-wide variables. We identify four main components of the problem: quality or product specifications, process operating conditions, production costs and production due-dates.

The flexible recipe model can be applied to a variety of scenarios. For instance, during batch process operation, processing times of some tasks may vary without set-point adjustment, thus affecting the properties (quality) of the products obtained in such tasks. Then, to meet customer requirements, another batch of the same product might be able to compensate for these effects. For example, let's assume a process in which A is converted into B; one batch with low conversion of A could be compensated for by another batch with a higher conversion, assuming that these two batches are going to be mixed afterwards, so that the final product quality corresponds to the customer and legal requirements. Otherwise, the processing time might be optimized without set-point adjustment by compensating for the quality within the same batch. For instance, a batch of product A that is first heated in one piece of equipment before reacting in another. A reduction in the processing time of the first task could be offset by a higher reaction time. Moreover, the processing time could be optimized with some set-point adjustment. In this situation, the properties of intermediates produced might be altered only at the expense of a higher operation cost. For instance, the reaction time could be reduced by increasing the reaction temperature, although this recipe modification would imply a higher operation cost. Which of the above-mentioned strategies should be applied in each case will depend on the specific process and on the available knowledge of the different tasks of the process. For example, such ways of operation might not be very suitable for highly restrictive processes, such as those found in the pharmaceutical industry, but they are probably convenient to specialty batch chemical production where customer requirements are defined simply by a set of product properties and not by the specific way the product has been produced.

The preceding discussion leads to the basic concept upon which the modeling of scheduling problems considering the flexible recipe is built.

6.3.1
Proposed Concept for the Flexible Recipe Model

The flexible recipe model is regarded as a constraint on quality requirements and on production costs.

In this approach, recipe items are classified into four groups:

- The vector of process operating conditions, $\boldsymbol{poc}_i$, of stages i of a recipe. It includes parameters like temperature, pressure, type of catalyst, batch size, etc.

- The product specification vector, ps_i, at the end of each process stage i of a recipe. It might include parameters like conversion of a reactant, purity or quality aspects.
- Processing time, PT_i, at each stage i of a recipe.
- Waiting time, TW_i, that is, the time between the end of a stage and the next stage start time.

Then, the product specifications vector of a batch stage will in general be a function, Ψ, of processing time, waiting time, process set-points, and product specifications at different stages $i*$ where the different inputs to stage i are produced. Moreover, within this model, product specifications, ps, and process operation conditions, poc, are subject to optimization within a flexibility region, α and Δ respectively.

A general algorithm representation of the flexible recipe model for short term scheduling is presented in Eq. (1). This model contains the nominal recipe and its capacity to accept modifications. The model adjusts the different recipe parameters for each individual batch performed in a specific production plan Θ, where Θ is the variable that permits integrating batch process scheduling with the recipe optimization procedure.

Each specific production plan Θ is defined when the specific orders to be delivered at a specific set of due dates, S, is specified and when the specific set of different plant resources, A, is assigned to each order. Besides this, each production plan has to meet some physical plant constraints, T, such as the multistage flowshop or jobshop batch plant topology constraints, T, operating with a set J of equipment units and a set R of process resources. Each production plan will be generated to meet the market constraints: set I of production orders in a given set DD of time horizon or due dates.

A performance criterion ϕ is also included. This criterion may vary from batch to batch and it may contain economic as well as process variables. The flexible recipe model validity constraints are considered in σ and Δ regions.

Optimize $\Phi(PT_i, TW_i, ps_i, poc_i, \Theta)$
  subject to recipe constraints,
  $$ps_i = \Psi(PT_i, TW_i, ps_i, poc_i),$$
  $$ps_i \subset \sigma,$$
  $$poc_i \subset \Delta,$$
  subject to production environment constraints,
  $$\Theta(S, A) \subset \Omega(T, J, R, I, DD)$$

(1)

The model may interact with the short-term scheduling level either offline or online.

6.4
Flexible Recipe Model for Recipe Initialization

At the start of a batch the initial conditions may differ from those prescribed by the master recipe, even to the extent of making successful completion unlikely. Examples are deviations in catalyst activity, available heat, raw material quality and equipment fouling, among others. In such cases, the flexible recipe concept makes it possible to alter the still-adjustable process conditions, so as to ensure the most successful completion of the run. Otherwise, because of scheduling requirements, it may be worthwhile modifying the processing time of a stage of the master recipe, by modifying some operating conditions, so as to debottleneck some piece of equipment or accomplish some product due-dates. The procedure of generating the initialized control recipe from the master recipe is called recipe initialization. This procedure also implies the need to specify in which specific equipment unit and in which product sequence will each stage of the recipe be carried out.

In general, the objective is to generate the best control recipe for different production scenarios. Specifically, the proposed framework adjusts the different parameters of a master control recipe model to deviations in prices or quality of delivered raw materials and in expected initial process conditions. For instance, in such a case where available steam pressure is lower than the nominal value at the beginning of a batch, recipe items will have to be initially adapted to this fact. Another aim of this framework is to adjust the different recipe items to the availability of plant resources and equipment units.

The inputs of the problem are the production master recipe for each product, that is, the different components that define each recipe, the available equipment units for each task, the list of common utilities, the market requirements expressed as specific amounts of products to be delivered at given instants, and others. The algorithm has to determine the optimal sequence of the tasks to be performed in each unit, the values of the different parameters that specify each recipe, that is, the *initialized control recipe* and the use of utilities as a function of time.

Specifically, the optimal schedule in each case is efficiently reached using the S-graph approach [27]. This approach implies a branch-and-bound algorithm. This algorithm proceeds from a root node corresponding to the nominal master control recipe. From this root, partial schedules (nodes of the tree) are built adding schedule-arcs to the preceding nodes. At each node, a flexible recipe model is solved to calculate a relaxation of the algorithm. The solution of this model at the end of a leaf gives the optimal timing, considering the flexible recipe, of the schedule associated to that leaf. The optimal schedule corresponds to the leaf with best objective function value. Hence, a model for schedule timing integrated with a flexible recipe is necessary. The proposed model is linear, simply to permit a rapid convergence of the algorithm.

6.4.1
Flexible Recipe Model for Schedule Timing

In addition to timing restrictions, two sorts of flexible recipe constraint have to be considered: product specifications and process operating conditions and their consequences on the production cost. The product specifications vector, **ps**, is a function, Ψ, of processing time, waiting time, process set-points, and other product specifications. The model adjusts these recipe parameters for each individual batch performed in a specific production plan, Θ, this plan being a function of orders to be satisfied, *S*, and of plant resources, *A*. Each production environment also has to meet some physical plant constraints, Ω, such as the plant topology, *T*, operating with a set *J* of equipment units and a set *R* of production resources. Each production plan is generated to meet the market constraints (set of production orders in a given set *DD* of time horizons or due dates). A performance criterion ϕ is also included.

Hence, two sort of flexible recipe constraints have to be considered to define the flexible recipe model Ψ: product specifications (quality of the final products) and process operation conditions (set points) and their consequences on the production cost.

6.4.2
Quality and Production Cost Model

Product specifications, $\boldsymbol{ps}_i$, might depend on processing time, waiting time, process operation conditions and product specifications at different stages $i*$ where different inputs to stage *i* are processed. At the first stage of a batch, $i*$ will represent the raw materials. It will also be assumed that, within a time interval, a linear model can be adjusted to predict small deviations from process specifications, $\boldsymbol{\delta ps}_i$, as a function of small deviations from the nominal values of PT_i, TW_i, $\boldsymbol{poc}_i$ and $\boldsymbol{ps}_{i*}$ (Eq. (2)).

$$\boldsymbol{\delta ps}_i = \boldsymbol{a}_i \delta PT_i + \boldsymbol{b}_i TW_i + \sum_{i*} \boldsymbol{C}_{i,i*} \boldsymbol{\delta ps}_{i*} + \boldsymbol{d}_i \delta \boldsymbol{poc}_i \qquad (2)$$

where $\boldsymbol{a}_i$ and $\boldsymbol{b}_i$, are the vectors that linearly correlate the effect of processing and waiting times of stage *i* on product specifications. $\boldsymbol{C}_{i,i*}$ is the matrix that linearly correlates the effect of the different product specification inputs to stage *i* from stage $i*$ on product specifications, and $\boldsymbol{d}_i$ the vector that correlates the effect of small deviations in process operation values on the product specifications.

For instance, consider the production of one batch of product A. The stage *i* of this process consists in heating A in equipment unit 1. Stage $i + 1$ constitutes the reaction of A to give B in equipment unit 2. The main important product specification at stage $i = 1$ is the temperature reached in unit 1 and at the second stage, the conversion of reactant A and the temperature at the end of this stage. Therefore, the vector $\boldsymbol{ps}_1$ will only contain one element (temperature at the end of the stage 1). The vector $\boldsymbol{ps}_2$ will have two elements, conversion of reactant and temperature. The vector $\boldsymbol{a}_1$ will consequently contain one element that will correlate the effect of small deviations in processing time of stage 1 on the temperature reached at stage 1. Similarly, $\boldsymbol{a}_2$ will have two elements, and each element will correlate the effect of processing time on each relevant product specification *j*, $\boldsymbol{ps}_{j,\,2}$. If waiting time has no effect on product specifications, the vector $\boldsymbol{b}_i$ is null. Otherwise, product specifications at stage 2 will clearly be affected by product specifications at stage 1. So, the matrix $\boldsymbol{C}_{2,\,i*}$ will be {1 $\times$ 2}. Its elements correlate the effect of small deviations in the temperature reached at stage 1 on the conversion and temperature at the end of stage 2.

Final products must meet some quality (product specifications) requirements. The model also considers the possibility of mixing different batches of the same product, produced within a fixed horizon, to be sold or used together. Therefore, the properties of the last task of each batch, or, in the case of some batches being mixed, the properties of the final products mixed, must meet such requirements, $\delta ps_p{}^o$. That is, only deviations up to a point will be permitted (Eq. (3)).

$$\sum_m B_m \delta ps_m \leq \delta ps_p^o \sum_m B_m \quad \forall p, \forall m \tag{3}$$

where B_m is the batch size of product p at stage m, and m belongs to the set of last recipe stages of product p batches that are mixed.

Process operation modification can have an influence on the operation cost. This fact is also considered in the flexible recipe model. Thus, within a time interval, the set-point modification is assumed to have a linear dependence with batch-stage cost (Eq. (4)).

$$\delta Cost_i = f_i \delta poc_i \tag{4}$$

6.4.3
Flexibility Regions

In Eq. (5), Δ and σ define the flexibility regions for poc_i and ps_i respectively. The width of these regions will basically depend on the accuracy of the model presented in the previous section. That is, the regions are defined in which the model deviates from reality by only a predetermined percentage value, ε. Assuming linearity, each of these regions can be described by a set of R^n hyper planes (Eq. (5)) where n will be number of variables considered or degree of flexibility of the batch process considered.

$$L_i \delta poc_i + l'_i \delta PT_i + l''_i \delta TW_i \leq M_i \quad \forall i \tag{5}$$

where L_i, l'_i and l''_i are the matrices that define the hyper planes bounding (M_i) the process flexibility to be considered within the linear model.

6.4.4
Integration with the Scheduling Tool

Within the S-graph framework, a partial schedule is obtained at each node of the branch-and-bound algorithm. That is, at each node some equipment units may be already scheduled and some others not. The problem is relaxed by solving the linear flexible recipe model. Therefore, if a node has a relaxation higher than the best bound, the branch corresponding to that node is cut. Figure 6.8 shows the Linear Progamming (LP) model to be solved at each node of the branch-and-bound algorithm procedure where the objective function contemplates a trade off between production makespan and production costs. Thus, the recipe is optimized as well as the timing of the partial schedule. Here TI_i and TF_i are the starting and ending times of task i respectively, S_i is the set of states that task i generates and S_i^* the set of states that feed task i^*.

Timing of the schedule constraints,

$$TI_i \geq 0$$
$$TF_i = TI_i + PT_i + TW_i \quad \forall i$$
$$TI_i = TF_i \quad \forall i, i'/\exists s \in \left\{ \overline{S}_i \cap S_{i*} \right\}$$
$$TW_i \leq TW_i^{max} \quad \forall i$$
$$MS \leq TF_i \quad \forall i$$

Flexible recipe model,

$$\delta ps_i = a_i \delta PT_i + b_i TW_i + \sum_{i*} C_{i,i*} \delta ps_{i*} + d_i \delta poc_i$$

Flexibility region,

$$L\delta poc_i + l_i' \delta PT_i + l_i'' \delta TW_i \leq M_i \quad \forall i$$

Performance criterion,

$$\sum_m B_m \delta ps_m \leq \delta ps_p^0 \sum_m B_m \quad \forall p, \forall m$$
$$\delta Cost_i = f_i \delta poc_i$$
$$\min \left(MS \cdot F^* + \sum_i \delta Cost_i \right)$$

Figure 6.8 Formulation for recipe initialization and multipurpose batch process schedule timing

6.4.5
Motivating Example

The proposed framework for recipe initialization integrated to production scheduling has been tested in the batchwise production of benzyl alcohol from the reduction of benzaldehyde through a crossed Cannizarro reaction. This reaction has been extensively studied by Keesman [28]. In that work, an input-output kind of black box model is developed in order to describe the behavior of the reaction phase of the recipe. The model predicts the reaction yield, $ps_{i, 1}$, as a function of the reaction temperature, $poc_{i, 1}$, reaction time, PT_i, amount of catalyst, $poc_{i, 2}$ and amount of one reactant in excess, $poc_{i, 3}$. Then, the model is used to optimize different recipe components analyzing the effects of model accuracy on the results. However, in that work only one batch phase of the recipe was considered. In the following study, the whole batch recipe train and a production environment are considered in order to fully exploit the potential of a more realistic batch process scenario.

The flexible recipe model, Ψ, for this reaction phase and given the linearity required by the model proposed in Section 6.4.2, becomes,

$$\delta ps_{i,1} = 4\delta PT_i + (4.4, 95, 95) \begin{pmatrix} \delta poc_{i,1} \\ \delta poc_{i,2} \\ \delta poc_{i,3} \end{pmatrix} \quad \forall i \in \{\text{Reaction phase}\} \tag{6}$$

The coefficients of Eq. (6) are the linear coefficients of the Keesman quadratic model. The flexibility of this batch stage, contained in $\boldsymbol{\Delta}$ and $\boldsymbol{\sigma}$ regions according to Eq. (5), is defined by the set of cutting planes (Eq. (7)) that bounds the deviation of $\sigma ps_{i,\,1}$ predicted by Eq. (6) and that predicted by the quadratic model.

For simplicity, it has been assumed that the hypervolume of $\mathbf{R}^4$ containing $\boldsymbol{\Delta}$ and $\boldsymbol{\sigma}$ is a hypercube. Equation 7 represents the hypercube of maximum volume that bounds the flexibility region with a tolerance of less than 1.5 % for the reactant conversion.

$$\begin{pmatrix} 1 & 0 & 0 & 0 \\ -1 & 0 & 0 & 0 \\ 0 & 1 & 0 & 0 \\ 0 & -1 & 0 & 0 \\ 0 & 0 & 1 & 0 \\ 0 & 0 & -1 & 0 \\ 0 & 0 & 0 & 1 \\ 0 & 0 & 0 & -1 \end{pmatrix} \begin{pmatrix} \delta poc_{i,1} \\ \delta poc_{i,2} \\ \delta poc_{i,3} \\ \delta poc_{i,4} \end{pmatrix} \leq \begin{pmatrix} 0.5\,°\text{C} \\ 0.7\,°\text{C} \\ 8.5\,\text{g} \\ 27\,\text{g} \\ 7.5\,\text{g} \\ 90\,\text{g} \\ 0.1\,\text{h} \\ 0.3\,\text{h} \end{pmatrix} \tag{7}$$

This reaction stage has been incorporated in the whole recipe. It is assumed that a preparation stage performed in equipment unit U1 and two separation stages carried out in equipment units U3 and U4 are also necessary to produce the alcohol. The reaction stage takes place in equipment unit U2. Reaction temperature at the second stage, $\delta poc_{i,\,1}$, depends on the temperature reached at the first one, $\delta ps_{i',\,2}$, as follows:

$$\delta poc_{i,1} = \delta ps_{i',2} \tag{8}$$

where i' corresponds to any preparation stage and i to any reaction stage of the alcohol recipe. The temperature reached at the preparation stage depends on the processing time according to:

$$\delta ps_{i',2} = 10\delta PT_{i'} \tag{9}$$

This recipe has been introduced into the production scenario given in Table 6.1. P1 represents the production of benzyl alcohol. The rest of products P2, P3, and P4 share equipment units and resources with product P1.

Table 6.1 Batch production environment

Products (N)	Equipment unit Processing Time (h)				Number of batches
P1	U1	U2	U3	U4	3
	0.5	1.75	2.0	0.5	
P2	U1	U3	U4	U6	1
	1.0	2.0	1.5	1.0	
P3	U7	U4	U6	U5	2
	2.0	1.0	1.0	1.0	
P4	U2	U3	U7	U5	1
	1.5	1.0	2.0	1.5	

Figure 6.9 shows the Gantt charts corresponding to the optimum production scheduling for the proposed case study when the fixed recipe at nominal operation conditions is contemplated and when recipe adaptation is considered. The resultant production makespan is 10.75 h for the fixed recipe environment. When the proposed flexible recipe framework is considered, the production makespan diminishes to 10.45 h (2.8% makespan reduction). Also, a different sequence of batches is obtained when it is imposed that the mixing of the three batches of alcohol has to meet the nominal reaction yield ($\delta ps_p^{\,o} = 0$). The optimal solution is obtained in 25.5 CPU seconds using an AMD-K7 Athlon 1 GHz.

The resultant process operating conditions of the three alcohol batches for the flexible recipe scenario are summarized in Table 6.2.

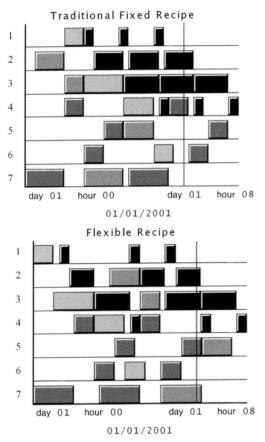

Figure 6.9 Optimal Gantt chart of batch production environment of Table 6.1 when considering the fixed recipe and the recipe adaptation respectively. The case study recipe is represented in **black**

Table 6.2 Formulation for recipe initialization and multipurpose batch process schedule timing

Batch	Temperature (°C)	Processing time (h)	Amount of KOH (g)	Amount of H₂CO (g)	Conversion (%)
1st	64.5	1.2	500	425	75
2nd	64.5	1.2	500	425	75
3rd	63.8	1.2	500	425	72

To see the effect of initial process deviations on the recipe, Keesman [28] limited the reaction temperature to 63 °C ($\delta poc_{i,1} = -1$). After optimizing the reaction stage alone, it is found that the reaction time has to be extended to 1.76 h so that the total amount of KOH reaches 528 g and the amount of formaldehyde goes to 475 g in order to keep the intended reaction yield. For these new nominal conditions, the resultant production makespan for the scenario described in Table 6.1 is 11.03 h, which means a reduction in productivity of 5.5 %. Otherwise, a better process performance can be achieved by applying the flexible recipe model to optimize the entire batch plant. The linear flexible recipe, Ψ, and the model validity constraints for these new nominal conditions are shown in Eqs. 10 and 11, respectively.

$$\delta ps_{i,1} = 3.75\delta PT_i + (101, 112.5) \begin{pmatrix} \delta poc_{i,2} \\ \delta poc_{i,3} \end{pmatrix} \quad \forall i \in \{\text{Reaction phase}\} \tag{10}$$

$$\begin{pmatrix} 1 & 0 & 0 & 0 \\ -1 & 0 & 0 & 0 \\ 0 & 1 & 0 & 0 \\ 0 & -1 & 0 & 0 \\ 0 & 0 & 1 & 0 \\ 0 & 0 & -1 & 0 \\ 0 & 0 & 0 & 1 \\ 0 & 0 & 0 & -1 \end{pmatrix} \begin{pmatrix} \delta poc_{i,1} \\ \delta poc_{i,2} \\ \delta poc_{i,3} \\ \delta PT_i \end{pmatrix} \leq \begin{pmatrix} -1\,°C \\ 1\,°C \\ 12\,g \\ 23\,g \\ 13\,g \\ 28\,g \\ 0.57\,h \\ 0.4\,h \end{pmatrix} \tag{11}$$

Now, the optimal production makespan becomes 10.61 h. Therefore, using the proposed framework, limiting the reaction temperature to 63 °C, only implies a 1.5 % reduction in process productivity. The new process conditions for the different batches of the alcohol production appear in Table 6.3.

Table 6.3 Optimal process operation conditions for three batches of alcohol after limiting reaction temperature

Batch	Temperature (°C)	Processing time (h)	Amount of KOH (g)	Amount of H₂CO (g)	Conversion (%)
1st	63	1.55	512	438	78.3
2nd	63	1.36	512	438	71.9
3rd	63	1.36	512	438	71.9

Notice that in this case study the cost of modifying different process variables has been considered negligible. Usually, nominal values should correspond to an economic optimum. Thus, altering such nominal conditions should result in overrunning this economic optimum in spite of an eventual increase in plant productivity. Obviously, a more realistic scenario should also consider the costs associated to deviations in process operation conditions from nominal values.

6.5
Flexible Recipe Model for Recipe Correction

The recipe initialization is performed at the beginning of the batch phase, taking into account known initial deviations. But other run-time deviations may arise. However, under certain circumstances it is possible to compensate for the effects of these unknown disturbances during the batch run, provided that continuous or discrete measurements are available.

The flexible recipe model is the relationship that correlates a batch process output as a function of the selected input items of the recipe. This model is regarded as a constraint on quality requirements and on production cost. Figure 6.7 shows the environment proposed here for real-time recipe correction.

While a batch process takes place, different online continuous process variables and discrete variables values, sampled at different times, are taken. From this information, a process state assessment is performed. This assessment gives information about how the batch process is being carried out to the flexible recipe model for recipe correction. The time at which process state assessment is performed, and so at which actions take place, might be different from the moment at which a deviation is detected. Interaction or integration of the flexible recipe model with production scheduling algorithms is necessary to account for the ultimate effect of recipe correction on overall plant capacity.

Three different kinds of models are identified:
- A prediction model that estimates the continuous and discrete sampled (at the sampling time) product specification variables, as a function of the actual control recipe that has already been established by the offline initialization tool. Then, the process state assessment consists of the evaluation of the batch-process run. The predicted product specification i, pvs_i^w, expected by the offline recipe initialization model, is compared with the actual variable observed at the wth process statement, ps_i^w. If this deviation observed is greater than a fixed permitted error, ε, some actions will be taken in order to offset this perturbation.
- A correction model for control recipe adjustments, which describes the ultimate effect of the values measured at the time of the process state assessment as well as of those run-time corrections made during the remainder of the processing time.
- A rescheduling strategy to adjust the actual schedule to the recipe modifications.

6.5.1
Rescheduling Strategy, Ω

The output of the flexible recipe model for recipe correction might give variations in processing time or resource consumption, which would make the existing plant resources schedule suboptimal or even infeasible. Therefore, in order to accommodate for these deviations in the actual plant schedule, a rescheduling strategy is to be used. There are two basic alternatives to update a schedule when it becomes obsolete: to generate a new schedule or to alter the initial schedule to adapt it to the new conditions. The first alternative might in principle be better for maintaining optimal solutions, but these solutions are rarely achievable in practice and require prohibitive computation times.

Hence, a retiming strategy is integrated into the flexible recipe model for recipe correction. At each deviation detected, optimization is required to find the best corrected control process recipe. From this, it is proposed to solve the LP shown in Eq. (12) along with a linear representation of the recipe correction model, to adjust the plant schedule to each recipe correction. In case of dealing with a multipurpose plant, it might happen that a given schedule becomes infeasible because of process disturbances. In such a situation further actions should be taken, like changing the order sequence or canceling a running batch:

$$
\begin{aligned}
& \min \left(\Phi(PT_i, TW_i, \boldsymbol{ps}_i, \boldsymbol{poc}_i, \boldsymbol{\Theta}) \right) \text{ subject to,} \\
& TI_{i,j} \geq 0 \quad \forall i,j \\
& TF_{i,j} = TI_{i,j} + PT_i + TW_{i,j} \forall i,j \\
& TI_{i,j} = TF_{i',j} \forall j, i, i' / \exists s \in \left\{ \overline{S}_i \cap S_{i'} \right\} \\
& TI_{i,j} \geq TF_{i,j-1} \forall i,j \\
& TW_{i,j} \leq TW_i^{\max} \forall i,j
\end{aligned}
\tag{12}
$$

correction flexible recipe model constraints

where $TI_{i,j}$, $TF_{.j}$, PT_i and $TW_{i,j}$ are the initial, ending, processing and waiting times of each stage i of a batch corresponding to the specific sequence j of the schedule. The sequence is assumed to be fixed. S_i is the set of stages that feed stage i. $S_{i'}$ is the set of stages fed by stage i'. ϕ is the performance criterion of the flexible recipe model.

6.5.2
Batch Correction Procedure

Within each batch-run, the algorithm of Fig. 6.10 is applied. This algorithm first predicts the expected deviations in process variables from the nominal values as a function of the corrections already taken. Then, the process state assessment verifies if there exist significant discrepancies between the observed variables and the predicted. If so, it freezes process variables of all batch-stages already performed and of the batch-stages that are currently being performed and are not the actual batch-stage

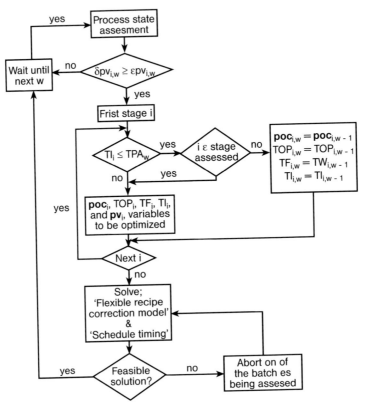

Figure 6.10 Batch correction procedure algorithm

of assessment and reoptimizes the actual recipe taking into account the effect on the schedule timing.

6.5.3
Application: Model-Based Advisory System for Recipe Correction and Scheduling

In this section, an integrated model-based advisory system is designed to give online support for batch process operation. This application integrates recipe modifications as well as modifications in the actual plant schedule-timing, thus breaking the traditional approach to disassociate recipe correction from plant-wide adjustment. It is based on the flexible recipe concept.

The application envisaged gives advice to plant operators and schedulers on how to react in the face of disturbances, so that different kinds of actions can be supported, for instance:

- correct recipe parameters to offset compensating disturbances to meet product specifications at the expense of processing time;

- allow batches to end on time, that is, finishing the batch below expected product quality;
- reduce processing times in order to fit a rush order;
- modify process operating conditions and processing time to partly compensate for disturbances, finishing the batch below product specifications, but not as low as if no action would have been taken.

Different elements form the integrated model advisory system presented here; The recipe adaptation set, where recipe flexibility is defined, the plant adaptation set, where plant schedule adaptation is included using different kinds of rescheduling alternatives, and finally, an integrated criterion, where recipe items modifications, product due-dates accomplishment and product specification deviations costs are included.

In this application, the flexible recipe concept is included in the recipe adaptation set. This set consists of a statistical process model optimized from historical process data and relevant model constraints. Variables in the process model may be classified into those that appear perturbed (P), those that may be tackled to compensate disturbances (M) and those that define the output of the batch process in terms of quality or yield (O). Hence, the flexible recipe model (Ψ), included at the recipe adaptation set is as follows,

$$O = \Psi(PT, \mathbf{M}, \mathbf{P})$$
$$O \subset \sigma \qquad (13)$$
$$\{\mathbf{M}, PT\} \subset \delta$$

where δ is the flexibility region for process operating conditions and σ is the flexibility region for product specifications.

The plant adaptation set, the other key concept of the advisory system, describes the plant resources management, including the relevant equipment information for scheduling, and defines penalties for due-date violations for the accepted orders. In this application, sequence of products is predefined and is assumed not to change. Hence, the plant adaptation set is described by Eq. (13).

When a deviation between the expected and the actual behavior during processing is observed, some advice on how to react is requested. The application presented here considers two different kind of perturbations: process disturbances on some input variable of the batch recipe stages, and rush orders, a new order to be satisfied at a specific due-date and to be fitted in the actual production plan. Figure 6.11 shows this advice system mechanism window.

The application being presented here has been simplified to just consider a linear flexible recipe model at the recipe adaptation set. Then, an LP formulation is used to calculate process adaptation effects.

As soon as a deviation is detected, the control recipe can be readjusted, and this readjustment may have an impact on the whole plant operation. A number of scenarios for recipe readjustment are considered, for instance;

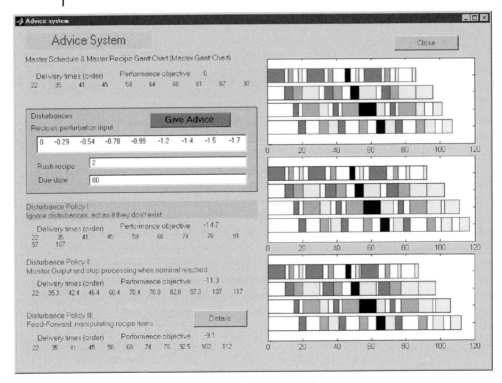

Figure 6.11 Integrated batch operation advice system results.

- The perturbation may be compensated for within the same batch stage, tackling manipulating variables with a consequence on operation cost.
- Other batches may have to be corrected, for instance reducing their processing time, to accommodate the impact of correcting a disturbance on a specific batch-stage (i.e., on its processing time).
- The timing of the schedule may have to change, imposing a delay on product delivery.

An integrated optimization criterion computing consequences of different scenarios coordinates the recipe adaptation set actions with the plant adaptation set ones in order to maximize overall batch plant performance. This architecture has been implemented in MATLAB 6.0.

When disturbances are encountered, the recipe adaptation set may decide to vary some processing times of some tasks of some recipes. This will have an impact on the actual production schedule, so some orders will have to be shifted forward (increased). The plant adaptation set is optimizing this order-shifting to minimize a function depending on delivery-date delays. That is, not all order delays will have the same (for instance, economic) impact on the overall objective function, but the plant adaptation set, tries to shift orders so that the overall impact is minimized. This prob-lem results in an LP formulation that is solved using MATLAB optimization toolbox.

6.5.4
Advisory System Case Study

The case study corresponds to a multiproduct batch plant, over an advising-span of 1 week. During this week, the plant is producing three different products with a specific given sequence. The recipes and sequence of the case study are shown at Tables 6.4 and 6.5, respectively.

Table 6.4 Master recipes of advisory system case study

Recipes	Processing time of stages (h)			
R1	10	5	3	4
R2	3	5	10	5
R3	2	10	5	6

Table 6.5 Master Schedule of advisory system case study

	Master schedule							
Sequence	R3	R1	R2	R3	R1	R2	R3	R3
Due dates	41	45	58	64	68	81	87	97

Recipe adaptation set (ras) variables are classified into a perturbed variable (P), manipulated variable (M), output variable (O) and processing time (PT). For simplification, it is assumed that there is only one (critical) variable of each. Besides, the relationship among these variables is considered to be defined by a linear model and it is considered that there is only one flexible task for each product recipe. Table 6.6 summarizes the recipe adaptation set (ras) parameters.

Table 6.6 Recipe adaptation set of case study

	Flexible	Coefficients for		
Recipe	phase	δM	δI	δPT
R1	2	2.5	2.0	1.5
R2	2	1.0	0.5	0.5
R3	3	2.0	1.0	0.5

A disturbance may be totally offset by modifying the manipulated variable and keeping the master processing time and master output (or quality). Or may be ignored, keeping the manipulated variable and processing time unmodified, so that the output variable (quality) will be affected, or, otherwise, a disturbance may be totally compensated by processing time, or partially by all variables.

The integrated advisory criterion computes the effect of modifying manipulated and output variables from the nominal values. Table 6.7 show the costs associated with modifying these variables. Also, the cost of delivery-dates delay from due dates for each order is shown at Table 6.8.

Table 6.7 Integrated advisory criterion 1

Recipes	Output deviation cost (u)	Manipulated variable deviation cost (u)
R1	1.0	0.1
R2	2.0	0.5
R3	0.5	0.7

Table 6.8 Integrated advisory criterion 2

Delivery-dates deviation cost (u)									
0.2	0.1	0.02	0.5	0.0	0.2	0.1	0.1	0.2	0.03

Two types of disturbances are considered; process disturbances and rush orders. From plant operation, recipe perturbation input variable is retrieved. In this case study, disturbances follow an exponential increase. This would be the case, for instance, of a catalyst being used for all product recipes whose activity is decaying at each use along the production makespan. In the face of disturbances (process disturbances and rush order) the application gives advice on how to react following three policies;

- Policy I. This policy ignores process disturbances, and therefore does not modify manipulated variables or processing times. This situation has a direct impact on the output variable, that is, quality of products.
- Policy II. Here, process disturbances are totally compensated for by modifying processing time (not modifying manipulated variable and keeping output variables equal to nominal values). This situation has a direct impact on delivery dates of products.
- Policy III This policy modifies all recipe items: manipulated variable, processing time and output variable. Within this situation, disturbances may have an impact on delivery dates as well as on product quality, depending on their weight on the overall objective function.

In all policies, a rush order is always accepted. In the case study shown, disturbances have a negative impact of −14.7 u. for policy I, −11.3 u. for policy II and of −9.1 u. for policy III. Policy III is the one containing more degrees of freedom to react in the face of disturbances, and therefore is showing the best performance. Figure 6.11 shows the application results window.

6.6
Final Considerations

The increasing interest recently observed in rationalizing batch/hybrid process operations is well-justified. The great advantage offered by batch process stages resides in their inherent flexibility, which may give an adequate answer to present uncertain product demand, variable customer specifications, uncertain operating conditions, market prices variations and so on [29]. In batch plants, there is no reason why the same product must be made every batch; there is the possibility of tailoring a product recipe specifically for a particular customer. In this chapter firstly a review of relevant approaches to represent and exploit this potential flexibility of batch/hybrid process has been given. A novel framework (flexible recipe) has been presented that allows further exploration of flexible manufacturing capabilities of such types of processes. This framework proposes a new philosophy for recipe management in batch process industries that includes the possibility of recipe adaptation in a real-time optimization environment.

Based on these novel concepts, a model-based integrated advisory system is presented. The system gives on-time advice to operators on how to react when process disturbances occur. This advice takes into account modification in recipe parameters (product quality, specifications, processing time, process variables) as well as modifications in the production schedule. A process state assessment module for evaluation of abnormal situations should advise when proper actions should be taken. The result is a user-friendly application for optimal batch process operation in industrial practice.

Acknowledgments

Financial support for this research received from the "Generalitat de Catalunya", FI program and project GICASA-D (No I-353) are fully appreciated. Also, support received in part by the European Community (project n° G1RD-CT-2001-00466-0466) is acknowledged. Funds were also received from Spanish MCyT (project n° DPI2002-00806). Enlightening discussions and suggestions received from Prof. Antonio Espuña and Prof. Verwater-Lukszo are thankfully appreciated.

Nomenclature

A	Assignment of different batch plant resources.
B_m	Batch size of product p at stage m.
C_{i,i^*}	Correlation matrix with the effect of product specifications inputs to stage i from stage i^*.
DD	Set of production horizon or due-dates.
l	Observed perturbed variable.
I	Set of production orders.

J	Set of equipment units.
M	Manipulable variable of advisory system.
O	Output variable of batch process.
P	Perturbed variable.
PT_i	Processing time of each stage i of a recipe.
$poc(t)i$	Process operation conditions vector as a function of time t of stage i.
ps_i^w	Observed product specification vector at the wth process state assessment moment of batch process stage i.
ps_{i*}	Observed product specification vector at the end of batch process stage $i*$.
pvs_i^w	Expected vector of product specification vector at the wth process state assessment moment of batch process stage i.
R	Set of process Resources.
S_i	Set of states generated by task i.
S_{i*}	Set of states that feed tasks $i*$.
S	Sequence of different batches.
T	Multistage flowshop or jobshop batch plant topology.
$TI_{i,j}$	Starting time of task i.
$TF_{i,j}$	Ending time of task i.
TPA_i^w	wth moment at which stage i of a batch is being assessed.
TW_i	Waiting time at stage i.
$T(P_i)$	Steam temperature condensation at pressure Pi.
Δ	Flexibility region for process operation conditions.
λ	Steam enthalpy.
Ω	Scheduling constraints.
ϕ	Performance criterion function of the Flexible Recipe model.
ψ^{pred}	Quality and production cost modelling function of prediction model.
$\psi^{correct}_w$	Quality and production cost modelling function of correction model of the wth process assessment moment.
Θ	A specific production plan.
σ	Flexibility region for product specifications.

References

1 *Reynold T. S. (1983) 75 years of Progress. A History of the American Institute of Chemical Engineers 1908–1983. American Institute of Chemical Engineers, New York*

2 *Parakrama R. Improving batch chemical processes. The Chem. Eng. 1985 p. 24–25*

3 *Reklaitis G. V. (1985) Perspectives for computer-aided batch process engineering. Chem Eng Prog, 8 (1985) p. 9–16*

4 *Reklaitis G. V. Sunol A. K. Rippin D. W. T. Hortacsu D. (1996) Batch Processing Systems Engineering. NATO ASI Series 143, Spinger-Verlag, Berlin*

5 *Stephanopoulos G. Ali S. Linninger A. Salomon E. AIChE Symp. Ser. 323 (2000) p. 46–57*

6 *Puigjaner L. Espuña A. Reklaitis G. V. (2002). Frameworks for discrete/hybrid production systems. In: Braunschwerg B. Gain R. (eds.) Software Architectures and Tools For Computer Aided Process Engineering. Computer-Aided Chemical Engineering, 11. 88 (No. 9) (1985) Elsevier, Amsterdam, pp 663–700*

7 *Engell S. Kowalenski S. Selmlz C Stursberg O. Continuous-discrete interaction in chemical processing plants. Proc. IEEE 88 (2000) p. 1050–1068*

8 *Barton P. I. Pantelides C. C.* Modeling of continued discrete/continuous processes. AIChE J. 40 (6) (1994) p. 966–979

9 *Kondili E. Pantelides C. C. Sargent R. N. H.* A general algorithm for short-term scheduling of batch operations – 1. Mixed integer linear programming formulation. Comput. Chem. Eng. 17 (1993) p. 211–227

10 *Pantelides C. C. Unfixed frameworks for optimal process planning and scheduling. Proceedings of the. 2nd Conference on Foundation of Computer-Aided Process. Operations. CACHE (1994) pp. 253–274, New York*

11 *Schilling G. Pantelides C. C.* A simple continuous time process scheduling fromulation and a novel solution algorithm. Comput. Chem. Eng. S20 (1996) p. S1221–S1226

12 *Reklaitis G. V. Mockus L.* Mathematical programming formulation for scheduling of batch operations based on non-uniform time discretization. Acta Chim. Esloven. 42 (1995) p. 81–86

13 *Mockus L. Reklaitis G. V.* Continuous time representation in batch/semicontinuous process scheduling-randomized heuristics approach. Comput. Chem. Eng. S20 (1996) p. S1173–S1178

14 *Zhang X. Sargent R. W. H.* The optimal operation of mixed production facilities – extensions and improvements. Comput. Chem. Eng. S20 (1996) p. S1287–S1292

15 *Shah N.* Single and multi-site planning and scheduling: current status and future challenges. AIChE Symp. Ser 320 (1998) p. 75–90

16 *Silver E. Ryke D. Peterson R.* (1998) Inventory Management and Production Planning and Scheduling. John Wiley and Sons, New York

17 *Graells M. Canton J. Peschaud B. Puigjaner L.* General approach and tool for the scheduling of complex production systems. Comput. Chem. Eng. 225 (1998) p. S395–S402

18 *Puigjaner L.* Handling the increasing complexity of detailed batch process simulation and optimization. Comput. Chem. Eng. 23S (1999) p. S929–S943

19 *Canton J.* (2003) Integrated Support System for Planning and Scheduling of Batch Chemical Plants. PhD Thesis Universitat Politchcnica de Catalunya

20 *Rijnsdorp J. E.* (1991) Integrated Process Control and Automation. Elsevier Amsterdam

21 *Verwater-Lukszo Z.* (1997) A Practical Approach to Recipe Improvement and Optimization in the Batch Processing Industry. PhD Thesis. Eindhoven Technische Universiteit, Eindhoven, The Netherlands

22 *Box G. E. P. Draper N. R.* (1969) Evolutionary Operation. Wiley, New York

23 *Graells M. Loberg E. Delgado A. Font E. Puigjaner L.* Batch production scheduling with flexible recipes: the single product case. AIChE Symp. Ser. 320 (1998) p. 286–292

24 *Romero J. Espuña A. Friedler F. Puigjaner L.* A newframework for batch process optimization using the flexible recipe. Ind. Eng. Chem. Res. 42 (2003) p. 370–379

25 *ANSI/ISA – S88.01 (1995) Batch Control. Part 1: Models and Terminology. American National Standards Institute, Washington D.C.*

26 *Verwater-Lukszo Z. Keesman K. J.* Computer-aided development of flexible batch production recipes. Prod. Planning Control 6 (1995) p. 320–330

27 *Sanmarti E. Holczinger T. Puigjaner L. Friedler F.* Combinatorial framework for effective scheduling of multipurpose batch plants. AIChE J. 48 (11) (2002) p. 2557–2570

28 *Keesman K. J.* Application of flexible recipes for model building batch process optimization and control. AIChE J. 39 (4) (1993) p. 581–588

29 *Rippin D. W. T.* Batch process systems engineering: a retrospective and prospective review. Comput. Chem. Eng. S17 (1993) p. S1–S13

7
Supply Chain Management and Optimization

Lazaros G. Papageorgiou

7.1
Introduction

Modern industrial enterprises are typically multiproduct, multipurpose and multisite facilities operating in different regions and countries and dealing with a global international clientele. In such enterprise networks, the issues of global enterprise planning, coordination, cooperation and robust responsiveness to customer demands at the global as well as the local level are critical for ensuring effectiveness, competitiveness, business sustainability and growth. In this context, it has long been recognized that there is a need for efficient integrated approaches to reduce capital and operating costs, increase supply-chain productivity and improve business responsiveness that considers various levels of enterprise management, plant-wide coordination and plant operation, in a systematic way.

A supply chain is a network of facilities and distribution mechanisms that performs the functions of material procurement, material transformation to intermediates and final products, and distribution of these products to customers. A definition provided by theSupplyChain.com (<urls>http://www.thesupplychain.com<urle>) is:

> "SCM is a strategy where business partners jointly commit to work closely together, to bring greater value to the consumer and/or their customers for the least possible overall supply cost. This coordination includes that of order generation, order taking and order fulfilment/distribution of products, services or information. Effective supply-chain management enables business to make informed decisions along the entire supply chain, from acquiring raw materials to manufacturing products to distributing finished goods to the consumers. At each link, businesses need to make the best choices about what their customers need and how they can meet those requirements at the lowest possible cost."

Computer Aided Process and Product Engineering. Edited by Luis Puigjaner and Georges Heyen
Copyright © 2006 WILEY-VCH Verlag GmbH & Co. KGaA, Weinheim
ISBN: 3-527-30804-0

A similar definition has also been given by Beamon (1998) by defining a supply chain as an integrated process with a number of business entities (i.e., suppliers, manufactures, distributors, retailers). A key characteristic of a supply chain is a forward flow of material from suppliers to customers and a backward flow of information from customers towards suppliers.

The supply-chain concept has in recent years become one of the main approaches to achieving enterprise efficiency. The terminology implies that a system view is taken rather than a functional or hierarchical one. Enterprises cannot be competitive without considering supply-chain activities. This is partially due to the evolving higher specialization in a more differentiated market. Most importantly, competition drives companies toward reduced cost structures with lower inventories, more effective transportation systems, and transparent systems able to support information throughout the supply chain. A single company rarely controls the production of a commodity as well as sourcing, distribution, and retail.

Many typical supply chains today have production that spans several countries and product markets are global. The opportunities for supply-chain improvements are large. Costs of keeping inventory throughout the supply chain to maintain high customer service levels (CSLs) are generally significant. There is wide scope to reduce the inventory while still maintaining the high service standards required. Furthermore, the manufacturing processes can be improved so as to employ current working capital and labor more efficiently.

It has widely been recognized that enhanced performance of supply chains necessitates (a) appropriate design of supply-chain networks and its components and (b) effective allocation of available resources over the network (Shah 2004).

In the last few years, there has been a multitude of efforts focused on providing improvements in supply-chain management and optimization. These efforts span a wide range of models, from commercial enterprise resource planning systems and so-called advanced planning systems to academic achievements (for example, linear and mixed-integer programming and multiagent systems).

- There are three main areas in supply-chain modeling research:
- supply-chain design and planning
- simple inventory-replenishment dynamics
- "novel" applications (e.g., optimization of taxation/transfer prices, cross-chain planning etc.).

The main aim of this chapter is to provide a comprehensive review of recent work on supply-chain management and optimization, mainly focused on the process industry. The first part will describe the key decisions and performance metrics required for efficient supply-chain management, while the second will critically review research work on enhancing the decision-making for the development of the optimal infrastructure (assets and network) and planning. The presence of uncertainty within supply chains will also be considered, as this is an important issue for efficient capacity utilization and robust infrastructure decisions. Next, different frameworks are presented which capture the dynamic behavior of the supply chains by establishing efficient inventory-replenishment management strategies. The subsequent section of this chapter considers management and optimization of supply chains involving other novel aspects. Finally, available software tools for supply-chain management will be outlined and future research needs for the process systems engineering community will be identified.

7.2
Key Features of Supply Chain Management

Management of supply chains is a complex task, mainly due to the large size of the physical supply network and inherent uncertainties. In a highly competitive environment, improved decisions are required for efficient supply-chain management at both strategic and operational levels, with time horizons ranging from several years to a few days, respectively. Depending on the level, one or more of the following decisions are taken:

- number, size and location of manufacturing sites, warehouses and distribution centres;
- production decisions related to plant production planning and scheduling;
- network connectivity (e.g., allocation of suppliers to plants, warehouses to markets etc.);
- management of inventory levels and replenishment policies;
- transportation decisions concerning mode of transportation (e.g., road, rail etc.) and also material shipment size.

In general, the supply chains can be categorized as domestic and international, depending on whether they are based in a single country or multiple countries, respectively (Vidal and Goetschalckx 1997). The latter case is more complex, as more global aspects need to be considered such as:

- different tax regimes and duties
- exchange rates
- transfer prices
- differences in operating costs.

It should be mentioned that effective application of suitable forecasting techniques are often critical to successful supply-chain management (see, for example, Makridakis and Wheelright (1989)). These quantitative forecasting techniques provide accurate forecasts (usually for product demands), which can then be used for planning purposes.

The quality of the efficiency and effectiveness of the derived supply-chain networks can be assessed by establishing appropriate performance measures. These measures can then be used to compare alternative systems or design a system with an appropriate level of performance. Beamon (1998) has described suitable performance measures by categorizing them as qualitative and quantitative. Qualitative performance measures include: customer satisfaction, flexibility, information and material flow integration, effective risk management and supplier performance. Appropriate quantitative performance measures include:

- measures based on financial flow (cost minimization, sales maximization, profit maximization, inventory investment minimization and return on investment);
- measures based on customer responsiveness (fill rate maximization, product lateness minimization, customer response time minimization, lead-time minimization and function duplication minimization).

7.3
Supply Chain Design and Planning

Supply-chain design and planning determines the optimal infrastructure (assets and network) and also seeks to identify how best to use the production, distribution and storage resources in the chain to respond to orders and demand forecasts in an economically efficient manner.

It is envisaged that large benefits will stem from coordinated planning across sites, in terms of costs and market effectiveness. Most business processes dictate that a degree of autonomy is required at each manufacturing and distribution site, but pressures to coordinate responses to global demand while minimizing costs imply that simultaneous planning of production and distribution across plants and warehouses should be undertaken. The need for such coordinated planning has long been recognized in the management science and operations research literature. A number of mathematical models have been presented with various features; steady-state, multiperiod, deterministic or stochastic.

Early research in this field was mainly focused on location-allocation models. Geoffrion and Graves (1974) present a model to solve the problem of designing a distribution system with optimal location of the intermediate distribution facilities between plants and customers. In particular, they aim to determine which distribution centre (DC) sites to use, what size DC to have at each selected site, what customer zones to serve and the transportation flow for each commodity. The objective is to minimize the total distribution cost (transportation cost and investment cost) subject to a number of constraints such as supply constraints, demand constraints and specification constraints regarding the nature of the problem. The problem is formulated as a mixed-integer linear programming (MILP) problem, which is solved using Benders decomposition. The model is applied to a case study for a supply chain comprising 17 commodity classes, 14 plants, 45 possible distribution centre sites and 121 customer demand zones.

The risks arising from the use of heuristics in distribution planning were also identified and discussed early on by Geoffrion and van Roy (1979). Three examples were presented in the area of distribution planning demonstrating the failure of "common sense" methods to come up with the best possible solution. This is due to the failure to enumerate all possible combinations, the use of local improvement procedures instead of global ones, and the failure to take into account the interactions in the system.

Wesolowsky and Truscott (1975) present a mathematical formulation for the multiperiod location-allocation problem with relocation of facilities. They model a small distribution network comprising a set of facilities aiming to serve the demand at given points. The model incorporates two types of discrete decisions, one involving the assignment of customers to facilities and the other the location of the nodes. They consider both steady-state and time-varying demands.

Williams (1983) develops a dynamic programming algorithm for simultaneously determining the production and distribution batch sizes at each node within a supply-chain network. The average cost is minimized over an infinite horizon.

Brown et al. (1987) present an optimization-based decision algorithm for a support system used to manage complex problems involving facility selection, equipment location and utilization, and the manufacture and distribution of products. They focus on operational issues such as where each product should be produced, how much should be produced in each plant, and from which plant product should be shipped to customer. Some strategic issues are also taken into account such as location of the plants and the number, kind and location of facilities (plants). The resulting MILP model is solved using a decomposition strategy. It is applied to a real case for the NABISCO Company.

A two-phase approach was used by Newhart et al. (1993) to design an optimal supply chain. First, a combination of mathematical programming and heuristic models is used to minimize the number of product types held in inventory throughout the supply chain. In the second phase, a spreadsheet-based inventory model determines the minimum safety stock required to absorb demand and lead-time fluctuations.

Pooley (1994) presents the results of a MILP formulation used by the Ault Foods company to restructure their supply chain. The model aims to minimize the total operating cost of a production and distribution network. Data are obtained from historical records; data collection is described as one of the most time-consuming parts of the project. Binary variables characterize the existence of plants and warehouses and the links between customers and warehouses.

Wilkinson et al. (1996) describe a continent-wide industrial case study. This involved optimally planning the production and distribution of a system with 3 factories and 14 market warehouses and over 100 products. A great deal of flexibility existed in the network which, in principle, enables the production of products for each market at each manufacturing site.

Voudouris (1996) develops a mathematical model designed to improve efficiency and responsiveness in a supply chain. The target is to improve the flexibility of the system. He identifies two types of manufacturing resources: activity resources (manpower, warehouse doors, packaging lines, etc.) and inventory resources (volume of intermediate storage, warehouse area). The activity resources are related to time while the inventory resources are related to space. The objective function aims at representing the flexibility of the plant to absorb unexpected demands.

Pirkul and Jayaraman (1996) present a multicommodity system concerning production, transportation, and distribution planning. Single sourcing is forced for customers but warehouses can receive products from several manufacturing plants. The objective is to minimize the combined costs of establishing and operating the plants and the warehouses to customers.

Camm et al. (1997) present a methodology by combining integer programming, network optimization and geographical information systems (GIS) for Procter and Gamble's North American supply chain. The overall problem is decomposed into a production (product-plant allocation) problem and a distribution network design problem. Significant benefits were reported with reconstruction of Procter and Gamble's supply chain (reduction of 20 % in production plants) and annual savings of $ 200m.

McDonald and Karimi (1997) consider multiple facilities that effectively produce products on single-stage continuous lines for a number of geographically distributed customers. Their basic model is of a multiperiod linear programming (LP) form, and takes account of available processing time on all lines, transportation costs and shortage costs. An approximation is used for the inventory costs, and product transitions are not modeled. They include a number of additional supply-chain related constraints such as single sourcing, internal sourcing and transportation times.

Other planning models of this type do not consider each product in isolation, but rather groups products that place similar demands on resources into families, and bases the higher level planning function on these families. More sophisticated models exist in the process systems literature. A model which selects processes to operate from an integrated network while ensuring that the network capacity constraints are not exceeded is described in Sahinidis et al. (1989). Means of improving the solution efficiency of this class of problems can be found in Sahinidis and Grossmann (1991) and Liu and Sahinidis (1995).

Uncertainty in demands and prices are modeled in Liu and Sahinidis (1996) and Iyer and Grossmann (1998) by using a number of scenarios for each time period, thus resulting in multiscenario, multiperiod optimization models. Computational enhancements of the above large-scale model have been proposed by applying projection techniques (Liu and Sahinidis 1996) or bilevel decomposition (Iyer and Grossmannn 1998). A potential limitation of these approaches is that they use expectations rather than a variability metric of the second-stage costs. Ahmed and Sahinidis (1998) resolved this difficulty by introducing a one-side robustness measure that penalizes second-stage costs that are above the expected cost. Similar measures based on expected downside risk have been developed by Eppen et al. (1989), and have recently been applied to capacity planning problems for pharmaceutical products at different stages in clinical trials (Gatica et al. 2003).

Applequist et al. (2000) focus on risk management for chemical supply-chain investments. They introduce the risk premium approach in order to determine the right balance between expected value of investment performance and associated variance. An investment decision is approved provided that its expected return is better than those in the financial market with similar variance. An efficient polytope integration procedure is described to evaluate expected values and variances.

Gupta and Maranas (2000) consider the problem of mid-term supply-chain planning under demand uncertainty. A two-stage stochastic programming approach is proposed with the first stage determining all production decisions (here-and-now) and all supply-chains decisions are optimized in the second stage (wait-and-see). This work is extended by Gupta et al. (2000) by integrating the previous two-stage framework with a chance constraint programming approach to capture the tradeoffs between customer demand satisfaction and production costs. The proposed approach was applied to the problem of McDonald and Karimi (1997).

Sabri and Beamon (2000) develop a steady-state mathematical model for supply-chain management by combining strategic and operational design and planning decisions using an iterative solution procedure. A multiobjective optimization procedure is used to account for multiple performance measures, while uncertainties in production, delivery and demands are also included.

A MILP model is proposed by Timpe and Kallrath (2000) for the optimal planning of multisite production networks. The model is multiperiod, based on a time-indexed formulation allowing equipment items to operate in different modes. A novel feature of the model is that it can accommodate different timescales for production and distribution of variable length, thus facilitating finer resolution at the start of the planning horizon. The above model was applied to a production network of four plants located in three different regions. A larger example is briefly described in Kallrath (2000), which demonstrates the use of an optimization model involving 7 production sites with 27 production units operating in fixed-batch mode.

Bok et al. (2000) present a multiperiod optimization model for continuous process networks with main focus on operational decisions over short time horizons (one week to one month). Special features of the supply chain are taken into account such as sales, intermittent deliveries, production shortfalls, delivery delays, and inventory profiles and job changeovers. A bilevel decomposition solution procedure is proposed to reduce computational effort and deal with larger scale problems.

Tsiakis et al. (2001) describe a multiperiod MILP model for the design of supply-chain networks. The model determines production capacity allocation among different products, optimal layout and flow allocations of the distribution network by minimizing an annualized network cost. Demand uncertainty is also introduced in the multiperiod model using a scenario-based approach with each scenario representing a possible future outcome and having a given probability of occurrence.

Papageorgiou et al. (2001) present an optimization-based approach for pharmaceutical supply chains to determine the optimal product portfolio and long-tem capacity planning at multiple sites. The problem is formulated as a MILP model, taking into account both the particular features of pharmaceutical active ingredient manufacturing and the global trading structures. Particular emphasis is placed upon modeling of financial flows between supply-chain components. A comprehensive review on pharmaceutical supply chains is given by Shah (2003).

Kallrath (2002) describes a multiperiod mathematical model that combines operational planning with strategic aspects for multisite production networks. The model is similar to the one presented by Timpe and Kallrath (2000) but allows flexible production unit-site allocation (purchase, opening, shutdown), and raw material purchases and contracts. Sensitivity analyses were also performed, indicating that the optimal strategic decisions were stable up to a 20 % change in demand.

Ahmed and Sahinidis (2003) propose a fast approximation scheme for solving multiscenario integer optimization problems, which is particularly relevant to capacity planning problems under discrete uncertainty.

Jackson and Grossmann (2003) describe a multiperiod nonlinear programming model for the production planning and distribution of multisite continuous multi-product plants where each production plant is represented by nonlinear process models. Spatial and temporal decomposition solution schemes based on Lagrangean decomposition are proposed, to enhance computational performance.

Ryu and Pistikopoulos (2003) present a bilevel approach for the problem of supply-chain network planning under uncertainty. The resulting optimization problem is then solved efficiently using parametric programming techniques.

Levis and Papageorgiou (2004) extend the previous work of Papageorgiou et al. (2001) to consider the uncertainty of outcome of clinical trials. They propose a two-stage, multiscenario MILP model to determine both the product portfolio and the multisite capacity planning, while taking into account the trading structure of the company. A hierarchical solution algorithm is proposed to reduce the computational effort needed for the solution of the resulting large-scale optimization models.

Neiro and Pinto (2004) present an integrated mathematical framework for petroleum supply-chain planning by considering refineries, terminals and pipeline networks. The problem is formulated as a multiperiod, mixed-integer nonlinear programming model, and essentially extends previous work (Pinto et al. 2000) for single refinery operations with nonlinear process models and blending relations. The case study solved represents part of a real-world petroleum supply-chain planning problem in Brazil involving four refineries, five terminals and pipeline networks for crude oil supply and product distribution.

7.3.1
MultiSite Capacity Planning Example

Consider a multisite pharmaceutical capacity planning example (Levis and Papageorgiou 2004) with seven potential products (P1–P7) subject to clinical trials, four alternative locations (A–D), where A and B are the sales regions, A is the intellectual property (IP) owner, and B, C and D are the candidate production sites.

The entire time horizon of interest is 13 years. In the first 3 years, no production takes place and the outcomes of the clinical trials are not yet known. Initially, there are two suites already in place at production site B. Further decisions for investing in new manufacturing suites are to be determined by the optimization algorithm. It is assumed that the trading structure is given together with the internal pricing policies as shown in Fig. 7.1.

Five out of seven potential products are selected in the product portfolio while the optimal enterprise-wide pharmaceutical supply chain is illustrated in Fig. 7.2 where location C is not chosen.

The investment decision calendar is illustrated in Fig. 7.3. Note that investment decisions for additional manufacturing suites are taken in the early time periods while the clinical trials are still on going. The proposed investment plans take into account the construction lead-time (2 and 3 years for nonheader and header suites, respectively) and safeguard the availability of the newly invested equipment right after the end of the clinical trials phase.

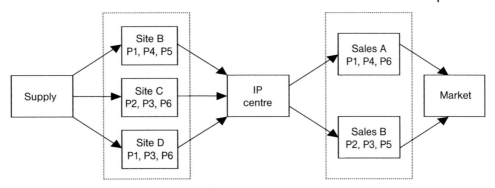

Figure 7.1 Trading structure of the company. *P1–P7* Potential products subject to clinical trials, *A–D* four alternative locations where *A* and *B* are the sales regions, *A* is the intellectual property (*IP*) owner, and *B*, *C* and *D* are the candidate production sites

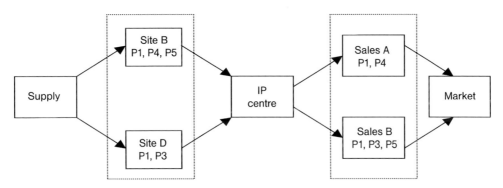

Figure 7.2 Optimal business network

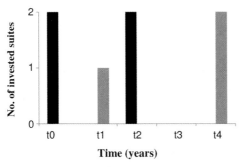

Figure 7.3 Investment decisions calendar. *Black*: site B; *grey*: site D

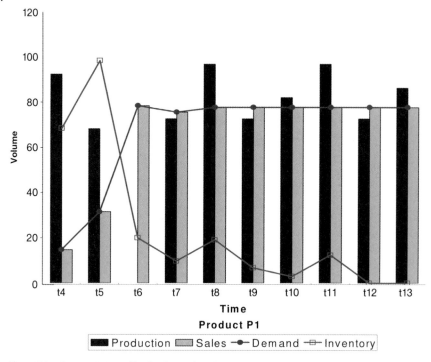

Figure 7.4 Characteristic profiles for the product P1

Operational variables (detailed production plans, inventory and sales profiles) for the selected products are also determined as illustrated in Fig. 7.4. In particular, the production of product P1 is taking place at both manufacturing sites B and D. Mainly due to the proposed investment plan and production policy, the total manufactured amount of P1 fully satisfies customer demand at all time periods.

7.4
Analysis of Supply Chain Policies

The operation of supply chains is a complex task, mainly due to the large physical production and distributions network flows, the inherent uncertainties and the dynamics associated with the internal information flow. At the operations level, it is crucial to ensure enhanced responsiveness to changing market conditions. In this section, different frameworks are presented which capture the dynamic behavior of the supply chains by establishing efficient inventory-replenishment management strategies.

Beamon (1998) provides a comprehensive review of supply-chain models and classifies them as analytical and simulation. Analytical models usually use an aggregate description of supply chains and optimize high-level decisions involving unknown configurations, while simulation models can be used to study the detailed dynamic operation of a fixed configuration under operational uncertainty.

In general, simulation is particularly useful in capturing the detailed dynamic performance of a supply chain as a function of different operating policies. Usually, these simulations are stochastic, thus deriving distributions of characteristic performance measures based on samples from distributions of uncertain parameters.

Gjerdrum et al. (2000) describe a procedure for modeling of the physical and decision-making business process aspects of a supply chain. A specialty chemical process with international markets, secondary manufacturing plants and primary manufacturing plants illustrates the model. The above procedure proposes pragmatic, noninvasive policy and parameter modifications (e.g., safety stocks) that improve performance measures such as average inventory levels, probability of stock-outs and customer service levels (CSLs) are identified. A stochastic simulation approach is then proposed using the above procedure and sampling from the uncertain parameters to assess future performance of the supply chain.

The above work has recently been extended by Hung et al. (2004) adopting an object-oriented approach to model both physical processes (e.g., production, distribution) and business processes of the supply chain. An efficient sampling procedure is also developed which reduces significantly the number of simulations required.

A model predictive control (MPC) framework for planning and scheduling problems is adopted by Bose and Pekny (2000). The framework consists of forecasting and optimization modules. The forecasting module calculates target inventories for future periods, while the optimization module attempts to meet these targets in order to ensure the desired CSL while minimizing inventory. Simulation runs are then performed to study the dynamics of a consumer goods supply chain focusing on promotional demand and lead time as the main control parameters. Different coordination structures of the supply chain are also investigated.

Van der Vorst et al. (2000) present a method for modeling the dynamic behavior of food supply chains by applying discrete-event simulation based on time colored Petri-nets. Alternative designs of the supply-chain infrastructure and operational management and control are then evaluated with the main emphasis being placed upon distribution of food products.

Perea-Lopez et al. (2001) describe a dynamic modeling approach for supply-chain management by considering the flow of material and information within the supply chain. The impact of different supply-chain control policies on the performance of supply chains is evaluated using a decentralized decision-making framework. This is demonstrated through a polymer case study with one manufacturing site, one distribution network and three customers.

Perea et al. (2003) extend their previous work (Perea et al. 2001) by proposing a multiperiod MILP optimization model within an MPC strategy. A centralized approach is adopted where the corresponding MILP model considers the whole supply chain, involving suppliers, manufacturing, distribution and customers simultaneously. The benefits of centralized over decentralized management are then emphasized with a case study with profit increases of up to 15%.

Agent-based techniques have recently been reported in the process systems litera-ture for the efficient management of supply-chain systems. Garcia-Flores et al. (2000) and Garcia-Flores and Wang (2002) present a multiagent modeling system for supply-chain management of process industries. Retailers, warehouses, plants and raw material suppliers are modeled as a network of cooperative agents. A commer-cial scheduling system is integrated in the multiagent framework, as plant schedul-ing usually dominates the supply-chain performance. A case study with a single mul-tipurpose batch plant of paints and coatings is then used to illustrate capabilities of the system.

A similar approach has also been reported by Gjerdrum et al. (2001a) to simulate and control a demand-driven supply-chain network system, with the manufacturing component being optimized through mathematical programming. A number of agents have been used, including warehouses, customers, plants, and logistics func-tions. The plant agent, which is responsible for production scheduling, is using opti-mization techniques while the other agents of the supply chain are mainly rule-based. The proposed system is then applied to a supply chain with two manufactur-ing plants by investigating the effect of different replenishment policies in the supply-chain performance.

Julka et al. (2002a,b) propose an agent-based framework for modeling, monitoring and management of process supply chains. A refinery application is considered for the efficient management of crude oil procurement business process by investigat-ing the impact of different procurement policies, demand fluctuation and changes in plant configuration.

7.4.1
A Pharmaceutical Supply Chain Example

A pharmaceutical supply chain example (Gjerdrum et al. 2000) is shown in Fig. 7.5. A primary manufacturing plant is situated in Europe. Secondary formulation sites in Asia and America receive AI from this plant and produce final products for the main warehouses in Japan and the US. There are two main SKUs in the Japanese market: products A and B. Also, in the US market there are two principal products, C and D.

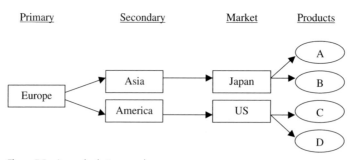

Figure 7.5 A supply chain example

There are also several other product SKUs that share secondary manufacturing resources (in Asia and America) which are handled by other (not explicitly modeled) warehouses e.g., in Europe.

The CSL deemed sufficient for this supply chain by decision-makers (called the target CSL) is 97 %. The low figure is due to large inventories held at external warehouses downstream in the supply chain, which will buffer against any temporary stock-outs. Occasional SKU stock-outs are therefore not considered to be harmful since the end consumer is not affected. Therefore, the aim of this case study is to reduce inventory while maintaining the target CSL.

The demand-management SKU data is presented in Table 7.1.

Table 7.1   Demand management SKU base case data.

	Product A	Product B	Product C	Product D
Safety stock (weeks)	6	6	6	6
MOQ (units)	10,000	5000	5000	20,000
Deviation from forecast (%)	25	20	25	25
Pack size (units)	12	30	30	30
Initial stock (units)	15,000	15,000	60,000	60,000

The production data are given in Tab. 7.2.

Table 7.2   Secondary manufacturing site base case data

	Asia	USA
Safety stock (kg)	50	150
AI Order quantity (kg)	30	130
Production capacity (h week–1)	60	60
Production rate (units h^{-1})	650	1200
Flexibility of production (%)	40	25
Initial AI stock (kg)	80	180

One of the most important supply-chain performance indicators is the amount of unutilized working capital in the chain. By closely tracking simulated inventory, substantial costs can be taken out of the supply chain. This must be done while maintaining required high CSLs, low probabilities of stock-outs and supply-chain efficiency in general. Therefore, we simulate the inventory levels as well as the more traditional customer-focused aspects of the supply-chain performance. Each simulation is repeated 100 times to ensure statistically trustworthy results.

CSL is defined as part-fill on-time. When there is inventory enough at hand, the sales equal demand. When inventory runs out, it is assumed that inventory that is there can be sold while the rest of the demand is left unfulfilled.

The probability of stock-out (PSO) is simply the number of times the inventory is zero divided by the total number of data points, i.e., the horizon length times the number of simulated simulation runs. Hence, this complete supply chain is dynamically simulated over a prespecified horizon of, e.g., 2 years. First, initialization of the model is performed with respect to fixed policy parameters, business processes and initial stock levels. At each time-step (e.g., each week) in the horizon, input parameters and model data from previous time steps are collected. Actual sales and machine breakdown are generated from stochastic distributions. Stock positions, current supply-chain orders and forecasts are then updated and evaluated. When any model variables violate the current procedures or policies (such as the safety stock) the model issues new directions of action (such as issuing new orders).

To obtain statistically significant results of the stochastic process, the simulation is repeated over a number of runs. At each run, informative data are collected. Finally, all the supply-chain simulation results are extracted and evaluated.

The experiments presented in Table 7.3 demonstrate the response of the supply chain to changes in internal and external parameters. These experiments are useful in that they provide information on whether current policies and external factors can be modified while maintaining strong supply-chain performance. INV is the horizon-average finished product (SKU) inventory. High demand variability represents a complex market to forecast. High order quantity shows what happens if a typical service level parameter is modified. Although these parameters obviously affect the performance levels, it seems that the demand management performance levels are not severely affected by the demand variability or the order quantity. The ramping and decreasing demands give conservative measures, since it is assumed that the forecast will not be updated during the horizon. The US market is able to handle a soaring demand better, due to fewer conflicting products.

Table 7.3 Supply-chain experiments. *CSL* Customer service level, *PSO* probability of stock-out, *INV* horizon-average finished product inventory

Experiment	Product A (Asia)			Product C (USA)		
	CSL (%)	PSO (%)	INV	CSL (%)	PSO (%)	INV
Base case	98.33	1.81	48,074	98.81	1.33	105,041
High demand variability	98.05	2.18	48,678	97.76	2.31	107,315
High order quantity	97.94	2.30	58,688	98.21	1.96	109,049
Soaring demand	95.52	4.54	43,997	98.73	1.22	112,554
Collapsing demand	99.83	0.19	54,645	99.51	0.75	111,789

In Fig. 7.6, the expected number of stock-outs for various policies of safety stock is shown for product C in the US market. The value represents the risk of a stock-out occurring during the simulated period.

In Fig. 7.7, the CSL defined as part-fill on-time is shown for different policies of weekly forward cover stock for product C. It can be seen that a policy of 4 weeks of forward cover will just about be sufficient to satisfy the target CSL of 97%.

In Fig. 7.8, the resulting mean average inventory values for various stock policies are shown for product C. The 4 weeks forward cover corresponds to an average inventory of about 70,000 units. This value can then be utilized to calculate the market inventory cost.

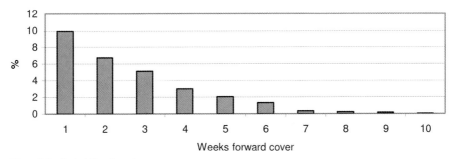

Figure 7.6   Probability of stock-out

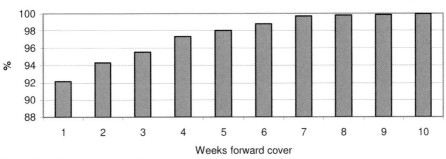

Figure 7.7   Customer service level

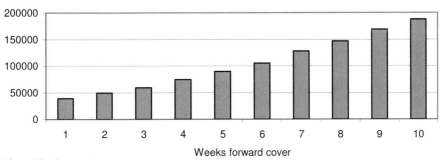

Figure 7.8   Average inventory

7.5
Multienterprise Supply Chains

In recent years, the concept of multienterprises has gained increasing attention, since it promotes all the benefits of extended multienterprise collaboration. The determination of policies that optimize the performance of the entire supply chain as a whole, while ensuring adequate rewards for each participant is crucial and the relevant work is very limited.

Conflicting interests in general extended multienterprise supply chains frequently lead to problems in how to distribute the overall value to each member of the supply chain. A simple approach to enhancing the performance of a multienterprise supply chain is to maximize the summed enterprise profits of the entire supply chain subject to various network constraints. When the overall system is optimized in this fashion, there is no automatic mechanism to allow profits to be fairly apportioned among participants. Solutions to this class of problems usually exhibit quite uneven profit distribution and are therefore impractical. They do however give an indication of the best possible total profit attainable in the supply chain as well as an indication of the best activities to carry out.

van Hoek (1998) raises the question of how to divide supply-chain revenues among the players in a supply-chain system when there is no leading player to determine how distribution of benefits should be handled. He states that supply-chain control is no longer based on direct ownership but rather on integration over interfaces of functions and enterprises. Traditional performance indicators limit the possibilities of optimizing the supply-chain network because the measures do not correctly address the wide opportunities for improvement.

Cohen and Lee (1989) deliver a model for making resource deployment decisions in a global supply-chain network and solve different scenarios using an extensive mixed-integer nonlinear programming (MINLP) model. They also discuss several "policy options" for plant utilization, supply and distribution strategies.

Pfeiffer (1999) describes transfer pricing in a supply chain consisting of procurement, manufacturing and selling units of one single company. His theoretical model handles one commodity at each node and does not include any capacity constraints. He proposes a transfer-price system governed by the headquarters, which fixes a specific transfer-price level. Each node optimizes its own decisions independently to maximize a given profit function, according to the price level fixed by the headquarters. After the decentralized optimization, headquarters evaluates and collects the overall results obtained and chooses a new transfer price that leads to a higher profitability.

Alles and Datar (1998) claim that cost-based transfer prices of the companies are usually based in their competitive environment. The enterprise may cross-subsidize products in order to increase their ability to increase prices. Often, transfer prices for relatively lower cost products are decreased. The authors give evidence that transfer prices are determined based on strategic decisions rather than on internal cost systems.

Jose and Ungar (2000) propose an approach to decentralized pricing optimization of interprocess streams in chemical industry companies. Their iterative auction method determines the prices of process streams so as to maximize an objective for a single chemical company, while each division within the company is constrained by its available resources. The approach is interesting in that each division conceals its private information from the other parties within the so-called micro supply chain. It normally takes several iterations for a model to converge, and the user has to define the limited amount of slack resources utilized. One of the main conceptual differences between their approach and the one presented in this paper is that they regard the channel members as adversarial and competitive for resources rather than cooperative. Also, they use a slack resource iterative auction approach, whereas in this paper the solution approach is to solve a noniterative separable MILP problem.

Ballou et al. (2000) stress the importance of common objectives in the supply chain. Unattainable improvements for single companies in terms of cost savings and customer service enhancements can be obtained by cooperative companies. The authors point out that problems arise if some of the firms benefit at the expense of the others. The conflict resolution between supply-chain partners must be of focal interest, and to keep the coalitions intact, the rewards of cooperation must be redistributed. They identify three means to achieve this:

1. *Metrics* could be developed to capture the nature if interorganizational cooperation to simplify benefit analysis.
2. *Information sharing* mechanisms could transfer information about the benefits of cooperation among the members in the supply chain.
3. *Allocation methods* could be developed that *fairly* distribute the rewards of cooperation between the members.

According to Pashigian (1998), multiproduct industries form a new market structure characterized by novel market relationships among companies. Collusion takes place when the firms in an industry join together to set prices and outputs. Such an agreement is said to form a cartel. However, game theorists insist that there is an inherent incentive for each firm to cheat on such an agreement, in order to gain more profits for itself.

Vidal and Goetschalckx (1997) present a nonconvex optimization model together with a heuristic solution algorithm for the management of a global supply by maximizing the after-tax profits of a multinational corporation. The model also considers transfer prices and the allocation of transportation costs as explicit decision variables.

Gjerdrum et al. (2001b, 2002) present a MINLP model including a nonlinear Nash-type objective function for fair profit optimization of an *n*-enterprise supply-chain network. The supply-chain planning model considers intercompany transfer prices, production and inventory levels, resource utilization, and flows of products between echelons. Efficient solution procedures for the above model are described by Gjerdrum et al. (2001b, 2002) based on separable programming and spatial branch-and-bound respectively. Computational results indicate profits very close to those obtained by simple single-level optimization (e.g., maximization of total profit), but more equitably distributed among partners.

Chen et al. (2003) propose a fuzzy decision-making approach for fair profit distribution for multienterprise supply-chain networks. The proposed framework can accommodate multiple objectives such as maximization of the profit of each participant enterprise, the CSL, and the safe inventory level.

7.6
Software Tools for Supply Chain Management

The coordination of operations on a global basis requires the implementation and use of software to support these decisions. State-of-the-art software requires the ability to perform constraint-based, multisite planning that can become extremely complex. Modern software supply-chain tools aim to integrate traditionally fragmented views of operations, and to provide a holistic view of the problem, rather than linking separate planning operations.

7.6.1
Aspen Technologies (<urls>http://www.aspentech.com<urle>)

Aspen MIMI supply-chain suite includes the Aspen Strategic Analyzer that can be used to identify strategic and operational options such as capacity addition, production constraints and distribution modes. Aspen is the leading provider of supply-chain solutions in the process industry, by market share.

7.6.2
i2 Technologies (<urls>http://www.i2.com<urle>)

Supply-Chain Optimization is a holistic solution and framework to help companies create a macrolevel model for the entire supply chain that controls an integrated workflow environment. The user may select to extend the capabilities and study in depth specific areas of the supply-chain problem, such as logistics, production, demand fulfilment and profit/revenue analysis.

7.6.3
Manugistics (<urls>http://www.manugistics.com<urle>)

Network Design and Optimization is a supply-chain design and operation package that is part of the Manugistics supply-chain suite. Among its capabilities is the design of a supplier, manufacturing site and distribution site network in the most effective way. The process considers inventory levels, production strategy, production and transportation costs, lead times and other user-specified constraints.

7.6.4
SAP AG (<urls>http://www.sap.com<urle>)

MySAP SCM (Supply-Chain Management) is designed to be a complete supply-chain solution. The supply network planning and deployment tool assists planners to balance supply and demand while simultaneously considering purchasing, manufacturing, inventory and transportation constraints. Integrated with other support tools, it aims to provide a complete optimization framework.

7.7
Future Challenges

It is clear that considerable research work has been done on process supply-chains especially in the areas of network design and planning. However, a number of issues provide interesting challenges for further research.

As many modern supply chains are characterized by their international nature, optimization-based decisions are required for various features such as taxes, duties, transfer prices, etc. Systematic integration of business/financial and planning models should be considered for efficient supply-chain management (see, for example, Shapiro (2003), Romero et al. (2003), Badell et al. (2004), Badell M., Romero J., Huertas R., Puigjaner L. Comput. Chem. Eng. 28 (2004) p. 45–61).

Significant effort has already been put into supply-chain modeling under uncertainty commonly related to product demands. The treatment of uncertainty still requires research effort to capture more aspects such as product prices, resource availabilities etc. In order to ensure that investment decisions are made optimally in terms of both reward and risk, suitable frameworks for the solution of supply-chain optimization problems under uncertainty are required. Most of the existing frameworks are suitable for two-stage problems, while there is a need for appropriate multistage, multiperiod optimization frameworks for supply-chain management.

As most of the resulting optimization problems, and predominantly cases under uncertainty, will be of large scale, there is great scope for developing efficient solution procedures. Aggregation and decomposition techniques are envisaged as such promising solution alternatives. It should be mentioned that it is quite important to maintain industrial focus for the successful development of such solution methods.

The analysis of supply-chain policies for process industries has recently emerged and this research area is expected to expand. Suitable frameworks seem to be the ones based on agents, MPC and object-oriented systems. A key issue here is the appropriate integration of business and process aspects (see, for example, Hung et al. (2004)).

Another emerging research area is the systematic incorporation of sustainability aspects within supply-chain management systems, necessitating the development of multiobjective optimization frameworks (see, for example, Zhou et al. (2000), Hugo and Pistikopoulos (2003)).

Finally, research opportunities are evident in the appearance of new types of supply chain, associated, for example, with hydrogen (fuel cells), energy supply, water provision/distribution, fast response therapeutics and biorefineries.

Acknowledgments

The author is grateful to Nilay Shah, Jonatan Gjerdrum, Panagiotis Tsiakis, Gabriel Gatica and Aaron Levis for useful discussions and contributions to this work.

References

1 *Ahmed S. Sahinidis N. V.* Ind. Eng. Chem. Res. 37 (1998) p. 1883–1892

2 *Ahmed S. Sahinidis N. V.* Oper. Res. 51 (2003) p. 461–47

3 *Alles M. Datar S.* Manage. Sci. 44 (1998) p. 451–461

4 *Applequist G. E. Pekny J. F. Reklaitis G. V.* Comput. Chem. Eng. 24 (2000) p. 2211–2222

5 *Ballou R. H. Gilbert S. M. Mukherjee A. A.* Ind. Market. Manage. 29 (2000) p. 7–18

6 *Beamon B. M.* Int. J. Prod. Econ. 55 (1998) p. 281–294

7 *Bok J. K. Grossmann I. E. Park S.* Ind. Eng. Chem. Res. 39 (2000) p. 1279–1290

8 *Bose S. Pekny J. F.* Comput. Chem. Eng. 24 (2000) p. 329–335

9 *Brown G. G. Graves G. W. Honczarenko M. D.* Manage. Sci. 33 (1987) p. 1469–1480

10 *Camm J. D. Chorman T. E. Dill F. A. Evans J. R. Sweeney D. J. Wegryn G. W.* Interfaces 27 (1997) p. 128–142

11 *Chen C. L. Wang B. W. Lee W. C.* Ind. Eng. Chem. Res. 42 (2003) p. 1879–1889

12 *Cohen M. A. Lee H. L.* J. Manuf. Oper. Manage. 2 (1989) p. 81–104

13 *Eppen G. D. Marti R. K. Schrage L.* Oper. Res. 37 (1989) p. 517–527

14 *García-Flores R. Wang X. Z.* OR Spectrum 24 (2002) p. 343–370

15 *Garcia-Flores R. Wang X. Z. Goltz G. E.* Comput. Chem. Eng. 24 (2000) p. 1135–1141

16 *Gatica G. Papageorgiou L. G. Shah N.* Chem. Eng. Res. Des. 81 (2003) p. 665–678

17 *Geoffrion A. M. Graves G. W.* Manage. Sci. 20 (1974) p. 822–844

18 *Geoffrion A. M. van Roy T. J.* Sloan Manage. Rev. Summer (1979) p. 31–42

19 *Gjerdrum J. Jalisi Q. W. Z. Papageorgiou L. G. Shah N.* 5th Annual International Conference of Industrial Engineering Theory – Applications and Practice, Hsinchu, Taiwan, 2000

20 *Gjerdrum J. Shah N. Papageorgiou L. G.* Prod. Planning Control (2001a) p. 12 81–88

21 *Gjerdrum J. Shah N. Papageorgiou L. G.* Ind. Eng. Chem. Res. 40 (2001b) p. 1650–1660

22 *Gjerdrum J. Shah N. Papageorgiou L. G.* Eur. J. Oper. Res. 143 (2002) p. 582–599

23 *Gupta A. Maranas C. D.* Ind. Eng. Chem. Res. 39 (2000) p. 3799–3813

24 *Gupta A. Maranas C. D. McDonald C. M.* Comput. Chem. Eng. 24 (2000) p. 2613–2621

25 *Hugo A. Pistikopoulos E. N.* Proceedings of the 8th International Symposium on Process Systems Engineering, 2003, A and B , 214–219

26 *Hung W. Y. Kucherenko S. Samsatli N. Shah N.* J. Oper. Res. Soc. 2004 55 (2004) p. 801–813

27 *Iyer R. R. Grossmann I. E.* Ind. Eng. Chem. Res. 37 (1998) p. 474–481

28 *Jackson J. R. Grossmann I. E.* Ind. Eng. Chem. Res. 42 (2003) p. 3045–3055

29 *Jose R. A. Ungar L. H.* AIChE J. 46 (2000) p. 575–587

30 *Julka N. Srinivasan R. Karimi I.* Comput. Chem. Eng. 26 (2002a) 1755–1769

31 *Julka N. Srinivasan R. Karimi I.* Comput. Chem. Eng. 26 (2002b) 1771–1781

32 *Kallrath J.* Chem. Eng. Res. Des. 78 (2000) p. 809–822

33 *Kallrath J.* OR Spectrum 24 (2002) p. 219–250

34 *Levis A. A. Papageorgiou L. G.* Comput. Chem. Eng. 28 (2004) p. 707–725

35 *Liu M. L. Sahinidis N. V.* Ind. Eng. Chem. Res. 34 (1995) p. 1662–1673

36 *Liu M. L. Sahinidis N. V.* Ind. Eng. Chem. Res. 35 (1996) p. 4154–4165

37 *Makridakis S. Wheelright S. C.* Forecasting methods for Management, 1989, Wiley, New York

38 *McDonald C. M. Karimi I. A.* Ind. Eng. Chem. Res. 36 (1997) p. 2691–2700

39 *Neiro S. M. S. Pinto J. M.* Comput. Chem. Eng. 2004 28 (2004) p. 871–896

40 *Newhart D. D. Stott K. L. Vasko F. J.* J. Oper. Res. Soc. 44 (1993) p. 637–644

41 *Papageorgiou L. G. Rotstein G. E. Shah N.* Ind. Eng. Chem. Res. 2001 40 275–286

42 *Pashigian B. P.* Price Theory and Applications 1998 McGraw Hill Boston

43 *Perea-Lopez E. Ydstie B. E. Grossmann I. E.* Comput. Chem. Eng. 27 (2003) p. 1201–1218

44 *Perea-Lopez E. Ydstie B. E. Grossmann I. E. Tahmassebi T.* Ind. Eng. Chem. Res. 40 (2001) p. 3369–3383

45 *Pfeiffer T.* Eur. J. Oper. Res. 116 (1999) p. 319–330

46 *Pinto J. M. Joly M. Moro L. F. L.* Comput. Chem. Eng. 24 (2000) p. 2259–2276

47 *Pirkul H. Jayarama V.* Transp. Sci. 30 (1996) p. 291–302

48 *Pooley J.* Interfaces 24 (1994) p. 113–121

49 *Romero J. Badell M. Bagajewicz M. Puigjaner L.* Ind. Eng. Chem. Res. 42 (2003) p. 6125–6134

50 *Ryu J. H. Pistikopoulos E. N.* Proceedings of the 4th International Conference on Foundations of Computer-Aided Process Operations,(2003) p. 297–300

51 *Sabri E. H. Beamon B. M.* Omega 28 (2000) p. 581–598

52 *Sahindis N. V. Grossmann I. E.* Comput. Chem. Eng. 15 (1991) p. 255–272

53 *Sahindis N. V. Grossmann I. E. Fornari R. E. Chathranthi M.* Comput. Chem. Eng. 13 (1989) p. 1049–1063

54 *Shah N.* Proceedings of the 4th International Conference on Foundations of Computer-Aided Process Operations (2003) p. 73–85

55 *Shah N.* Escape–14 conference 2004

56 *Shapiro J.* Proceedings of the 4th International Conference Conference on Foundations of Computer-Aided Process Operations (2003) p. 27–34

57 *Timpe C. H. Kallrath J.* Eur. J. Oper. Res. 126 (2000) p. 422–435

58 *Tsiakis P. Shah N. Pantelides C. C.* Ind. Eng. Chem. Res. 40 (2001) p. 3585–3604

59 *van der Vorst J. G. A. J. Beulens A. J. M. van Beek P.* Eur. J. Oper. Res. 122 (2000) p. 354–366

60 *van Hoek R. I.* Supply Chain Manage. 3 (1998) p. 187–192

61 *Vidal C. J. Goetschalckx M.* Eur. J. Oper. Res. 98 (1997) p. 1–18

62 *Voudouris V. T.* Comput. Chem. Eng. 20S (1996) p. S1269–S1274

63 *Wesolowsky G. O. Truscott W. G.* Manage. Sci. 22 (1975) p. 57–65

64 *Wilkinson S. J. Cortier A. Shah N. Pantelides C. C.* Comput. Chem. Eng. 20S (1996) p. S1275–S1280

65 *Williams J. F.* Manage. Sci. 29 (1983) p. 77–92

66 *Zhou Z. Y. Cheng S. W. Hua B.* Comput. Chem. Eng. 24 (2000) p. 1151–1158

Section 4
Computer-integrated Approaches in CAPE

Section 4 presents a review on actual trends and shows new advances in the integration of software tools and process data. The material in this section is organized in five chapters.

Chapter 1 sets the goals for an integrated process and product design, possibly including the product application process in the analysis. Functions to be met by the product are the specifications. However, identifying a feasible chemical product is not enough, it needs to be produced through a sustainable process. Chapter 1 defines the general integrated chemical product-process design problem. The important issues and needs are identified with respect to their solution and illustrated through examples. Any CAPE method/tool needs to organize the scales and complexity levels so that the events at different scales can be described and understood: from property prediction at the nanoscale to phenomena occurring at the equipment scale. Integrated product-process design where modeling and supply chain issues play an important role is also highlighted.

Chapter 2 shows where, why, and how models of various types are used throughout the life of an industrial or manufacturing process. This justifies the need for tool and data integration across the process life cycle. Modeling addresses a diversity of goals, and relies on a range of forms, approaches, and tools. Models differ widely in the detail level, time, and length scale. Several industrial case studies help illustrate the challenges of modeling throughout the life cycle, since there is a huge range of models used to help answer vital sociotechnical questions through the life cycle of the process or product.

The last chapter of Section 3 has already presented the elementary principles and systematic methods of supply chain modeling and optimization. Chapter 3 shows how the practical implementation of supply chain management software suffers from several deficiencies, due to a limited focus or a lack of integration. To overcome these deficiencies and respond better to industrial demands facing a more dynamic environment, there is a need to explore new strategies for supply chain management. Integrated solutions are required for the next generation of software tools given the number and complex interactions present among main components in the global supply chain: financial flows, negotiation and environmental aspects need to be considered simultaneously along with a number of operating and design constraints. Agent-based systems are considered as a promising architecture for integration.

Reliable and consistent thermophysical property data for pure components and mixtures are essential for CAPE calculations. Chapter 4 reviews the data needed and the quality requirements. Major sources for physical properties and phase equilibrium data collections are compared. The text also provides up to date references to information sources available on the Internet.

The major issue in CAPE tools integration is to ensure software component interoperability and to allow seamless data exchange between tools. This issue is discussed in Chapter 5 *that focuses on operational standards in the domain of CAPE, namely the CAPE-OPEN. Promising software interoperability standards, leading to service-oriented architectures and the emerging Semantic Web are also described. The organizational and economic consequences of the trend towards interoperability and standards in CAPE are also shortly described.*

1

Integrated Chemical Product-Process Design: CAPE Perspectives

Rafiqul Gani

1.1
Introduction

Chemical process design typically starts with a general problem statement with respect to the chemical product that needs to be produced, its specifications that need to be matched, and the chemicals (raw materials) that may be used to produce it. Based on this information, a series of decisions and calculations are made at various stages of the design process to obtain first a conceptual process design, which is then further developed to obtain a final design, satisfying at the same time, a set of economic and process constraints. The important point to note here is that the identity of the chemical product and its desired qualities are known at the start but the process (flow sheet/operations) and its details are unknown.

Chemical product design typically starts with a problem statement with respect to the desired product qualities, needs and properties. Based on this information, alternatives are generated, which are then tested and evaluated to identify the chemicals and/or their mixtures that satisfy the desired product specifications (qualities, needs and properties). The next step is to select one of the product alternatives and design a process that can manufacture the product. The final step involves the analysis and test of the product and its corresponding process. The important points to note here are that (1) the identity of the chemical product is not known at the start but the desired product specifications are known, and (2) process design can be considered as an internal subproblem of the total product design problem in the sense that once the identity of the chemical product has been established, the process and/or the sequence of operations that can produce it is determined. Note also that after a process that can manufacture the desired chemical product has been found, it may be necessary to evaluate not only the product but also the process in terms of environmental impact, life cycle assessment and/or sustainability.

From the above descriptions of the product and process design problems, it is clear that an integration of the product and process design problems is possible and that such an integration could be beneficial in many ways. For example, in chemical

Computer Aided Process and Product Engineering. Edited by Luis Puigjaner and Georges Heyen
Copyright © 2006 WILEY-VCH Verlag GmbH & Co. KGaA, Weinheim
ISBN: 3-527-30804-0

product design involving high value products where the reliability of the chemical product is more important than the cost of production, product specifications and process operations are very closely linked. In pharmaceutical products, there is a better chance to achieve success the first time with respect to their manufacture by considering the product-process relations. In the case of bulk chemicals or low-value products, the use of product-process relations may be able to help obtain economically feasible process designs. In all cases, issues related to sustainability and environmental constraints (life cycle assessment) may also be incorporated.

As pointed out by Gani (2004a), integration of the product and process design problems can be achieved by broadening the typical process design problem to include at the beginning, a subproblem related to chemical product identification and to include at the end, subproblems related to product and process evaluation, including, lifecycle and/or sustainability assessments. Once the chemical product identity has been established, Harjo et al. (2004) proposes the use of a product-centric integrated approach for process design. Giovanoglou et al. (2003), Linke and Kokossis (2002) and Hostrup et al. (1999) have developed simultaneous solution strategies for product-process design involving manufacture of bulk chemicals, while Muro-Sune et al. (2004) have highlighted the integration of chemical product identification and its performance evaluation. In all cases, integration is achieved by solving simultaneously some aspects of the individual product and process design problems. Recently, Cordiner (2004) and Hill (2004) have highlighted issues related to product-process design with respect to agrochemical products and structured products, respectively. Issues related to multiscale and chemical supply chain have been highlighted by Grossmann (2004) and Ng (2001).

The objective of this chapter is to provide an overview of some of the important issues with respect to integrated product-process design, to highlight the need for a framework for integrated product-process design by employing computer-aided methods and tools, and to highlight the perspectives, challenges, issues, needs and future directions with respect to CAPE/PSE related research in this area.

1.2
Design Problem Formulations

In principle, many different chemical product-process design problems can be formulated. Some of the most common among these are described in this section together with a brief overview of how they can be solved.

1.2.1
Design of a Molecule or Mixture for a Desired Chemical Product

These design problems are typically formulated as, given the specifications of a desired product, determine the molecular structures of the chemicals that satisfy the desired product specifications, or, determine the mixtures that satisfy the desired product specifications (see Fig. 1.1).

In the case of molecules, techniques known as computer-aided molecular design (CAMD) can be employed, while in the case of mixtures, techniques known as computer aided mixture-blend design (CAM^{b}D) can be employed. More details on CAMD and CAM^{b}D can be found in Achenie et al. (2002) and Gani (2004a,b). These two problems are also typically known as the reverse of property prediction as product specifications defined in terms of properties need to be evaluated and matched to identify the feasible alternatives (molecules and/or mixtures). This can be done in an iterative manner by generating an alternative molecule or mixture and testing (evaluating) its properties through property estimation. This problem (molecule design) is mainly employed in identifying chemicals that are added to the process, such as solvents, refrigerants and lubricants that may be used by the process to manufacture a chemical product. In the case of mixture design, petroleum blends and solvent mixtures are two examples where the product is designed without process constraints.

1.2.1.1
Examples of Solvent Design
Consider the following process-product design problems where solvent design has an important role. The production of an active ingredient in the pharmaceutical industry needs the addition of a new solvent to an existing solvent-solute (reactant) mixture such that solubility is increased, and thereby the conversion is achieved. The new solvent must be totally miscible with the existing solvent (water) and must also be inert to the reactant. First determine all compounds that are totally miscible with water (use either a database or predict water miscibility). Next, screen out those that may react with the solute (this can be checked through the calculation of the chemical equilibrium constant). Next, identify those that will most likely dissolve the solute (this can be checked through the calculation of solubility). For this problem, alcohols, ketones, glycols are likely candidates.

Consider also the following problem: it is necessary to design/select an alternative solvent to remove oleic acid methyl ester, which is a fatty acid used in treatment of textile, rubber and waxes. The alternative solvent must be better than diethyl ether and chloroform in terms of safety and environmental impact while having solubility properties as good as the known solvents. Also, it must be liquid at temperatures between 270 K to 340 K. Searching of a compound that is acyclic (and containing only C, H and O atoms) that has a melting point below 270 K and boiling point above 340 K, has a Hildebrand solubility parameter that is ± 16.95 MPa$^{1/2}$, and an octanol partition coefficient less than 2 generates the following candidates: 2-heptanone, 3-hexanone, methyl isobutyl ketone, isopryl acetate and many more.

Interesting examples of application of CAMD in the agrochemicals, materials and pharmaceutical industries can be found in the edited monograph by Reynolds, Holloway and Cox (1995).

1.2.2
Design of a Process

These design problems are typically formulated, given the identity of a chemical product plus its specifications in terms of purity and amount and the raw materials that should be used, and determine the process (flow sheet, condition of operations, etc.) that can produce the product (see Fig. 1.1).

This is a typical process design problem, which now can be routinely solved (see for example, textbooks on chemical process design) with CAPE methods/tools when the chemical product is a low-value (in terms of price) bulk chemical. The optimal process design for these chemical products is usually obtained through optimization and process/operation integration (heat and mass integration) in terms of minimization of single or multiparametric performance function. For high-value chemical products, however, a more product-centric approach is beneficial, as pointed out by Harjo et al. (2004), Fung and Ng (2003) and Wibowo and Ng (2001).

1.2.2.1
Examples of Product-Centric Process Design

Harjo et al. (2004) have recently developed a systematic method for product-centric process design and illustrated the application of their method through the design of processes for the manufacturing of phytochemical (plant-derived chemical) products. Harjo et al. (2004) provide a general structure of phytochemical manufacturing processes (Fig. 1.2) and provide a list of heuristic rules for constructing flow sheet alternatives (see Appendix).

As examples of application, Harjo et al. (2004) have considered the manufacturing of carnosic acid, which is known to have a powerful antioxidant activity and can be recovered from popular herbs such as rosemary and sage. One of the flow sheet alternatives generated and evaluated by the authors is presented in Fig. 1.3. The solution of the problem required a number of methods and tools, some of which needed to be developed (property and unit operation models) while others needed to be adopted (such as heuristics for flow sheet generation, flow sheet simulation and evaluation, etc.). Since solvents also play an important role in these processes, methods for solvent search are also needed.

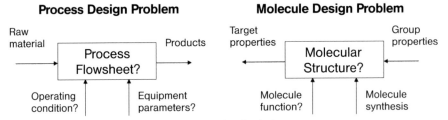

Figure 1.1 Differences between process design and molecule design problems. The question marks indicate what is unknown (needs to be determined) at the start of the design problem solution

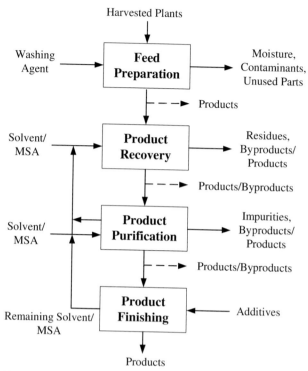

Harvested Plants

Washing Agent → **Feed Preparation** → Moisture, Contaminants, Unused Parts

- - → Products

Solvent/ MSA → **Product Recovery** → Residues, Byproducts/ Products

- - → Products/Byproducts

Solvent/ MSA → **Product Purification** → Impurities, Byproducts/ Products

- - → Products/Byproducts

Remaining Solvent/ MSA → **Product Finishing** ← Additives

Products

Figure 1.2 The main processing steps for manufacture of phyto-chemicals (from Harjo et al. 2004, reproduced with permission from IChemE)

1.2.3
Total Design of a New Chemical Product

In these design problems, given, the specifications (qualities, needs and properties) of a desired product, the objective is to identify the chemicals and/or mixtures that satisfy the given product specifications, the raw materials that can be converted to the identified chemicals and a process (flow sheet/operations) that can manufacture them sustainably, while satisfying the economic, environmental and operational constraints.

As illustrated in Fig. 1.4, solution of this problem could be broken down into three subproblems: a chemical product design problem that only identifies the chemicals (typically formulated as a molecule or mixture design problem), a process design part that determines a process that can manufacture the identified chemical or mixture (typically formulated as a process design problem) and a product-process evaluation part (typically formulated as product analysis and/or process analysis problems). In principle, mathematical programming problems can be formulated and solved to simultaneously identify the product and its corresponding optimal sustainable process. The solution of these problems are however not easy, even if the necessary

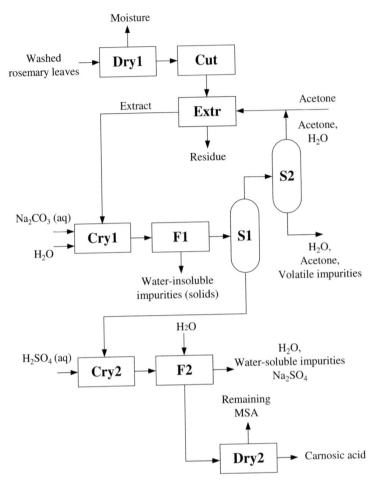

Figure 1.3 Generated process flow sheet for the manufacture of carnosic acid (from Harjo et al. 2004, reproduced with permission from IChemE)

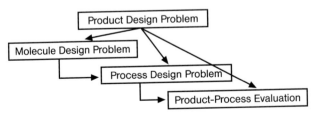

Figure 1.4 Product design problem includes the molecule design and the process design problems

models are available (Gani 2004a). Cordiner (2004) and Hill (2004) also provide examples of problems (formulations and structured products) of this type and the inability of current CAPE methods and tools to handle them.

Numerous examples of new alternatives for the production of known chemical products can be found in the open literature and have been successfully addressed by the CAPE/PSE community. Examples of complete product-process design for new high-value chemical products may not be easy to find because of reasons of confidentiality. Interesting examples of some well known high-value chemical products from the pharmaceutical and specialty chemical industries can however be found, e.g., design and manufacturing of penicillin (Queener and Swartz 1979), and production of intracellular protein from recombinant yeast (Kalk and Langlykke 1985).

1.2.4
Chemical Product Evaluation

In these problems, given a list of feasible candidates, the objective is to identify/select the most appropriate product based on a set of product-performance criteria.

This problem is similar to CAMD or CAM^{b}D except for the step for generation of feasible alternatives. Also, usually the product specifications (quality, needs, and properties) can be subdivided into those that can be used in the generation of feasible alternatives and those that can be used in the evaluation of performance. A typical example is the design of formulated products (also known as formulations) where a solvent (or a solvent mixture) is added to a chemical product to enhance its performance. Here, the feasible alternatives are generated using solvent properties while the final selection is made through the evaluation of the product performance during its application. Consider the following problem formulations:

- Select the optimal solvent mixture and the paint to which it must be added by evaluating the evaporation rate of the solvent when a paint product is applied (Klein et al. 1992).
- Select the pesticide and the surfactants that may be added by evaluating the uptake of the pesticide when solution droplets are sprayed on a plant leaf (Munir 2005).
- Select the active ingredient (AI) or drug/pesticide product and the microcapsule encapsulating it by evaluating the controlled release of the AI (Muro-Sune et al. 2005).
- Select solvent mixtures for crystallization of a drug or active ingredient (Karunanithi et al. 2004).

In all the above design problems, the manufacturing process is not included but instead, the application process is included and evaluated to identify the optimal product.

Consider the following product evaluation problem from the agrochemical industry. A pesticide product consisting of an active ingredient and a surfactant and other additives need to be evaluated in terms of which surfactant can be added to the system to enhance the uptake of the AI into the plant from the water droplets sprayed on the leaf surface. Solution of this problem requires property models that can predict the solubility of the AI in the water plus surfactants mixture, the diffusion of the AI through the leaf and into the plant, the evaporation of the water and many more. A modeling framework able to generate the necessary model for evaluation of the specified problem has been proposed recently by Muro-Sune et al. (2005). Through an integrated set of methods, models, and tools it is possible to not only evaluate the formulated product through its performance (in terms of uptake rate) but also find the best combination of AI, surfactant and plant that provides an improved product.

1.2.5
Chemical Process Evaluation

These problems are formulated typically, given the details of a chemical product and its corresponding process, to perform a process evaluation to improve its sustainability.

To perform such analysis, it is necessary to have a complete design of the process (mass balance, energy balance, condition of operations, stream flows, etc.) as the starting point and new alternatives are considered only if the sustainability indices are improved. Here, the design (process evaluation) problem should also include retrofit design. Uerdignen (2003) and Jensen et al. (2003) provide examples of how such analysis can be incorporated into an integrated approach by exploiting product-process relations. Note that the choice of the product and its specifications, the raw materials, the process fluids (for example, solvents and heating/cooling fluids), the by-products, conditions of operation, etc., affect the sustainability indices.

1.3
Issues and Needs

Three issues and needs with respect to integrated product-process design are considered in this chapter, namely, the issue of models and the understanding of the associated product-process complexities, the issue of integration, and the issue of problem definition.

1.3.1
The Need for Models

According to Charpentier (2003), over fourteen million different molecular compounds have been synthesized and about one hundred thousand can be found on the market. Since however, only a small fraction of these compounds are found in nature, most of them will be deliberately conceived, designed, synthesized and manufactured to meet human needs, to test an idea or to satisfy our quest for knowledge. The issue/need here is the availability of sufficient data to enable a systematic study leading to the development of appropriate mathematical models. This is particularly true in the case of structured products where the key to success could be to first identify the desired end-use properties of a product and then to control product quality by controlling microstructure formation. Another feature among product-process design problems is the question of systems with different scales of time and size.

For integration of product-process design, it is necessary to organize scales and complexity levels in order to understand and describe the events at the nano- and microscales and to better convert molecules into useful products. The relation between length and time scales is very nicely illustrated through Fig. 1.5, which is adapted from Charpentier (2003). Figure 1.6 highlights the relationship between scales (related to the product) and events (phenomena, operation, application, etc., related to the process).

Examples of multiscale modeling for product-process design can be found in structured products and their manufacture, such as polymers by polymerization and solid crystals by crystallization. In polymerization, the nanoscale is used in kinetics, the microscale for mass and energy transport, the mesoscale for particle-particle and particle-wall interactions, the macroscale for global polymerization reactor behaviour, and the megascale for reactor runaway analysis and energy consumption analy-

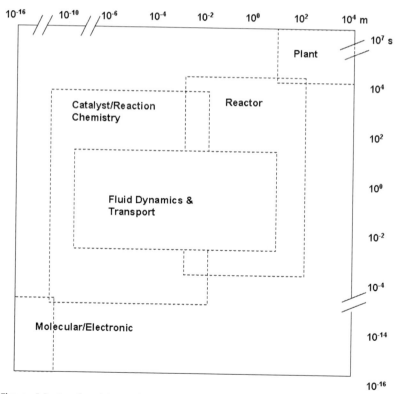

Figure 1.5 Length and time scale covered by multiscale approach

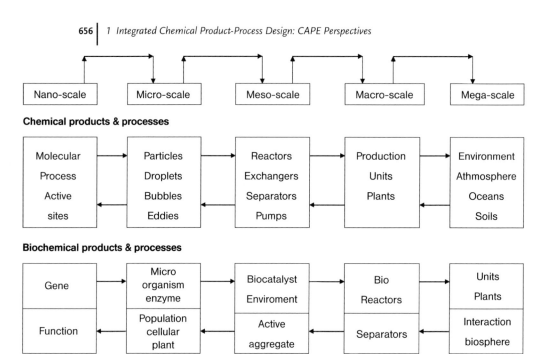

Figure 1.6 Scales and complexity in chemical and biochemical product-process engineering

sis. For a biochemical process, the nanoscale may be used for molecular and genomic processes and metabolic transformations, the microscale is used for enzyme and integrated enzyme systems, the mesoscale is used for the biocatalyst and active aggregates, macroscale and megascales are used for bioreactors, units and plants involving interactions with the biosphere (Charpentier 2003).

1.3.2
The Need for Integration

According to IMTI (2000), integrated product/process development is the concurrent and collaborative process by which products are designed with appropriate consideration for all factors associated with producing the product. To make product right the first time and every time, product and process modeling must support, and be totally integrated with the design function, from requirements capture through prototyping, validation and verification, and translation to manufacture.

Another issue/need is the increasing complexity of new chemical products and their corresponding technologies, which provide opportunities for the CAPE community to develop/employ concurrent, multidisciplinary optimization of products and processes. Through these methods and tools, the necessary collaborative interaction between product designers and manufacturing process designers early in the product realization cycle can be accomplished. Few collaborative (design) tools are available to help turn ideas into marketable products, and those that are available are technology and product-specific. Optimizing a product design to meet a set of requirements or the needs of the different production disciplines remains a manually intensive, iterative process whose success is entirely dependent on the people involved.

1.3.3
Definition of Product Needs

Good understanding of the needs (target properties) of the product is essential to achieve "first product correct", even though it may be difficult to identify the product needs in sufficient details to provide the knowledge needed to design, evaluate and manufacture the product.

1.3.4
Challenges and Opportunities

Based on the above discussion, a number of opportunities have been identified by IMTI (2000), which are summarized below:

Definition of Product-Process Needs (Design Targets)
- Provide knowledge management capability that captures stakeholder requirements in a complete and unambiguous manner.
- Provide modeling and simulation techniques to directly translate product goals to producibility requirements for application to product designs.

Methods/Tools for Product-Process Synthesis/Design
- Provide a first principles understanding of materials and processes to assure that process designs will achieve intended results.
- Provide the capability to automatically create designs from the requirements data and from the characterization of manufacturing processes.
- Provide the capability to automatically build the process plan as the product is being designed, consistent with product attributes, processing capabilities, and enterprise resources.
- Create and extend product feasibility modeling techniques to include financial representations of the product as an integral part of the total product model.

Modeling Systems and Tools
- Provide a standard modeling environment for integration of complex product models using components and designs from multiple sources/disciplines, where any model is completely interoperable and plug compatible with any other model.

Integration
- Provide the capability to create and manipulate product/process models by direct communication with the design workstation, enabling visualization and creation of virtual and real-time prototyped product.
- Provide simulation techniques and supporting processing technologies that enable complex simulations of product performance to run orders of magnitude faster and more cost-effectively than today.
- Provide the capability to simulate and evaluate many design alternatives in parallel to perform fast tradeoff evaluations, including automated background tradeoffs based on enterprise knowledge (i.e., enterprise experience base).
- Provide integrated, plug and play toolset for modeling and simulation of all life cycle factors for generic product types (e.g., mechanical, electrical, chemical).

1.4
Framework for Integrated Approach

The first step to addressing the issues/needs listed above could be to define a framework through which the development of the needed methods and tools and their application in product-process design can be facilitated. Integration is achieved by incorporating the stages/steps of the two (product and process) design problems into one integrated design process through a framework for integration. This framework should be able to cover various product-process design problem formulations, be able to point to the needed stages/steps of the design process, identify the methods and tools needed for each stage/step of the design process and finally, provide efficient data storage and retrieval features. In an integrated system, data storage and retrieval are very important because one of the objectives for integration is to avoid duplication of data generation and storage. The design problems and their connec-

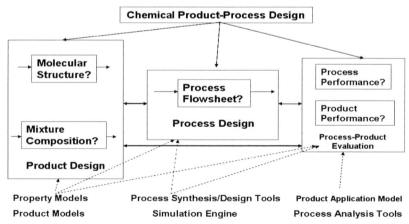

Figure 1.7 Integration of product-process design (see Table 1.1 for data flow details)

tions are illustrated through Fig. 1.3. The various types of design problems described above can be handled by this framework by plugging the necessary methods and tools into it. One of the principal objectives of the integration of product-process design is to enable the designer to make decisions and calculations that affect design issues related to the product as well as the process. The models and the types of models needed in integrated product-process design are also highlighted in Fig. 1.7. In the sections below, the issues of models and data flow and workflow in an integrated system are briefly discussed.

1.4.1
Models

One of the most important issues and needs related to the development of systematic computer-aided solution (design) methodologies are the models. For integrated product-process design, in addition to the traditional process and equipment model, product models and product-process performance models are also needed. A product model characterizes all the attributes of the product while a product-performance model simulates the function of the product during a specific application. Figure 1.8 illustrates the contents and differences among the process, product and product-performance models. Constitutive (phenomena) models usually have a central role in all model types.

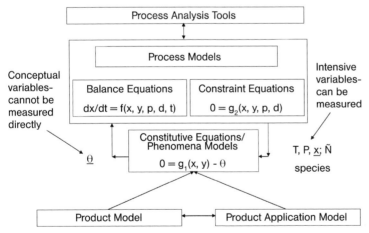

Figure 1.8 Different types of models and their connection to each other

1.4.2
Data Flow and Workflow

The data flow related to the framework for integrated product-process design is highlighted through Table 1.1, where the input data and output data for each design (sub)problem is given.

As highlighted in the problem formulation section, the workflow for various types of design problems is different and needs to be identified. In general terms, however, the following main steps can be considered (note that, as discussed above, some of these steps may be solved simultaneously):

- Define product needs in terms of target (design) properties.
- Generate product (molecule, mixture, formulation, etc.) alternatives.
- Determine if process considerations are important.
  - If yes, define the process design problem and solve it.
  - If no, go directly to the product evaluation (analysis) step.

Table 1.1 Data flow for each design problem

Input data	Problem type	Output data
Building blocks for molecules, target properties and their upper/lower bounds and/or goal values	Molecular design (CAMD)	Feasible molecular structures and their corresponding properties
List of candidate compounds to be used in the mixture, target properties and their upper/lower bounds and/or goal values at specified conditions of temperature and/or pressure	Mixture design (CAMbD)	List of feasible mixtures (compounds and their compositions) and their corresponding properties
Desired process specifications (input streams, product specifications, process constraints, etc.)	Process design/synthesis (PD)	Process flow sheet (list of operations, equipments, their sequence and their design parameters)
Desired separation process specifications (input streams, product specifications, process constraints, etc.) and desired (target) solvent properties	Process solvent design	Process flow sheet (list of operations, equipments, their sequence and their design parameters) plus list of candidate solvents
Details of the molecular or formulated product (molecular structure or list of molecules and their composition and their state) and their expected function	Product evaluation	Performance criteria
Details of the process flow sheet and the process (design) specifications	Process evaluation	Performance criteria, sustainability metrics

Table 1.2 List of methods/algorithms and tools/software that may be used for each problem (design) type

Problem type	Method/Algorithm	Tools/Software
Molecular and mixture design (CAMD)	Molecular structure generation Property prediction and database Screening and/or optimization	ProCAMD
Process design/synthesis (PD)	Process synthesis/design Process simulation/optimization Process analysis	ICAS (PDS, ICAS-sim, PA)
Process solvent design	CAMD methods/tools Process synthesis/design Process simulation/optimization Process analysis	ICAS (ProPred, ProCAMD, PDS, ICAS-sim, PA)
Product evaluation	Property prediction and database Product-performance evaluation model Model equation solver	ICAS (ProPred, ICAS-utility, MoT)
Process evaluation	Process synthesis/design Process simulation/optimization Process analysis	ICAS (ICAS-sim, ICAS-utility; MoT, PA)

- Analyze the process in terms of a defined set of performance criteria.
- Analyze the product in terms of a defined set of performance criteria.

In Table 1.2, the methods/algorithms and their corresponding tools/software are listed. Under tools/software, only tools developed by the author and coworkers have been listed (see the tools and tutorial pages at www.capec.kt.dtu.dk/Software/).

Examples of application of the tools listed above are not given in this chapter but can be found in several of the referenced papers.

1.4.3
Simultaneous Molecular and Flow Sheet Design

Design of chemical products is often described as the design of molecules and their mixtures with desired (target) properties and specific performance, as drugs, pesticides, solvents or food products. Molecules likely to match the target properties and performance are identified, usually in experiment-based trial and error solution approaches. In product-centric process design, it is necessary to match a set of target performance criteria for the process, usually through process simulation. For process design, alternative process flow sheets can be generated through simulation-based synthesis/design methods, where simulation is mainly used to evaluate and test alternatives. Systematic methods for generation of process alternatives are either rule-based or mathematical optimization-based.

Group contribution methods, which provide the basis for molecule and mixture design, can also be applied in process design. That is, in the same way functional groups are defined to represent molecules and to estimate their properties, process groups are also defined to represent process flow sheets and to estimate their operational properties. Therefore, if a table of process groups representing a wide range of operations can be established, the technique of CAMD can be adapted to computer-

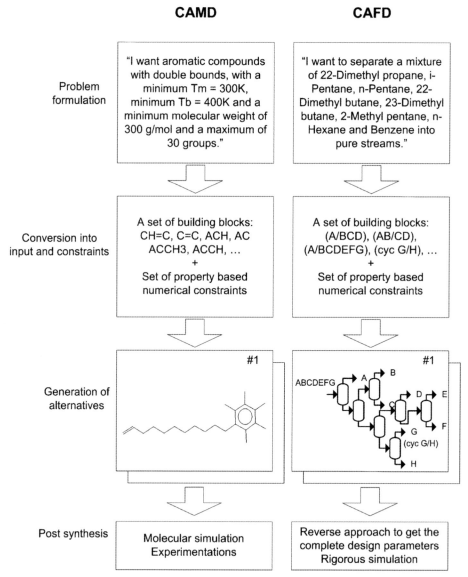

Figure 1.9 Common framework overview (from d'Anterroches et al. 2005, reproduced with permission from IChemE)

aided flow sheet design (CAFD), so that CAMD and CAFD can both be used for modeling, synthesis and design. Also, since CAMD can generate and evaluate thousands of molecules within few seconds of computer time, CAFD would also be able to generate numerous process alternatives without any loss of accuracy or application range. d'Anterroches et al. (2005) has developed a group contribution-based method for simultaneous molecular and mixture design. Figure 1.9 illustrates the features of a common framework for CAMD and CAFD.

1.5
Conclusion

As pointed out by Cordiner (2004), Hill (2004), and the referenced papers by Ng, even though the primary economic driver for a successful chemical product is speed to market, this does not mean that process design is not strategically important to these products. The important questions to ask (to list a few) pertain to the chemical product, how they will be manufactured, how sensitive is product quality related to cost and production, where they will be used or applied, how their performance will be evaluated and, how long a period will they be sustainable? Obviously, the answers to these questions would be different for different products and consequently, the methods and tools to be used during problem solution will also be different. Many opportunities exist for the CAPE/PSE community to develop systematic model-based solution approaches that can be applied to a wide range of products and their corresponding processes. It is the study through these model-based solution approaches that will point out under what conditions the process or operational issues become important in the development, manufacture and use of a chemical product. Successful development of model-based approaches will be able to reduce the time to market for one type of products, reduce the cost of production for another type of product, reduce the time and cost to evaluate another type of product. The models, however, need to be developed through a systematic data collection and analysis effort, before any model-based integrated product-process tools of wide application range can be developed. Finally, it should be noted that to find the magic chemical product, these computer-aided model-based tools will need to be part of a multidisciplinary effort where experimental verification will have an important role and the methods/tools could be used to design the experiments.

References

1 *Achenie L. E. K. Gani R. Venkatasubramanian V.* 2002 Computer-Aided Molecular Design: Theory and Practice, CACE-12, Elsevier Science, Amsterdam

2 *Charpentier J. C.* The future of chemical engineering in the global market context: market demands versus technology offers, Kem Ind 52(9) (2003) p. 397–419

3 *Cordiner J. L.* Challenges for the PSE community in formulations, Comput Chem Eng 29(1) (2004) p. 83–92

4 *d'Anterroches L. Gani R. Harper P. M. Hostrup M.* 2005 CAMD and CAFD (computer-aided flow sheet design), paper presented at World Chemical Engineering Conference, Glasgow, Scotland, July 2005

5 *Fung K. Y. Ng K. M.* Product centered processing: pharmaceutical tablets and capsules, AIChE J 49(5) (2003) p. 1193

6 *Gani R.* Chemical product design: challenges and opportunities, Comput Chem Eng 28(12) (2004a) p. 2441–2457

7 *Gani R.* Computer-aided methods and tools for chemical product design, Chem Eng Res Des 82(A11) (2004b) p. 494–1504

8 *Giovanoglou A. Barlatier J. Adjiman C. S. Pistikopoulos E. N. Cordiner J. L.* Optimal solvent design for batch separation based on economic performance, AIChE J 49 (2003) p. 3095–3109

9 *Grossmann I. E.* Challenges in the new millennium: product discovery and design, enterprise and supply chain optimization, global life cycle assessment, Comput Chem Eng 29(1) (2004) p. 29–39

10 *Harjo B. Wibowo C. Ng K. M.* Development of natural product manufacturing processes: phytochemicals, Chem Eng Res Des 82(A8) (2004) p. 1010–1028

11 *Hill M.* Product and process design for structured products: perspectives, AIChE J 50 (2004) 1656–1661

12 *Hostrup M. Harper P. M. Gani R.* Design of environmentally benign processes: integration of solvent design and process synthesis, Comput Chem Eng 23 (1999) p. 1394–1405

13 *IMTI* First product correct: visions and goals for the 21st century manufacturing enterprise, Integrated Manufacturing Technology Initative Report USA 2000

14 *Jensen N. Coll N. Gani R.* An integrated computer aided system for generation and evaluation of sustainable process alternatives, Clean Technol Environ Pol 5 (2003) p. 209–225.

15 *Kalk J. Langlykke A.* Cost estimation for biotechnology projects, ASM Manual of Industrial Microbiology and Biotechnology, ASM Press, Washington, D.C. 1986

16 *Karunanithi A. Achenie L. E. K. Gani R.* A computer-aided molecular design framework for crystallization solvent design, Chem Eng Sci 61 (2006) p. 1243–1256

17 *Klein J. A. Wu D. T. Gani R.* Computer-aided mixture design with specified property constraints, Comput Chem Eng 16 (1992) p. S229

18 *Linke P. A. Kokossis* Simultaneous synthesis and design of novel chemicals and chemical process flow sheets, in J. Grievink and J. van Schijndel (Eds.), ESCAPE-12, CACE-10, Elsevier Science, Amsterdam, (2002) pp. 115–120

19 *Ng K. M.* A multiscale-multifaceted approach to process synthesis and development, in R. Gani and S. B. Jørgensen (Eds.), ESCAPE-11, CACE-9, Elsevier Science, Amsterdam (2001) pp. 41–54

20 *Queener S. Swartz R.* Penicillins; biosynthetic and semisynthetic, in Economic Microbiology 3, Academic Press, New York (1979) pp. 35–123

21 *Reynolds C. H. Holloway M. K. Cox H. K.* 1995 Computer-aided molecular design: applications in agrochemicals, materials and pharmaceuticals, ACS Symposium Series, 589, Washington, D.C., USA

22 *Munir A.* Pesticide product and formulation design, Dissertation, CAPEC, Department of Chemical Engineering, DTU, Lyngby, Denmark 2005

23 *Muro-Sune N. Gani R. Bell G. Shirley I.* 2005 Model-based computer aided design for controlled release of pesticides, Comput Chem Eng, 30 (2005) p. 28–41

24 *Uerdingen E. Gani R. Fischer U. Hungerbühler K.* A new screening methodology for the identification of economically beneficial retrofit options in chemical processes, AIChE J 49 (2003) p. 2400–2418

25 *Wibowo C. Ng K. M.* Product-oriented process synthesis and development: creams and pastes, AIChE J 47(2) (2001) p. 2746

Appendix

Heuristics for constructing flow sheet alternatives (from Harjo et al. 2004, reproduced with permission from IChemE).

Feed Preparation
1. Consider reducing the size of the plant material to 2–5 mm to obtain good performance in industrial scale S-L extraction.
2. Consider using particle size bigger than 0.25 mm in S-L extraction to avoid clogging of the filter.
3. If the plant material is hard and abrasive, consider using size reduction by ball mill, fluid jet mill, or hammer mill.
4. If the plant material is soft and tough or fibrous and woody, consider using size reduction by cutting mill, disk mill, or hammer mill.
5. If the plant material is brittle or crystalline, consider using size reduction by fluid jet mill, hammer mill, or roller crusher.
6. If the plant material is tough, fibrous, and very heat-sensitive, consider using cryogenic size reduction by cutting mill, disk mill, fluid jet mill, or hammer mill.
7. Consider reducing the moisture content of harvested plants to about 10% for a safe storage.

Product Recovery
8. Consider using disk press for mechanical pressing of fibrous materials.
9. Consider using immersion type S-L extraction equipment if the target compounds are in low concentrations, strongly bound, and/or slowly diffusing.
10. Consider using percolation type S-L extraction equipment if the target compounds are in high concentrations, loosely bound, or slightly soluble in the solvent.

Product Purification
11. Whenever possible, consider using the same MSA as in the product recovery step.
12. For heat-sensitive materials, consider separations using L-L extraction, chromatography, or crystallization.
13. Consider using adsorption to separate natural pigments.
14. Consider using chromatography and/or crystallization for the separation of chiral molecules or when multiple single-compound products are desired.
15. Consider using pH swing crystallization for separation of compounds with acidic or basic groups.
16 Consider using large polarity differences between the mobile and stationary phases in reversed-phase chromatography to achieve high selectivity.
17. When handling liquid systems with little density difference, easily emulsified, or short contacting time is required, consider using centrifugal L-L extractors and separators.
18. When handling liquid systems containing suspended solids, easily emulsified, or large capacity, consider using reciprocating-plate L-L extractor columns.

19. When handling liquid systems with high viscosity or large capacity, consider using mixer-settler L-L extractors.
20. If the material has a very steep solubility curve (e.g., very sensitive to temperature), consider using cooling-type crystallizers.
21. If the material has a normal or moderate solubility curve, consider using evaporative-cooling, surface-cooling, or isothermal-evaporative crystallizer.
22. For batch and relatively low capacity processes, or feed with viscous solutions, consider using either plate-and-frame filter presses or leaf filters.
23. For continuous and large capacity processes, consider using either continuous rotary-vacuum-drum filter or continuous rotary-disk filter.

Product Finishing
24. If the feed to be dried is in liquid, suspension, or slurry solution forms, consider using either drum or spray dryers.
25. If the wet granular solids are to be dried, consider using either rotary or tray dryers.
26. Use freeze-drying only for heat-sensitive materials which may not be heated in the ordinary drying or when the loss of flavor and aroma must strictly be avoided.

2
Modeling in the Process Life Cycle

Ian T. Cameron and Robert B. Newell

This chapter deals with the important issues of where, why and how models of various types are used throughout the life of an industrial or manufacturing process. The chapter does not deal specifically with the modeling of the life-cycle process but concentrates on the use of models to address a plethora of important issues that arise during the many stages of a process' life, from the cradle to the grave.

In this chapter we first discuss the life-cycle concept in relation to a cradle-to-the-grave viewpoint and then in subsequent sections consider specific issues related to the modeling goals and realizations. Some important issues are discussed which surround model development, reuse, integration, model documentation and archiving. We also consider the future needs of such modeling approaches and the important implications of life-cycle modeling for corporations.

Throughout this chapter we refer to several specific industrial case studies that help illustrate the importance of modeling throughout the life cycle as well as the challenges of doing so. What is evident in the following sections is that there is a huge range of modeling used to help answer vital sociotechnical questions through the life cycle of the process or product.

It is important to appreciate that process and product engineering have vital links to social and human factors within a holistic approach to modeling. Major infrastructure projects continually reinforce a more complete view than that which is often taken by process and product engineers. In this chapter we expand the vision of modeling within the process or product life cycle to see just what has been achieved and where the challenges lie for the future.

2.1
Cradle-to-the-Grave Process and Product Engineering

Cradle-to-the-grave is a concept that now pervades most industrial operations, driven by concerns for safety, health and the environment (SHE). Sustainability and global environmental issues also figure highly in the drive for life-cycle analysis. Those con-

Computer Aided Process and Product Engineering. Edited by Luis Puigjaner and Georges Heyen
Copyright © 2006 WILEY-VCH Verlag GmbH & Co. KGaA, Weinheim
ISBN: 3-527-30804-0

cerns have been heightened by community pressures on government and industry regarding the impact of process and manufacturing developments on ecosystems as well as associated social impacts. This is largely driven by past disastrous events that have had major impacts on local communities, either directly through such events as fires, explosions or toxic releases or through more sinister chronic impacts or severe land contamination. Much of the risk and environmental management legislation and regulations in Europe, the United States and Australasia have now focused on cradle-to-the-grave concepts to control industrial impacts and analyze economics over the complete life cycle of the process. These important concepts have been expressed in numerous international standards such as the ISO14000 series and ISO15288 [1]. The life-cycle stages in these standards involve:

- concept
- development
- production
- utilization
- support
- retirement.

Process and product related activities define similar stages to the generic standard as shown in Section 2.1.2.

These holistic life-cycle concepts are particularly noticeable in such industries as vehicle manufacture, aluminum production and subsequent recycling, newspaper, glass production and recycling, plastic consumer articles and their reuse, to name just a few. The concepts are becoming increasingly common in the process and manufacturing industries, driven by tough environmental impact assessment regimes that demand in-depth analysis of the sociotechnical aspects of all major developments as well as facility expansions, well before any implementation. Life-cycle analysis is part of the responsible care program promoted by the International Council of Chemical Associations and in place since 1988 [2] after its initial start in Canada in 1985.

The following sections outline in broad terms the key concepts that undergird the life-cycle modeling activities and put these into a broader sociotechnical context beyond the mere process perspective. In this way, the life-cycle concept is seen as a much more holistic activity.

2.1.1
The Life-Cycle Concept

The process life cycle is characterized by several chronological stages as illustrated in Fig. 2.1. Accompanying the process life-cycle phases are certain activities associated with each stage. Of prime importance throughout the life-cycle perspective will be the issues of raw materials, wastes and emissions as well as energy consumption, generation and reuse. These issues are necessarily part of an integrated framework [3].

These stages involve:

- strategic planning
- research and development activities
- conceptual design of product and process
- detailed engineering designs
- installation and commissioning
- operations and production
- decommissioning of process
- remediation of process related facilities.

It is worth saying something about the phases of process life cycles to set the scene for a more in-depth analysis of the key issues from a modeling perspective.

Strategic Planning Phase

Here, the initial ideas of resource utilization or new product development have their genesis. This phase is driven by new business opportunities, perceived market needs or market push through the introduction of novel products, processes or markets. Modeling must incorporate uncertainty and is broad in scope and intent, as is to be expected in strategic and scenario planning.

Research and Development Phase

Following an initial strategy phase is often a more focused phase of research and development for those options identified as having the best potential. From the product viewpoint this might involve the modeling of market responses to product ideas to gain understanding of acceptance. This is common in the food and consumer products sectors.

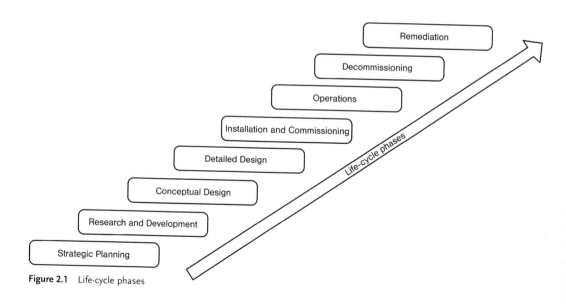

Figure 2.1   Life-cycle phases

From the process perspective this phase can often involve very lengthy and intense research that covers such areas as product qualities, reaction kinetics, product yields and effects of operating conditions on product spectrums. In pharmaceutical developments it can involve significant computer-aided drug design and possibly the use of animal experimentation.

This phase can also involve the analysis of environmental impacts through the treatment of process wastes such as solids, gases and liquids. Treatment options might be sought to eliminate or ameliorate impacts. Energy and utilities utilization for product creation will also be of importance.

Here, the modeling can involve a wide range of time and length scales, using such tools as quantum chemistry, molecular simulations and specialized model development at one end of the scale, and producing partial models for use at macroscale levels that deal with equipment and plant design at the other end of the scale. Other efforts at this phase can concentrate on issues of risk management and the choice of particular processing paths which apply inherently safer design and operations concepts. An important issue at this phase is the development and validation of physicochemical prediction models for a range of phase equilibria applications.

Conceptual Design Phase

For process technologies, conceptual design will lead to the development of input-output process models whereby overall economics can be better estimated. Flow sheeting in the traditional industries such as minerals, petroleum and chemicals is extensively used to generate these insights. Further, the use of these tools based on generic plant unit models leads to the development of the internal structure of the process system. Often the models are simply mixers, stream splitters, yield/stoichiometric reactors and component splitters. The aim will be to develop alternative flow sheet configurations to assess economics, product and by-product production plus environmental and risk impacts. The modeling is typically concerned with steady-state behavior.

Structural optimization using simple process models may also take place in this phase in order to consider optimal processing structures.

Of importance in this phase are the environmental impacts from air, noise, water and solid wastes. Extensive air-shed modeling is often required to assess potential impacts. Similar modeling may be required to assess impacts on groundwater and receiving waters such as rivers and bays. Particularly hazardous or noxious wastes may require modeling of their treatment options including destruction by incineration, landfill, chemical conversion or biological treatment. These can be significant modeling exercises in their scope and depth as well as time.

Detailed Design Phase

The detailed design phase leads inevitably to such outputs as the definitive engineering flow sheet as well as the piping and instrumentation (engineering line) diagrams. It also covers such project outputs as detailed specifications for procurement.

Modeling practice at this phase often involves the creation of equipment or unit specific models, which might take significant time to develop. The models might be integrated into larger software systems and dynamics can often be a significant issue driven by concerns for startup, shutdown, emergency response and regulatory control. As well, unit and plant-wide optimization modeling may take place at this point in order to consider best operational modes. Often, steady state assumptions move into continuous, dynamic considerations and then into hybrid (continuous-discrete) modeling environments.

Installation and Commissioning

The installation phase of process and product plants is a key area involving project planning and related models. Here, critical path models, dynamic resource allocation and control models play a vital role in this life-cycle phase.

Operations Phase

The operations phase includes such aspects as the construction and commissioning stages of process development. It clearly involves the day-to-day operations of the process under its intended operations policy over the useful life of the process or product. This operation period can vary widely depending on the industrial manufacturing sector being considered. Again it can be seen from the perspective of time and distance scales.

Within the traditional process industries that make use of major natural resources such as oil, gas and mineral deposits, the time scale can be of the order of decades. In the consumer product area it can be of the order of months, where market forces demand quick time to market and flexibility of manufacture and delivery modes. The computer and electronics industries as well as the food sector provide well-known examples of these 'quick-to-market' sectors.

Here, in the operations phase, modeling can focus on debottle-necking of existing processes for retrofit or incremental improvements. It can also address issues of effective supply chain design and operation under a dynamic market environment. It can also involve detailed risk modeling for improved design and operations as the external environment such as government regulations change. As well, the effect of changing resource characteristics can lead to significant process modeling that involves the assessment on the process of changing raw materials and product quality demands.

Modeling of routine plant maintenance through techniques such as critical path procedures or risk-based inspection (RBI) or maintenance (RBM) strategies is a key area of production operations. They require modeling of the system in terms of predicted risk of failure and the need for preventative maintenance procedures on varying time scales.

Decommissioning Phase

Most processes and products have a 'use by' date and inevitably come to a natural or in some cases, dramatic end. Decommissioning of the process or the product and product-line is now an important consideration in the life cycle. It can mean a very lengthy process of assessment and action. Much needs to be considered in the decommissioning phase, with a significant amount of work associated with the management of risk in achieving the outcomes.

In particular the decontamination of plant for disposal may require specialized treatment and this can lead to in-depth modeling of the processes. In the case of defunct nuclear facilities the process of decommissioning can take decades as seen from recent activities within Europe and elsewhere.

Remediation and Rehabilitation Phase

This is a phase of the life cycle which can involve significant financial resources, often in the past borne by government but now more likely by the operating companies if they remain financially viable. In many cases, specialized modeling and chemical experimentation is necessary to consider ways of achieving remediation of land and the environment. In many cases, where mining is carried out a remediation plan is activated, which is an integral part of the operations phase of the process. Modeling of mining operations and the remediation phase provides input to environmental management plans.

2.1.2
Process Modeling Within the Life Cycle

Modeling within the process life cycle is now an extremely important activity for all new industrial initiatives and expansions of existing facilities. However, what do we mean by the word 'modeling'? This seems an obvious question and for many there are simplistic answers. However, the modeling concept is far beyond the simple idea of a set of equations to be solved within a flow-sheeting package or numerical solver. In the context of life-cycle modeling, we do well to consider the generic definition of Minsky [4] who defined a model in the following terms: "A model (M) for a system (S) and an experiment (E) is *anything* to which E can be applied to answer questions about S."

As such, this definition captures a wide variety of models often used within the process life cycle, from physical models to mathematically-based models. A nonexhaustive review of modeling applications and modeling forms or approaches is given in Table 2.1. This expands the life-cycle phases of Fig. 2.1 by introducing some intermediate activities often seen in industrial projects.

Table 2.1 illustrates the diversity of modeling activities throughout the life cycle. There are many forms and approaches used in the phases, with significant reuse or integration of models – either directly or indirectly through data transfer from one model to another. The over-riding challenge in this area is unity of models, data and documentation.

The principal characteristics can be summarized as:

- a diversity of modeling goals throughout the life cycle phases encompassing such purposes as:
  - assessing market potential or response;
  - generating basic data or property relations for use in later phases;
  - estimating economic potential;

Table 2.1 Model use and characteristics for process life cycle.

Process life-cycle phase	Modeling applications	Modeling forms or approaches
Strategic planning	Market potential Basic economics Resource assessment	Purpose, goal, mission models Issue-based planning Scenario models Self-organizing models
Research and development	Resource characterization Basic chemistry Reaction kinetics Catalyst activity life Physicochemical behavior Pilot plant design and operation	Reaction systems models Catalyst deactivation models PFR, CSTR reactor models Elementary flow sheet packages Fluid-phase equilibria models Physical property models Molecular simulation Quantum chemistry models
Initial process feasibility	General mass and energy balances Alternate reaction routes Alternate process routes Input-output economic analysis Preliminary risk assessment	Flow-sheeting packages Semi-quantitative risk models Financial analysis models
Conceptual design	Mass and energy balances Plant/site water balances Initial environmental impact Detailed risk assessment Economic modeling	Flow-sheeting packages Environmental impact (air, noise, water, solid wastes) Social impact assessment models Risk consequence and frequency models Computational fluid dynamics (CFD)
Detailed design	Detailed mass and energy balances Vessel design and specifications Sociotechnical risk assessment Risk management strategies Project management	Flow-sheeting packages Dynamic simulation for plant and units CFD modeling and simulation Mechanical simulation (finite element methods and variants) 3D plant layout models Fire, explosion, toxic release models Fault tree and event tree models Air-shed models for dispersion of gases and particulates Noise models
Commissioning	Startup procedures Shutdown procedures Emergency response	Grafcet and ladder logic models Safety instrumented assessment models Risk assessment models
Operations	Process optimization Process batch scheduling Supply chain design and optimization	Scheduling models Unit and plant-wide optimization models (LP, NLP, MILP, MINLP) Queuing models

Table 2.1 Model use and characteristics for process life cycle (continued).

Process life-cycle phase	Modeling applications	Modeling forms or approaches
Operations (continued)	Real-time expert system models Neural nets and variants Empirical models (ARMAX, BJ) Maintenance models (CPN, RBI/M)	
Retrofit	Debottle-necking studies Redesign	Flow-sheeting packages Detailed dynamic simulation Computational fluid dynamics (CFD)
Decommissioning	Disposal processes and strategies Decontamination of equipment, approaches and optimal policies	Specialized models for the processes involved
Remediation and restoration	Geotechnical Contaminant extraction options	3D physical extraction pilot plants Soil processing models for decontamination

- evaluating system dynamics;
- predicting environmental impacts;
- optimizing product and process performance;
- designing products and processes (structure and units);
- planning production cycles or routine maintenance;
- improving viability, process performance or risk management factors;
- enhancing inherently safer designs and operations;
- a diversity of model forms and approaches to address the modeling goals such as:
  - social, economic, human factors and technical models;
  - mechanistic, empirical, stochastic and deterministic models;
  - physical and mathematical modeling approaches;
- a granularity in the model representations across the life cycle, typically increasing in detail as the life cycle phases progress chronologically;
- a diversity of time and length scales being captured in the models, representing the multiscale nature of product and process engineering;
- the current independence of much of the life-cycle modeling that is undertaken, typically by a range of external consultants, in-house company groups and government agencies;
- a diversity of tools to accomplish the modeling tasks including:
  - proprietary software products such as flow-sheeting packages;
  - purpose built models that are essentially standalone items in languages such as C, Fortran or Java;
  - specific models in commonly available software such as MS Excel or Matlab;
- a diversity of solution approaches to the models in the life cycle, for purposes of prediction, design, estimation or identification, spanning computation times of seconds to weeks for large-scale CFD computations.

2.1.3
The Multiscale Nature of Modeling During the Life Cycle

In Section 2.1.2, there is a wide range of time and length scales involved in modeling across the life cycle. This applies across the life cycle phases as well as within the individual phases. The rise in interest in multiscale modeling approaches and the integration and solution of composite models built from several partial models is driven primarily by product engineering where the nano or microscale characteristics are seen as vital to 'designer' products. Combined with the meso and macro-scales, typical of issues at the equipment and plant level, the emphasis on multiscale representations will continue to grow.

The chapter in this current book on multiscale process modeling gives a comprehensive overview of this important area. Within the modeling practice across the life cycle, Marquardt et al. [5] have also mentioned its importance.

2.2
Industrial Practice and Demands in Life-Cycle Modeling

The following sections deal with some of the important issues facing those carrying out modeling at stages of the process life cycle. It emphasizes the fact that modeling is far more than just developing a set of equations to be solved by a simulation tool [6, 7].

2.2.1
Modeling Methodology and Workflow

One of the key underlying concepts is methodology within modeling. This is closely related to workflow concepts in carrying out any modeling activity. Figure 2.2 illustrates a particular modeling methodology [8] which possesses generic character and is applicable to many occasions when modeling is performed. It shows how various stages of the modeling cycle use and refine other stages. There are seven steps shown in this workflow scheme, each having an important part to play. It is an iterative process which demands clear understanding of each task and the need to specify conditions of modeling cycle termination.

- *Goal set definition and decomposition.* This refers to the initial phase of asking: what is the purpose of the model? Here, the key goals are established with regard to model application area (control, design, optimization, etc.) and the desired outcomes from the modeling activity must be stated, leading to formal specifications. This includes a wide range of outcomes as mentioned in Section 2.1.3. This aspect of modeling is generally poorly done but is essential for establishing the termination conditions for the modeling cycle. It still remains a complex, difficult area with little guidance or techniques in how overall or canonical goals are decomposed into subgoals as represented in a goal tree or goal graph and then subsequently to define the subsystems.

- *Model conceptualization.* Here, the model form or multiple forms have been chosen and the conceptualization takes place. In mechanistic models this is related to defining balance volumes for mass, energy, momentum or population conservation, together with convective and diffusive streams connecting the balance volumes. Other objects and attributes within balance volumes relate to reaction, physical properties, spatial distribution and the like.

  In the case of empirical model building, a selection of potential model forms is initially needed. The selection can be based on physical insight or through the use of information criteria which seeks to balance model complexity, usually the number of parameters, against the quality of the model fit to the data. This approach might cover completely black-box models with arbitrary structure or grey-box models that incorporate some predefined structure.

- *Modeling data.* This refers to both the type of data needed for calibration or parameter estimation and model validation as well as physicochemical data necessary for the building of mechanistic models. Here, experimental design can play an important role and cover classic factorial designs to such techniques as Latin hypercube sampling for Monte Carlo models.

- *Model building and analysis.* This brings us to the task of actually constructing the physical or mathematical representation based on the chosen approach. In most cases, significant use of computer-aided tools can assist in this task. Nevertheless, the construction of mathematical models can be a time-consuming, error-prone task.

  Analysis of the resultant model set is absolutely necessary for reasons of degrees of freedom, model index and potentially observability, controllability and identifiability.

- *Model verification.* This relates to the systematic construction of solution code typically in the form of a computer program to solve the model. It requires such concepts as algorithm design, modular code construction and software verification tools. It seeks to ensure that the model solved is the model that was defined.

- *Model solution.* Solving the model can be a trivial task taking a few seconds of computer time to weeks of execution time on even the largest, distributed computing devices. Appropriate numerical methods are essential for minimizing computation time as well as ensuring solutions that are credible.

- *Model calibration and validation.* The final phase to be tackled in most industrial modeling is the need to calibrate the model against good plant or process data. This is nontrivial as 'good' data is often hard to obtain and requires significant effort and determination to get it. It leads to key parameter estimates being obtained.

  Model validation which asks: is the model prediction accuracy adequate for the task? This is another area that requires perseverance and in-depth statistical analysis – something often glossed over in industrial modeling.

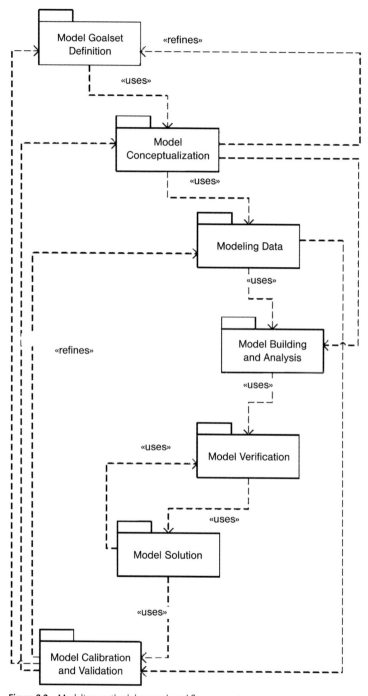

Figure 2.2 Modeling methodology and workflow concepts

2.2.2
Modeling Goals and Model Types

In Section 2.1.2 we discussed the various activity or application models commonly seen in industrial modeling practice. These were based on model outcomes in specific application areas. Such models as reaction kinetic models can have many forms, ranging from linear to nonlinear empirical models. Physical property models can have an enormous range of model forms. However, we can consider a more fundamental classification of process related models based on characteristics of the models being developed and used in particular applications. Figure 2.3 provides a simple class diagram for a range of typical model forms, which include most of those seen in the life-cycle process. Other variants of these model types are possible. The basic classification relates to steady state and dynamic models with actual behavior being captured as continuous, discrete or hybrid (continuous-discrete). Further classifications lead to empirical, mechanistic, stochastic and deterministic models, with their various subclasses. Each of these models has a particular mathematical form that relates to the model type and which requires specific solution methods to be applied – in most cases this involves some form of numerical computation.

Models are built for a purpose or 'goal' and that means they exist to answer questions about the reality they represent. As such, the idea of goal definition becomes an increasingly important idea that is not often explicitly stated or used in driving the modeling activities. There has been some work done in this area, using functional modeling approaches that are mainly driven by applications in fault diagnosis [9, 10].

Recent work [11] has been directed towards the concept of modeling goal development and evolution. This uses structured goal definitions to help drive the modeling with the intent of shortening the iterative nature of the modeling cycle and to improve modeling outcomes as shown in Fig. 2.2.

2.2.3
Model Ingredients

Industrial modeling requires a number of key ingredients to be effective. The following highlights some of these ingredients and comments on them.

Assumptions
Key assumptions that relate to the conceptualization need to be recorded and then used to check consistency with the resultant model. It is vital that all assumptions relevant to the model development are available in the modeling life cycle. This will include such items as assumed balance volumes, mechanisms, details of species in the system or the lumping of species into larger classes for tractability of modeling. It will include issues related to key flows in the system, whether convective or diffusive as well as sources and sinks of energy and component masses.

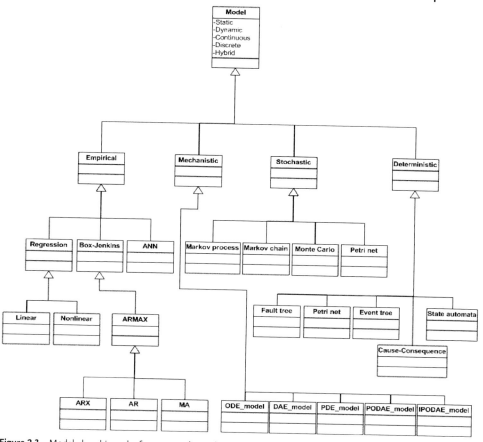

Figure 2.3   Model class hierarchy for commonly used models

Documentation

Documentation still remains one of the biggest challenges in industrial modeling. The adequacy of documentation in most modeling exercises is usually very poor, leading to difficulties in model reuse and much repetition of effort by future generations of engineers that need to revisit historical work.

Decision making and arguments concerning the positions taken on various modeling decisions can be documented in structured ways such as the IBIS system [12-14]. This consists of several elements such as issues to be addressed, positions to be taken in addressing those issues, arguments for and against the various positions. This is typically performed through hypertext and graphical descriptions.

The generation of readable model descriptions and final reports as part of the modeling exercise is vital for future reuse. In general, reports are not stored as part of the modeling exercise but remain separate items. Better integration is necessary for reuse and utilization of efforts.

Data

As mentioned in Section 2.2.1, modeling data – either physicochemical or plant data – is a vital ingredient for modeling. It requires careful analysis and archiving with the other inputs and outputs of the various phases of the modeling cycle. This aspect of modeling is one of the biggest challenges, with little if any facility in modern CAPE software tools to adequately store validation and other data together with the model.

2.2.4
Support Technologies

Modeling requires support technologies. This is a huge area and what follows is a brief mention of some of the key items often encountered in life-cycle modeling.

Model Definition and Building

There is a plethora of model-building environments available for each life-cycle modeling phase. In the engineering area, many allow concepts and relationships to be defined at a higher conceptual level using GUIs. Such systems as Model.La, ModKit, ICAS-ModDev and SCHEMA are in this category [8]. Other environments such as gPROMS, ASPEN Custom Modeler, Modelica and Daesim allow direct model descriptions in terms of equations in various forms that can be stored as model members. Matlab provides facilities for equation solution of ordinary differential equations (ODE) and differential-algebraic equations (DAE). There are also dedicated toolboxes that cover specialized areas such as neural network modeling, partial differential equation systems and finite element applications.

Simulation

Again, simulators abound in most phases of the process life cycle. Process simulation systems such as ASPEN, PROII and HYSYS have dominated the petroleum and petrochemicals sectors with many other similar products available, that vary in their application areas and capabilities. Generic simulation tools such as gPROMS, ModKit, ICAS, Daesim Dynamics and other simulators are available for general solution of models in the form of differential-algebraic equations that may also be hybrid in nature. Several also handle partial differential equation systems and a few allow simulation of integro-differential systems. These tools need to be chosen carefully for the task being tackled.

At the more fundamental chemistry level there is a wide range of tools for molecular simulation, quantum chemical calculations and the like. Other tools such as discrete element simulations can handle the dynamics of particulate systems and find wide use in minerals processing, bulk fertilizer and pharmaceutical applications.

Environmental simulation tools are easily available for such applications as neutral, buoyant and heavy gas dispersion estimates as well as prediction of groundwater flows, river, estuarine and ocean impact assessments. Significant adaptation of finite element or finite volume methods has been applied to such application areas. Many are made available by national environmental protection agencies such as the US EPA.

Data Analysis

Again, there are numerous tools for analyzing large data sets from process plant. The key issues here are good visualization facilities, ability to handle very large data sets (many millions or tens of millions of data points) and data reduction capabilities to handle multivariable data. The visualization of process data in real-time through the use of Web browser tools across the complete corporation is an increasingly important development. In many multinational corporations, the monitoring and analysis of real-time data from their world-wide operations is now commonplace. Typical of the systems in place for data capture is Honeywell's Process History Database (PHD), which can be linked to other applications such as MS Excel to view archived data.

Accompanying these issues is the ability to transform and manipulate data and to analyze its characteristics so that it is suitable for calibration and validation studies. Filtering, handling or reconstruction of missing data points, spectral analysis and time series analysis capabilities should be available to help tackle the data analysis problem.

2.2.5
Modeling Integration

Model integration is a major issue in industrial modeling practice. Model integration is poorly understood and practiced in the manufacturing and process industries. Much duplication of effort is required to integrate models across the process-product life cycle. In particular, the inputs and results from particular models are often transferred inefficiently between modeling applications. For example, the predictions of gaseous discharges from multiple sources in a process flow sheet are often required in area source dispersion models. The linked models are often not within the same corporation and even if they are much manual effort is often expended in data transfer. The initiative leading to pdXi data exchange protocols based on XML has not penetrated the process and related industries very well, if at all.

Some progress has been made in integrating disparate process models into proprietary software simulation systems through the use of CAPE-OPEN and Global CAPE-OPEN standard interfaces. This also addressed physical property routines, numerical solvers and continues to be extended.

The problem of data and model integration across the whole life cycle still remains as an unmet challenge.

2.3
Applications of Modeling in the Process Life Cycle: Some Case Studies

In this section we illustrate some aspects of life-cycle modeling on a recent, large-scale industrial development of an alternative fuel source. This is based on oil shale. This section shows some of the extensive and diverse modeling that takes place during the life cycle of such a process, and comments on the importance and challenges in such modeling.

2.3.1
Shale Oil Processing

Shale oil has a long history around the world as an alternate source of hydrocarbon-based fuels. Plants have operated in many countries, either as demonstration plants or for commercial production. Countries such as China, Estonia and Brazil currently produce shale oil. Several demonstration plants operated in the US, mainly in Colorado. In Australia oil shale was processed in the early 1840s to produce 'kerosene' from the kerogen content of the shale. In the 1930s shale oil was produced in crude, batch retorts which pyrolized shale that had been suitably mined and crushed. It was an expensive and inefficient process but provided alternative fuels in a time when normal petroleum fuels were in short supply. Much research and development work was done around the work during the 1970s to 1990s to enhance processing and recoverability of products as well as understand the potential environmental impacts.

Processing of oil shale into hydrocarbon products involves a combination of mining technologies, minerals processing and conventional hydrocarbon processing. Figure 2.4 shows a typical block diagram flow sheet representing a commercial operation located on the central coast of Queensland, near to the city of Gladstone [15]. This shows the complex nature of the process from mining and shale processing, which is heavily slanted towards solids processing technologies through to liquid-vapor systems typical of petroleum processing.

2.3.1.1
Research Modeling
One of the key research issues in oil shale processing is the understanding of product yields under various conditions of pyrolysis. Other research issues relate to the drying characteristics of crushed oil shale prior to the retorting of shale in the processor. Much academic and industrial research has sought to address these issues [16-18].

In particular, models for the kinetics of oil shale pyrolysis, which were represented by nonlinear models were important for reactor design [19]. These were based on extracting rate constants through parameter estimation techniques. Other work related to understanding the effect of particle size on retorting and drying.

2.3.1.2
Conceptual Design
Some selected aspects of conceptual design are mentioned here.

Preliminary Process Design
Modeling at the conceptual stage of the process was complemented by significant pilot plant studies that gave specific data from actual feedstock material, so that ini-

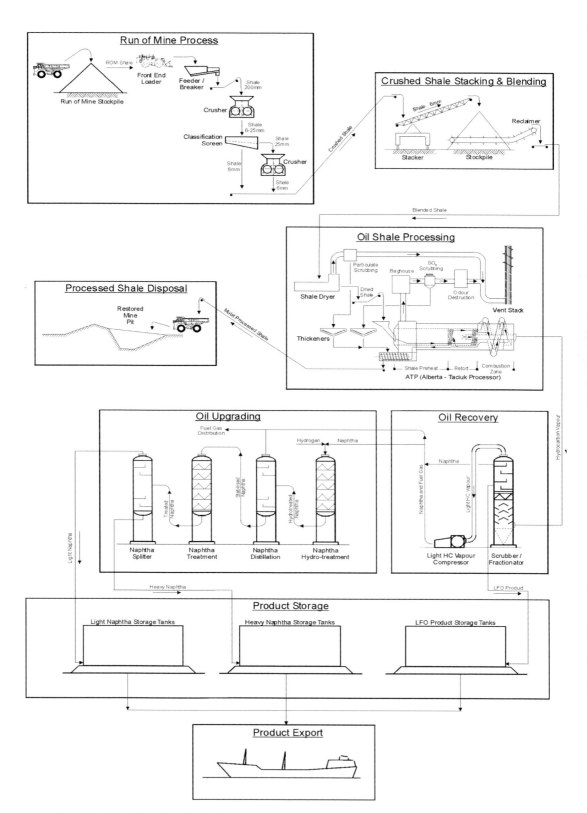

Figure 2.4 Overall flow sheet of oil shale processing

tial mass and energy balances were possible through the use of minerals processing and petrochemical flow-sheeting packages.

Targetted modeling of key items such as the Alberta Taciuk Processor (ATP) were needed to ensure a good design basis for such a large piece of rotating equipment. Initial modeling used simplified multiple, lumped-parameter models to represent the distributed nature of the processor operations. The key parameters such as heat transfer coefficients and heats of combustion being obtained from pilot plant measurements.

Environmental Aspects

Oil shale processing environmental issues continue to be one of the priority areas and as such a significant amount of modeling is required to answer questions related to environmental impacts. In particular the geographical location of such plants within complex air sheds poses particular challenges as to the impact of gaseous emissions. Complex dispersion models that can handle multiple sources such as the Industrial Source Complex version 3 (ISC3) [20] are needed to estimate such potential impacts of a wide range of gaseous species such as NO_2 and SO_2. Figure 2.5 shows the prediction of SO_2 levels (micrograms per cubic meter) from the planned stage 2 process. These showed that expected ground-level concentrations were well within accepted guidelines. It also showed potential areas of concern, which were very dependent on the geographical and atmospheric characteristics of the area.

Other modeling involved such areas as:
- Leaching models for spent shale to ascertain extraction rates of specific chemical substances from percolation of rain water. This involved the development of one-dimensional partial differential equation models with appropriate constitutive relations for the extraction rates.
- permeation modeling of leachate into surrounding land and potentially into ground waters [21-23];
- Total site water management modeling to help design and develop 'zero discharge' operating policies. This covers water segregation, reuse and effects of atmospheric conditions such as cyclone conditions as well as sediment control issues close to the Great Barrier Reef Marine Park.
- premining and postmining flow modeling of principal creeks in the areas surrounding the operations.

2.3.1.3
Detailed Engineering

Detailed engineering inevitably involves the design of complex process equipment.

In the case of oil shale processing, the ATP is a very complex device with four operational zones carrying out functions of shale preheating, spent shale cooling, retorting of dried shale and combustion of spent shale for energy recovery purposes. The zones are maintained at different pressures for operational reasons. Figure 2.6 shows a schematic of the ATP vessel and Fig. 2.7 shows the retort zone end of the demonstration pilot plant vessel, which is approximately 50 m in length, 10 m in diameter and has a loaded weight of over 2500 tonnes.

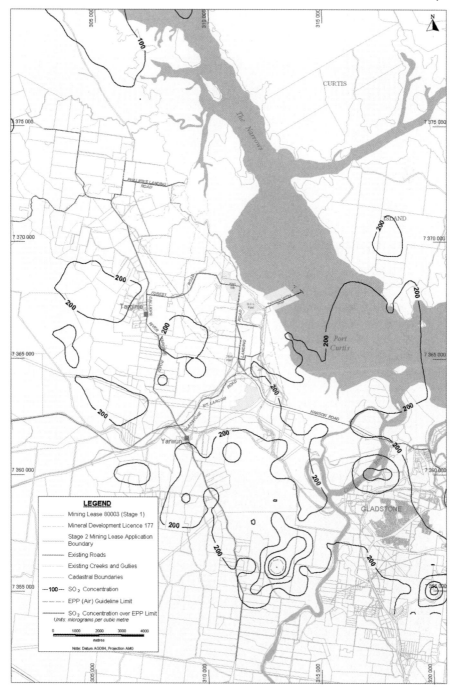

Figure 2.5   Predicted SO$_2$ levels for stage 2 process from complex
dispersion modeling

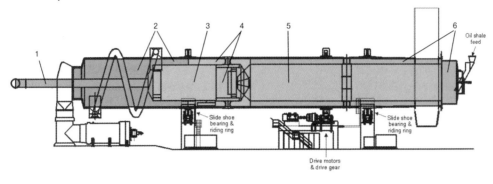

1 Vapor Tube   4 Retort Seal

2 Combustion Zone (750°C)   5 Preheat Zone (100–250°C)

3 Retort (500°C)   6 Cooling Zone

Figure 2.6   **ATP shale processor schematic**

Modeling aspects of detailed process design involved:

- Substantial flow-sheet modeling of the back-end process, which consists of fractionation, hydrogenation and product separation to produce a range of liquid and gaseous hydrocarbons products.
- Detailed process modeling of the ATP vessel using steady state models to compute the mass and energy balances. This involved the use of standalone programs to consider a range of scenarios for changing feed rates and pressure changes.
- Finite element stress modeling of the structural aspects of the ATP are used to ascertain adequacy of the internal mechanical design for vessel integrity.

Figure 2.7   ATP retort and combustion zones

2.3.1.4
Operations Modeling

Operational phase modeling considers the following issues:

- Initial modeling of the ATP dynamics to consider the effect of feed flow rate changes and moisture into the ATP vessel. This allowed the development of an initial tool for improving operator training and considering alternate control strategies for both set-point changes as well as disturbance rejection. This considered conventional PID loops as well as the potential for model-based control strategies.
- empirical modeling through planned step tests on the ATP to establish appropriate models for model-based control applications;
- improved flow-sheet modeling of the process to consider better operating policies for the hydrocarbon processing units of the plant;
- Development of a complex distributed parameter model of the preheat-cooling section of the ATP in order to answer questions about internal heat transfer design and potential hot gas bypassing. This involved the consideration of solids flow in rotating drums as a function of rotation speed and particle properties, the effect of particle size distributions on internal heat transfer and the effect of pressure driven flows in various flow zones within the device. A full distributed parameter model was developed to help answer the key questions.

2.3.1.5
Risk Modeling

Risk modeling that includes both financial and environmental risk is a crucial factor in modern process design and operations. In the case of shale oil production, there are a number of hazardous materials and conditions in the process. A full quantitative risk assessment was needed, which involved modeling all major events such as low and high pressure releases of gases and liquids, pool and jet fires as well as explosions. This led to estimates of iso-risk contours for potential fatalities as well as injury levels. Figure 2.8 shows predicted risk levels for radiation impacts at 4.7 and 23.0 kW m^{-2}, which are key legislative limits for nearby residential and industrial land-uses. In this case, risk levels were well within acceptance levels, since nearby industrial sites were located hundreds of meters away and residential areas were several kilometers away.

2.3.1.6
Socioeconomic Impact Analysis

One of the key issues that is amenable to modeling is the impact that the proposed operations will have on social and economic aspects of the region. Regional economic impacts were measured using the Queensland Multi-Regional Input-Output Model that takes into account linkages between regions. This provides a tool to consider alternative strategies for regional development.

This type of modeling can provide useful data on regional effects throughout the life cycle of the operation covering planning, conceptualization, construction and operational phases. The outputs can be linked to issues of employment levels, potential housing needs and the like.

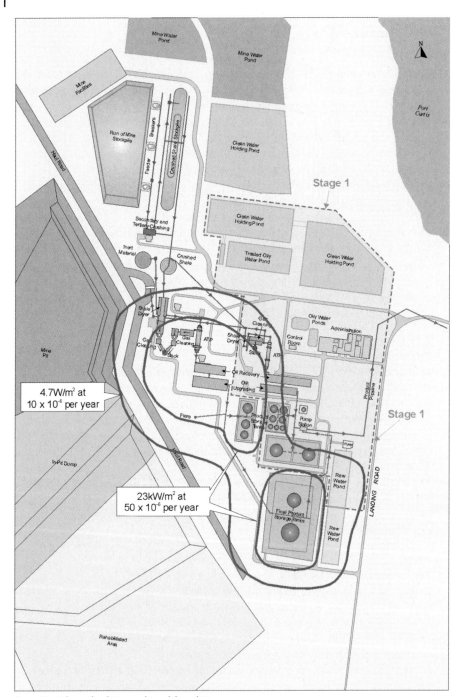

Figure 2.8 Thermal radiation risk model predictions

2.3.2
Utility and Inventory Management Applications

Two applications that illustrate less traditional or less recognized modeling are utility management in a brewery and inventory management in a warehouse.

Brewery Utility Management

Beer production is a deceptively simple process involving brewing (basically extraction of sugars from malted grain), fermentation (partial biological conversion of sugars to alcohol), filtration and packaging. It is complicated by the range of products generated by variations in ingredients and in the brewing and fermentation processes and by traditional processing in batches.

A key economic driver to remain competitive is to reduce and manage the consumption of utility streams, about six in total. Utility management is based on the prediction of current consumption rates and the prediction of consumption over short periods into the future.

Utility consumption during individual steps in the batch processing is modeled empirically by a piecewise-linear profile based on processing time and sometimes batch size. The model needs to know what processing steps are active and when the steps commenced and then it simply performs interpolations and summations based on simple utility flow sheets to predict utility rates at various real and virtual sensor locations. This information is then used by operations and technical personnel for monitoring and diagnosis.

Warehouse Inventory Management

Warehouses are also deceptively simple operations. The basic model is simply a dynamic mass balance that tracks inflows and outflows to predict inventories. A model validation is called a stocktake. Again there are many complicating factors: a large number of products, discrete inflows and outflows in terms of packs, pallets and truck loads of various sizes, spatial factors governing the placement of material in the warehouse, product 'use by' dates, real-time orders and just-in-time delivery and manufacture.

The economic drivers are minimizing inventories and wastage and increasing the ability to meet orders often with very small lead times.

The model must not only predict current and future inventories but also advise the warehouse staff on where products are to be found to meet orders by careful tracking in the spatial domain as well as the time domain.

2.4
Challenges in Modeling Through the Life Cycle

This section outlines a number of challenges currently being addressed that help to tackle some of the more intractable aspects of life-cycle modeling for process and product systems. Most are to do with information management which is at the heart of these challenges.

2.4.1
Model Management

It is clear that during the life cycle of a process or manufacturing plant there are a large range of models and associated data that have been gathered and developed by many different people with an often enormous associated cost over the life cycle.

While there are well developed procedures for the life-cycle management of computer software, little is done to manage models, resulting in much loss of information and knowledge and much duplication of effort. This often occurs over relatively short periods in the operations phase largely due to changes in personnel and computer systems, let alone between life-cycle phases which frequently occur within different organizations. In many cases there is much process and operating knowledge in models developed in the strategic planning, research and development and design phases that is never utilized in the operations phase.

Recent research into the tracking and representation of design information has led to several systems that seek to tackle this issue. The recording and utilization of modeling assumptions [7, 8, 33] and the transfer of software life-cycle management to modeling may one day address the model management issue more fully.

2.4.2
Model Repositories and Reuse

There have been efforts to address the issue of model repositories. It still remains a major challenge in process modeling and even more so in sociotechnical modeling, which presents a far greater challenge due to its diversity of applications and model forms. It is especially a problem when modeling over various life-cycle phases is carried out by a wide range of organizations not necessarily part of the corporation.

Many model repositories are simply files within a directory that were used to run a specific simulation similar to those in current flow-sheeting packages. More sophisticated repositories that are 'simulator neutral' are not very common. Amongst the attempts to produce such repositories are languages such as system for chemical engineering model assembly (SCHEMA) [24, 25], that seek to store models and model families in a natural language form that includes the underlying assumptions in the model as well as a description that can be used to generate simulation code suitable for a specific simulation tool. In this case, the advantage of not being tied to a specific simulation package is clear when industrial companies make corporate decisions to change simulation platforms. This constitutes a major problem in maintaining process models.

Other approaches have been proposed and prototyped. These include the repository of a modeling environment (ROME) system [26], which enables neutral models to be stored in terms of fundamental modeling objects. The ROME system provides import and export capabilities for various application software. It thus provides a means of storing, retrieving and using models over the life cycle. The wider problem of handling model storage across organizations is still largely unaddressed and unsolved. It is often made more difficult by privacy and nondisclosure issues related to specific modeling systems or models that have significant commercial value.

Along with this concept is the idea of exchanging models between different modeling environments such as gPROMS, Modelica or AspenPlus using a model exchange language like CapeML [27]. This provides a neutral exchange mechanism for model use across application packages. This was part of the larger program of CAPE-OPEN and Global CAPE-OPEN [28] that seeks to address mobility and exchange of modeling across diverse application platforms by setting open standard interfaces for computer-aided engineering applications. A number of major software vendors such as Aspen Technology, PS Enterprise, Fantoft Process and academic members support the Global CAPE-OPEN initiative.

2.4.3
Data Representation and Use

Data is at the heart of modeling and data representation is crucial to effective model development and use. In this context, significant work has been done to address the structuring of data across the design cycle. Development of the conceptual life-cycle model (CLiP) [29-31], has provided a framework in chemical engineering for design and model development and reuse.

The CLiP development, in theory, covers sociotechnical systems but is yet to be expanded and developed to a point where it can adequately cover the range of industrial modeling activities common to major industrial developments. It includes some concepts covering technical, social and material systems that act as upper level metamodels to instances of these classes.

One current development is the extension of the CLiP data model into an ontological representation called OntoCAPE [32], which seeks to put these concepts in the form of an ontology which can be used for reasoning about the domains covered by the concepts. This provides the possibility of building intelligent software agent systems that can help practitioners perform modeling and design tasks.

Data exchange in the area of process engineering has also been of major concern leading to such initiatives as the Process Data Exchange Institute (pDXi) which was initiated in 1989 by the American Institute of Chemical Engineers and numerous organizations within the US and Europe. It is however difficult to assess the actual usage of the standard.

More recently, initiatives such as the Standard for Exchange of Product Model Data (STEP) within the industrial automation and integration standard, ISO 10303 will provide extensive specifications for chemical engineering related equipment, processes assembly and design [34].

2.4.4
Documentation

Documentation in a corporation is a major challenge, given the enormous amounts of reports, figures, drawings, memos, letters, consulting documents and the like that are generated throughout the process or product life cycle. A number of large commercial systems exist to address this issue, such as Documentum [35], that provide enterprise content management (ECM). Recent developments provide collaborative workspaces that give facilities to share ideas and information within the corporation and beyond. Challenges still exist in being able to effectively link important documents to other technical systems such as hazard and risk registers or plant level systems, so that documents can be retrieved in a timely fashion for decision making purposes.

Acknowledgements

The authors would like to acknowledge the help given by Mr. Jim Schmidt, managing director (SPPD) and to Southern Pacific Petroleum (Development) for permission to quote details of the shale oil process referred to within this chapter. We also acknowledge permission from Unilever and CUB Limited to quote on warehouse and utility applications.

References

1 British Standards 2004
   http://bsonline.techindex.co.uk
2 *ICCA*, 2004, International Council of Chemical Associations, www.icca-chem.org/
3 *Rosselot K. S. Allen D. T. Life Cycle Concepts, Product Stewardship, and Green Engineering,* in D.T. Allen and D. Shonnard (eds.) *Green Engineering: Environmentally Conscious Design of Chemical Processes, ch. 13,* Prentice Hall PTR, Upper Saddle River, NJ 2002
4 *Minsky M.* Matter, Minds and Models in Semantic Information Processing, MIT Press, Boston, USA 1968
5 *Marquardt W. von Wedel L. Bayer B.* Perspectives on Lifecycle Process Modeling, FOCAPD, 5th International Conference on Computer-Aided Process Design, AIChE Symposium Series 323 96 (200) p. 192–214
6 *Virkki-Hatakka T. et al.* Modeling at Different Stages of Process Life Cycle, European Symposium on Computer-Aided Process Engineering (ESCAPE)-13, Elsevier Science, Amsterdam (2003) pp. 977–982

7 *Foss B. A. Lohmann B. Marquardt W.* A Field Study of the Industrial Modeling Process, J. Process Control , 8(5–6) (1988) p. 325–338
8 *Hangos K.M. Cameron I. T.* Process Modeling and Model Analysis, Academic Press, London 2001
9 *Lind M.* Modeling Goals and Functions of Complex Industrial Plant, J. Appl. Artificial Intell. 8 (1994) p. 259–283
10 *Modarres M. Cheon S.* Function Centered Modeling of Engineering Systems Using the Goal Tree Success Tree Technique and Functional Primitives, Reliab. Eng. Syst. Safe. 64 (199) p. 181–200
11 *Cameron I. T. Fraga E. S. Bogle I. D. L.* Goal Set Development and Evolution in the Conceptualization of Process Models, CAPE Centre, University of Queensland, Internal Report 2004
12 *Rittel H. Kunz W.* Issues as Elements of Information Systems, Tech Report Working Paper 131, Institute of Urban and Regional Development, University of California, Berkeley, USA 1970

13 *Bañares-Alcantará R. Lababidi H. M. S.* Design Support Systems for Process Engineering, Comput. Chem. Eng. 19 (1995) p. 267–301

14 *Bañares-Alcantará R. King J. M. P.* Design Support Systems for Process Engineering, Comput. Chem. Eng. 21 (1997) p. 263–276

15 *SPP-CPM* Stuart Oil Shale Project, Draft Environmental Impact Statement, Southern Pacific Petroleum (Development) Pty Ltd, Sinclair Knight Merz, Australia 1999

16 *Litster J. D. Bell P. R. F. Newell R. B. White E. T.* The Role of Kinetics in Oil Shale Retorting, in Proceedings of CHEMECA 82, Sydney, Australia, August 1982, pp. 35–39

17 *Do D. D. Litster J. D. Peshkoff E. Newell R. B. Bell P. R. F.* Pyrolysis of Queensland Oil Shale in a Fluidized Bed: Modeling and Experimental Studies, in Proceedings 1st Aust. Workshop on Oil Shale, Lucas Heights, Australia, May 1983, pp. 131–134

18 *Litster J. D. Rogers M. J. Newell R. B.* Modeling Fluid Bed Drying and Retorting of Rundle Oil Shale, in Proceedings of 2nd Australian Workshop on Oil Shale, Brisbane, Australia, December 1984, pp. 206–211

19 *Fincane D. George J. H. Harris H. G.* Perturbation Analysis of Second Order Effects in Kinetics of Oil Shale Pyrolysis, Fuel 56(1) (1977) p. 65–69

20 *United States Environmental Protection Agency (USEPA)* www.weblakes.com/lakeepa1.html 2004

21 *Connell L. D. Bell P. R.* Modeling Moisture Movement in Revegetating Waste Heaps :I Finite Element Model for Liquid and Vapor Transport Water Resources Res. 29(5) (1993) p. 1435–1443

22 *Connell L. D. Bell P. R. Haverkamp R.* Modeling Moisture Movement in Revegetating Waste Heaps: II Application to Oil Shale Wastes, Water Resources Res. 29(5) (1993) p. 1445–1455

23 *Syamsiah S. Krol A. Sly L. Bell P. R. B.* Adsorption and Microbial-Degradation of Organic Compounds in Oil-Shale Retort Water Fuel 72(6) (1993) p. 855–861

24 *Williams R. P. B. Keays R. McGahey S. Cameron I. T. Hangos K. M.* SCHEMA: an Object Oriented Modeling Language for Continuous and Hybrid Process Models, Asia Pacific Conference on Chemical Engineering (APC-ChE), Paper #922, 2002, Christchurch, New Zealand 2002

25 *McGahey S. Cameron I. T.* Transformations in Model Families, ESCAPE12, Comp. Chem. Eng., The Hague, The Netherlands, 26–29 May 2002

26 *Von Wedel L. Marquardt W.* ROME: A Repository to Support the Integration of Models over the Lifecycle of Model-Based Engineering Processes, in S. Pierucci (ed.) European Symposium on Computer-Aided Process Engineering, 10, Elsevier, Amsterdam, (2000) pp. 535–540

27 *Von Wedel L.* CapeML: A Model Exchange Language for Chemical Process Modeling Tech Report LPT-2002-16 Lehrstuhl für Prozesstechnik, RWTH Aachen University, Germany 2002

28 http://zann.informatik.rwth-aachen.de:8080/opencms/opencms/COLANgamma/index.html

29 *Bayer B.* Conceptual Information Modeling for Computer-Aided Support of Chemical Process Design, Dissertation Nr. 787, Lehrstuhl für Prozesstechnik, RWTH Aachen University, Germany 2003

30 *Bayer B. Krobb C. Marquardt W.* Data Model for Design Data in Chemical Engineering – Information Models, Tech Report LPT-2001-15, Lehrstuhl für Prozesstechnik, RWTH Aachen University, Germany 2002

31 *Schneider R. Marquardt W.* Information Technology Support in the Chemical Process Design Lifecycle, Chem. Eng. Sci. 57(10) (2002) p. 1763–1792

32 *Yang A. et al.* Principles and Informal Specification of OntoCAPE: COGENTS Project, Information Society Technologies (IST) Programme, IST-2001-34431, European Union 2003

33 *Bogusch R. Lohmann B. Marquardt W.* Computer-Aided Process Modeling with ModKit, Technical Report #8, RWTH Aachen University of Technology 1996

34 *International Standards Organization (ISO)* www.iso.org 2004

35 *Documentum Inc.* www.documentum.com 2004

3

Integration in Supply Chain Management

Luis Puigjaner and Antonio Espuña

3.1
Introduction

An introductory chapter (Section Three, Chapter 7) on the supply chain (SC) network has already presented the elementary principles and systematic methods of supply chain modeling and optimization. Here, the need for and integrated management of the SC is further emphasized and challenging solutions are presented. As seen, supply chain management (SCM) comprises the entire range of activities related to the exchange of information and materials between costumers and suppliers involved in the execution of product and/or service orders in an extremely dynamic environment (Fig. 3.1).

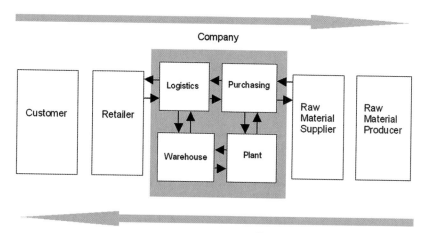

Figure 3.1   Flow of supply chain management information and materials

Computer Aided Process and Product Engineering. Edited by Luis Puigjaner and Georges Heyen
Copyright © 2006 WILEY-VCH Verlag GmbH & Co. KGaA, Weinheim
ISBN: 3-527-30804-0

A successful management of the supply chain management requires direct visibility of the global results of a planning decision in order to include this global perspective. This requires significant integration across multiple dimensions of the planning problem for nonconventional manufacturing networks and multisite facilities over their entire supply chain [1]. Objectives such as resources management, minimum environmental impact, financial issues, robust operation and high responsiveness to continuous needs must be simultaneously considered along with a number of operating and design constraints [2, 3].

Almost all the currently available SCM software suffers from several of the following demerits: product availability focus; reactive rather than proactive; long lead times; uncertainty treatment throughout; lack of flexibility in systems; performance measured functionally; poorly defined management process; no real partnership; insufficient performance measurement [4]. To overcome the above deficiencies and respond better to industrial demands facing a more dynamic environment (greater uncertainty of demand, shorter product life cycle, financial and environmental issues, fewer warehouses, new cost/service balance, globalization, channel integration and so on), there is a need to explore new strategies for supply chain management.

Moreover, an integrated solution is required for the next generation of SCM given the number and complex interactions present among supply chain main components in the global supply chain (Fig. 3.2).

This new scenario requires departing from current approaches that consider Supply chain optimization in a static way. Production-distribution-inventory systems that are now being used in manufacturing companies are static information systems. Periodically (typically weekly or monthly), all data (demand forecasts, available machine capacity, current inventory levels, desired inventory levels, etc.) are collected and fed into a huge optimization system (typically some type of linear programming system). After hours of computation a company-wide plan is obtained: this plan rep-

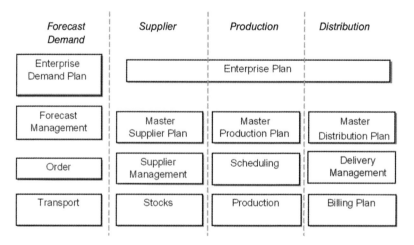

Figure 3.2 Current supply chain planning components

resents next week's (or month's) production schedule, the inventory levels at the various facilities and the distribution of the final goods. However, due to demand fluctuations or other unpredictable situations, the plan does not exactly meet the company's best course of action in length of changing conditions and data. So, production and plant managers adjust the plan to the best of their ability, given the frequent lack of data and their inability to compute the "best" course of action. Real-time adjustments in view of changing data are made manually and *ad hoc*. This situation is largely recognized and manifested in recent specialized forums of debate [5].

Therefore, new SCM solutions should be provided with the following main characteristics: interoperability, scalability, with open and flexible infrastructure; Web-oriented interface; autonomous, capable of self-organization and reconfiguration, coordination and negotiation, with optimization and learning mechanisms, so to evolve and adapt to the dynamic market environment; adaptive process modeling, rapid prototyping; involving production planning and scheduling; capable of making forecasts accurately and to incorporate e-commerce and w-commerce.

In this section, an open, modular-integrated solution is discussed that implements real-time supply chain optimization. The technology involved uses a network of cooperative and auto-associative software agents (smart agents) that constitute a decision support system for managing the whole supply chain in a real-time environment.

This chapter is organized as follows: First, a review of recent approaches to SCM integration is presented and the requirements for next generation of SCM integrated solution are identified. Then, a specific environment is presented that encompasses the SCM characteristics identified above. Finally, the integration of negotiation, environmental, forecasting and financial decisions is reported as an example of the new technology that may lead to better, fully integrated, easier to use and more comprehensive tools for SCM. A brief description of the architecture and functionalities of the solution implemented is also given.

3.2
Current State of Supply Chain Management Integration

Supply chain management (SCM) has been extensively studied in recent years. Lee and Billington [7] consider inventory handling as the central part of supply chain management integration. This work considers three areas of potential conflict:

- problems related to information and management of the SC;
- operational related problems;
- strategic and design problems.

Geoffrion and Powers [8] studied design aspects of the distribution network focusing on the storage location problem. The integration of production and distribution in the SC is examined in the work of Erengunc et al. [9]. The authors point out the necessary tradeoff between flexibility and quality on the one hand and product cost on the other in the SCM.

As the level of integration in supply chain management increases, the complexity of the resulting model limits some approaches to contemplate small academic examples. Some representative methodologies to represent the supply chain are summarized.

3.2.1
Methods Based on Deterministic Analytical Models

Heuristics methods were used by Williams [10] for planning and distribution of supply chain networks. The objective is to find the optimum production plan that satisfies the demand at minimum cost. Different models are used and compared with dynamic programming. In a later work, Williams [11] developed a dynamic programming model to determine simultaneously production and distribution lot sizing for each echelon in the chain. Inventory and operation costs associated to each mode of the chain are minimized.

A deterministic model was used by Ischii [12] to obtain the inventory levels and dead times associated to the solution of an integrated supply chain considering finite horizon and linear demand.

Cohen and Lee [13] presented a mathematical programming formulation (mixed-integer nonlinear programming) that minimizes net profit of manufacturing and distribution centers under management constraints (production resources) and logistics constraints (flexibility, availability and demand limitations). This work was later extended [14] to minimize fixed and transport costs along the chain subject to supply, capacity, assignment, demand and raw material constraints.

A combination of mathematical programming and heuristic models is used by Newhart et al. [15] to minimize the number of products in inventory through the network. A second step investigates the minimum inventory required to absorb demand delays and fluctuations.

Arntzen et al. [16] develop a "global" model for the supply chain (GSCM) that implies a formulation of the type mixed-integer linear programming to determine: (1) the number and location of distribution centers, (2) client assignment to distribution centers, (3) number of echelons in the SC, and (4) product assignment to production plants. The objective is to minimize the weighted sum of total costs (production, inventory, transport and other fixed costs) and active days.

The efficiency and response capacity of the SC can be improved by increasing its flexibility [17]. Here, flexibility is measured as the sum of the instantaneous difference between capacity and utilization of two types of resources: inventory and capacity resources. Given the necessary resources for manufacturing each product and bill of materials (BOM) information, the transport plan and delivery of each product is obtained and optimum inventory levels are achieved.

Camm et al. [18] developed an integer mathematical programming (IP) model to determine the best location for distribution centers of Procter and Gamble and its proper assignment to clients grouped by zones. The model developed uses a simple transport model and the allocation problem does not consider capacity constrains.

More recently, the design of multiechelon supply chain networks under demand uncertainty has been modeled mathematically as a mixed-integer linear programming optimization problem [19]. The decisions to the determined include the number, location and capacity of warehouses and distribution centers to be set up, the transportation links that need to be established in the network and the flows and production rates of materials. The objective is the minimization of the total annualized cost of the network, taking into account both infrastructure and operating costs.

3.2.2
Methods Based on Stochastic Models

A detailed presentation of uncertainty sources present in the supply chain that affects its operation performance can be found in Davis [20]. In the same work a method is developed to treat the uncertainty associated to the supply chain of Hewlett-Packard. According to Davis [20], three different sources of uncertainty can be found in the SC: (1) suppliers, (2) production, and (3) clients. Chain suppliers are characterized by their performance and their response can be predicted. Uncertainty problems related to production can be solved by reliability analysis maintenance techniques. Finally, client demand uncertainty can be dealt with specialized forecasting methods.

Stochastic models incorporate uncertain aspects of the supply chain and focus on certain parameters relative to its operation. For instance, in the work of Cohen et al. [21] a model is developed to establish a materials supply policy for each echelon in the SC. Four submodels are developed based on different costs for the control of materials, production, finished products storage and distribution. There are two probability distributions which are determined by the SC interactions, namely the materials' demand in manufacturing plants and clients demand in distribution centers.

Svoronos and Zipkin [22] consider a multiechelon SC distribution and estimate the average inventory level and the number of unfilled orders for a given base level of inventory. With these approximations, the authors build an optimization model to determine the inventory base level that implies minimum cost.

A mathematical programming model for three echelons SC (one product, one factory, one distribution center and a retailer) is developed by Pyke and Cohen [23]. The model minimizes the total cost subject to a service level constraint. A later work by Pyke and Cohen [24] considers the same network but with multiple types of products.

Lee et al. [25] present a mathematical model that describes the "bullwhip" effect (variance distortion along SC upstream). Although it is often impossible to know exactly the probability distribution functions of products demand, it will always be possible to specify a set of demand scenarios with high probability of occurrence. Scenario-based planning permits the capture of uncertainly by defining a number of possible future scenarios [26]. Thus, the objective consists of finding solutions that behave satisfactorily under all scenarios. Mobasheri et al. [27] describe a number of scenarios as possible states from the actual state. The authors claim that this is avoided forecasting, which is less reliable. The same approach is used to formulate and solve operational problems, like environmental impact along the chain [28, 29].

The midterm supply chain planning under demand uncertainly is addressed in a recent work with the objective to safeguard against inventory depletion at the production sites and excessive shortage for the customer [30]. A chance constraint programming approach in conjunction with a two-stage stochastic programming methodology is utilized for capturing the tradeoff between customer demand satisfaction and production costs.

The design of multiproduct, multiechelon supply chain networks under demand uncertainty is considered by Tsiakis et al. [31]. Compared to previous models, the model integrates three distinct echelons of the supply chain within a single, mathematical-based formulation. Moreover, it takes into account the complexity introduced by the multiproduct nature of the production facilities, the economics of scale in transportation and the uncertainty inherent in the product demands.

In a recent work [32], an optimization model is developed for the supply chain of a petrochemical company operating under uncertain demand and economic conditions. The proposed two-stage stochastic model succeeds in determining the optimum production volumes that maximize the volumes of products shipped for each scenario. However, the need for further investigation to study the dynamics of the petrochemical supply is recognized.

A novel approach to increase the supply chain competitiveness has been presented very recently [33]. The proposed strategy helps to coordinate the production/distribution tasks of the orders embedded in a SC by integrating the plant production scheduling with the transport scheduling to final markets. Uncertainty in the demand is considered and the problem is formulated as a two-stage stochastic optimization approach. The mathematical model looks for the detailed global schedule (production and transports) that maximizes the expected benefits.

3.2.3
Methods Based on Economic Models

The current trend of advanced planning and scheduling tools (APS) is to incorporate tools to model and change the current position of the financial and process managers during complex interconnected decision making in chemical process industries. Cash management models were considered in the supply chain following basically two stochastic approaches. Baumol's model [34] had an inventory approach assuming certainty. Cash was treated similarly as holding inventory and payments were assumed at a constant rate. On the contrary, the Miller and Orr cash management model [35] was based on the fact that perfect forecasts of cash were virtually impossible because the tuning of inflows depend on payments to customers. In consequence, lower and upper bounds of cash were calculated to create a safety stock. In an attempt to model the client-provider behavior, Christy and Grout [36] develop a supply chain economic model based on games theory. The integration of budgeting models into scheduling and planning models is also considered in a recent work [37]. A cash flow and budgeting model is coupled with an advance scheduling and planning procedure within the decision-making process for increased revenues across the supply chain.

A further step in integrating levels of decision making in the SC is contemplated in the work of Badell et al. [38]. This work considers business decisions and their impact in the SCM. It addresses the implementation of financial cross functional links with the SC operation and investment activities at the factory level when scheduling and budgeting in short term planning in batch processing industries. The target is to obtain tradeoff solutions preserving at most the profit and liquidity while satisfying customers.

3.2.4
Methods Based on Simulation Models

The fast development of new products and the increasing competitiveness of market agents have turned the SC system into a rapidly changing environment. Therefore, it becomes necessary to capture and characterize the dynamic behavior of enterprise systems and develop systematic procedures for decision-making support under these circumstances.

Although the interest in integrated dynamic approaches for SCM is recent, some studies were clearly reported during the last four decades. Forrester [39] performs dynamic analysis and simulation of industrial systems by means of discrete dynamic mass balances and linear and nonlinear delays in the distribution channels and manufacturing sites. Although this work contemplated small academic examples, it permitted the identification of the aforementioned demand amplification problem. Later, Towhill [40] reported some effects to control SCs based on a transfer function analysis and classical feed-forward control.

A changing environment is contemplated in the work of Back et al. [41]. They propose a strategy to cater for the dynamics of the environment and the disturbances, as well as for the dynamics of operations of the business. The same approach is used in the work of Perea-López et al. [42], but taking a step forward by considering a consumer-driven operation. They analyze the impact of heuristic control laws on the performance of the SC integrated by multiproduct multistage distribution networks and manufacturing sites. A more recent model, model predictive control (MPC), applied to the supply chain problem was reported by Brown et al. [43]. A comparison between these two control strategies can be found in Mele et al. [44].

A different approach is presented in the work of Mele et al. [45]. Here, a dynamic approach for SCM based on the development of a discrete event-driven system model of the SC contemplating several entities is reported. The interaction between these entities is explored through simulation techniques. The results obtained provide information about the tradeoff found in real systems and give valuable insight into SC dynamics. Thus, the proposed framework becomes a useful tool for decision-making support in real scenarios.

Present analytic approaches to decision making have severe limitations when dealing with the amount of computations, probabilities and nonanalytic knowledge. Thus, there is an increasing interest in decision theory with artificial intelligence tools. It is used to address important tasks such as planning, diagnosis, learning, and serves as the basis for the new generation of "intelligent" software known as normative systems. An emerging area is the utilization of multiagent systems, since decision making is the central task of artificial agents [46].

This chapter focuses on a multiagent viewpoint for SCM and design. The proposed approach is to consider each possible configuration and action advice as an independent agent provided with autonomous, interactive, cooperative, adaptive and proactive capabilities. This approach permits new levels of integration and additional functionalities in SCM (environmental issues, human factors, financial decisions) as it will be unveiled in the next sections.

3.3
Agent-based Supply Chain Management Systems

Since SCM is essentially concerned with coherence among multiple, globally-distributed decision makers, a multiagent modeling framework based on explicit communication between constituent agents (such as manufacturers, suppliers retailers, and customers) seems very attractive. Agents can help to transform closed trading partner networks into open markets and extend such applications as production, distribution, and inventory management functions across the entire SC, spanning diverse organizations [46].

Agents are autonomous pieces of software that are designed to handle very specific tasks. In the case of SCs, where one has to deal with thousands of products, numerous requirements in production quality control, and many types of interactions, no single agent can be designed to handle this overall task; therefore, we will have to design multiple specialized agents to guide the SC in its entirety. Multiagent systems may be regarded as a group of agents, interacting with one another to collectively achieve their goals. By drawing on other agents' knowledge and capabilities, agents can overcome their own limits of intelligence. Otherwise, knowledge is distributed among several intelligent entities, the agents.

Autonomous agents and multiagents represent a new way to designing, analyzing and implementing complex software systems [47]. A multiagent system uses cooperative agents towards a common goal. The agent is informed about the environment and/or can act on it. The agent has control of its own actions and internal state in a very flexible way, interacting when appropriate with other agents. Therefore, a multiagent system can emulate the behavior of distributed systems – like real world distributed supply chains – at a logical level, thus providing a resource for control of the real physical distributed systems.

3.3.1
Multiagent System Architecture

A multiagent system (MASC) built on an open, distributed, flexible, collaborative, and self-organizing architecture has been recently proposed for SCM [48]. Retailers, warehouses, plants and raw material suppliers are modeled as a flexible network of cooperative agents, each performing one or more SC functions following a client-server paradigm in an object-oriented fashion.

An agent must use all its knowledge about the SC to determine the most convenient values of each attribute that must be negotiated. According to the resulting set of interests and capabilities of the network, the agent translates the value of each attribute into its value of satisfaction. Since it may be expected that not all attributes are equally important for the agent, each attribute has a different weight according to the agent's scale of priorities. The total satisfaction that an agent obtains from a set of attributes is calculated taking into account the satisfaction given by each attribute and their respective weights. This final function, the utility function, gives an abstract value of the offers and counter-offers generated by both supplier and customer. Since objects of negotiation may be interdependent, the tradeoff between each pair of attributes considered is defined in a compensation matrix. This matrix is the element that enables the negotiation of all attributes at the same time, making the whole process faster and more similar to a human negotiation.

The steps of the negotiation are: an agent receives a message of his opponent and evaluates how much this offer satisfies his expectations. From this initial vector of satisfaction, the agent generates an improved counter-offer using his compensation matrix. Then the agent can generate several counter-offers with the same utility using its weights. The opponent must do all the same steps after evaluating all the counter-offers and choosing the one with the highest applicable utility. The negotiation process finishes when an agent launches two consecutive offers that are nearly equal. Obviously, different negotiation policies can be considered and compared.

Two types of agents are to be distinguished: the physical agent and management agent (Table 3.1). The physical agent represents the system's physical entities (distribution center, warehouse, plant, etc.) and is capable of simulating the behavior and decision making of the corresponding entity, while information handling between entities is carried out by the management agents. A central agent, which is a management agent, is also responsible for the overall network management optimization.

Table 3.1 Classification of the different entities in the MASC

Entities		
Agents	**Modules**	**Users**
Physical	Forecasting	
Management	Scheduling ...	

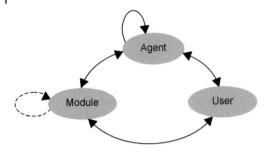

Figure 3.3 Communication types between entities

Module components (forecasting, scheduling, etc.) are information tools that act as servers to specific queries from other entity (agent, module, and user). These three types of entities define five types of communications as seen in Fig. 3.3 for illustration purposes.

The proposed MASC which contemplates a central agent (CA) as management agent offers a flexible architecture allowing representing real SCs with different policies of information control and decentralized decision making. The simplified structure of the MASC can be seen in Fig. 3.4.

The architecture proposed contemplates physical agents at each node of the supply chain network (client, warehouse, factory) while a central management agent communicates with all the other agents. Other management agents are also considered that may be subagents of these already enumerated. This flexible architecture should permit an easy adaptation to represent any real SC with its own level information sharing, from a centralized system to a wholly decentralized one. For instance, in a decentralized case every physical agent (manufacturer A and B, retailer) takes decisions on internal variables and negotiates with other physical agents within its own

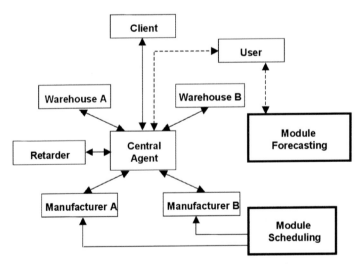

Figure 3.4 Basic multiagent system showing constituent entities (physical and management agents, modules and user)

SC. Here, the central agent plays the role of information handling without decision-making capacity. Otherwise, when a centralized control system is contemplated, the central agent is the only one to have the overall SC information and to make decisions, while every physical agent sends and receives information to and from him. In this case, the central agent is provided with improvement/optimization algorithms.

A fundamental aspect of this architecture is the traceableness required for a proper SC operation. Namely, all transactions carried out through the SC must be clearly registered to facilitate reproduction of the results obtained by simulation of a particular scenario.

The second important element of the framework is the agent modeling. A brief description of different physical and management agents contemplated is given below:

Physical Agents

As mentioned previously, physical agents represent the system's physical entities. Specifically, the following physical agents are considered:

- *Client agent.* The client agent initiates the system's information flow by placing an order to the central agent. Basically the information transmitted is related to the product type, the amount of it, the delivery date and acceptable price. The central agent receives this information and transmits it to potential suppliers (manufacturer/warehouse agents). Following their own internal logic, the potential supplier makes an offer to the central agent which is transmitted to the client. The client internal logic (amount of product, delivery, date and price) negotiates the selected supplier through the central agent. Final confirmation of the order by the supplier is received though the CA (Fig. 3.5).
- *Warehouse agent.* This agent models the material-handling of different kinds (raw materials, intermediates, final products) to be distributed through the SC. Clients of the warehouse may be manufacturers, other warehouses and product end-users. This agent mechanism has already been described (client agent), The internal logic for making an offer obeys to the following issues: (1) delivery date depending on distance, transport and preparation time; (2) available amount of

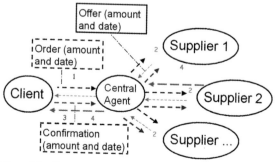

Figure 3.5 Client-system interaction

product that will depend on the stock and eventual delivery time, and (3) the product selling price which is dependent on production and storage cost plus warehouse expected profit. The warehouse behaves similarly to the client. However, complex inventory policies must be modeled that have to include demand uncertainty. Moreover, since the inventory control policy will influence the cost of downstream echelons in the SC, optimum negotiation at this point become very essential.

- *Manufacturer agent.* This agent models the actual manufacturing facility behavior in the SC producing intermediate and/or final products. It receives orders from clients/warehouses and makes an offer in terms of products amount, due date and price, which is calculated by using a production scheduling module. Manufacturers also operate like clients regarding raw materials supply. Planning tools are used to determine the amount of raw material needed. Production scheduling models will have into account the process type (continuous, batch, hybrid). Information provided by these models will be used at upper levels of manufacturing decision making (MRP, financial modules).

Management Agents

This type of agents does not represent a physical entity of the SC, but it rather simulates the SC operation in these aspects that are not necessary related to a physical entity of the SC. They could optimize the overall performance of the SC by modeling specific parameters associated to the other agents.

- *Central agent.* Coordination of the agent's network is achieved by means of the central agent. It essentially supervises and analyses the information flow between the other agents. As a result of information analysis, they may also have an active role by modifying adequate parameters looking for the optimum overall performance of the SC.
- *Other agents.* The architecture envisaged contemplates the existence of subagents operating within the agents already described. Namely, a manufacturer/warehouse agent contains coordinated subagents to simulate its real behavior. Typical examples of subagents are the sales/buy agents that simulate the corresponding department in a factory. These subagents negotiate transactions between physical agents (client, warehouse, manufacturer agents) and perceive their consequences. They follow the client-supplier logic described before. The architecture developed may consider the "multiowner" case where one SC competes with other SC. In that case, each SC has a partial view of the whole situation and therefore cannot manipulate variables belonging to a SC of a different "owner". Thus, negotiation between SCs is necessary; being the central agent of each SC the responsible for interchain negotiations mechanism to reach an agreement on multiple conditions (Fig. 3.6). Management agents are adaptive. Namely, they are able to learn from the operation of the SC. Otherwise, they must be provided with appropriate tools (modules) for internal optimization (vertical integration) and external optimum negotiation (horizontal integration).

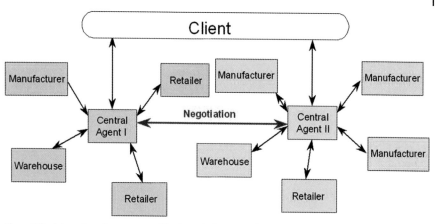

Figure 3.6 Interactions between two SCs through their central agents

Modules

Modules are not considered agents strictly speaking, but software tools needed to realize certain functionalities within the multiagent system. For instance, the warehouse physical agent model of inventory control may require demand forecasting tools to estimate the amount of future supply. Therefore, the warehouse agent will have to interact with the forecasting module to achieve proper inventory control.

Available modules are: forecasting, negotiation, planning and scheduling, financial, optimization (multiobjective), environmental and diagnosis. Other plug-in modules can be added in the future to contemplate further functionalities. The next sections focus on three of them (environmental, financial, negotiation) that provide a challenging insight into the level of integration achieved for the whole SC performance optimization.

3.4
Environmental Module

Environmental considerations in the SC are necessary because industrial products most often reach the client through a variety of steps that are subject to strict environmental regulations. Moreover, these requirements migrate upwards through the SC and create a need for flow of environmental information. An adequate methodology to systematize this information and provide a vehicle for environmental impact minimization is the life-cycle assessment (LCA).

The LCA approach has been adapted for the SC environmental assessment and improvement in this module [49]. The methodology used is summarized next.

Let us consider the elementary SC shown in Fig. 3.7. It contains all the basic constituents of a generic SC in a simplified way, but although simple permits the representation of a variety of scenarios. The assumptions made in this base case study will

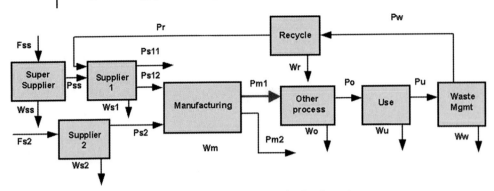

Figure 3.7 Elementary supply chain representation utilized in the environmental module base case study

give an insight into the characteristics of the model contained in the environmental module.

It should be observed that a high degree of aggregation is assumed in this SC representation so that energy and material streams are reduced to a minimum. This assumption implies that this SC representation is the result of a detailed modeling at the individual agent in the network, for instance, the manufacturer agent.

Now let us consider Pm1 as the selected product of interest for LCA evaluation. This product leaves the factory as seen in Fig. 3.7. Then, the purpose of this study is to obtain an eco-label for this product in terms of environmental burden emissions associated to it following the LCA methodology guidelines (goal and scope, inventory analysis, impact assessment, integration phases) applied to the SC system.

The first phase of LCA identifies the functional unit (product or process). This "functionality" can always be expressed as an equivalent product amount (in kg or MJ, according to the nature of the product) that will facilitate later calculations. The system boundaries are indicated in Fig. 3.7 (although in some cases it will be necessary to go deeper inside these boundaries to perform internal calculations to find environmental values for the global streams across the boundaries). Next, the inventory phase of manufacturing (the source block of product Pm1) takes place, where the input streams Ps12 and Ps2 data, the emissions represented by Wm and data related to the products Pm1 and Pm2 are tabulated. Additionally, inputs and emissions downstream the manufacturing agent (Wo, Wu and Ww) are also considered.

A key issue in inventory calculation is to establish the allocation policy for environment load associated to each product in each SC echelon. If causal relationship between inputs, outputs and emissions are known with certainty, inventory calculation can be easily done without need of an allocation procedure. Otherwise, the following general expression can be used for allocation at each SC echelon:

$$P_k \cdot v_k = f_k \left(-W \cdot v_w + \sum_s F_s \cdot v_s \right) - \sum_{p \neq k} f_{0p} \cdot P_p \cdot v_p \qquad (1)$$

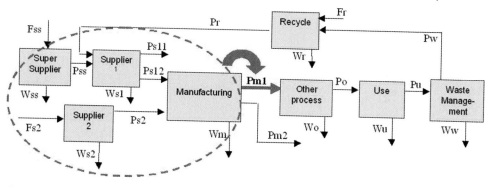

Figure 3.8 Forward allocation

where P_k represents the stream of product k and v_k is the corresponding eco-vector associated to product k. $W \cdot v_w$ is the waste stream weighted by its eco-vector, $F_s \cdot v_s$ is an input stream multiplied by its eco-vector and $P_p \cdot v_p$ is the corresponding output stream weighted by its eco-vector v_p. Finally, f_k and f_{op} are allocation factors depending on the allocation policy (e.g., mass allocation, energy allocation).

Allocation to the chain left side of manufacturing (super supplier, supplier 1, supplier 2) is also analyzed in the same way as for the manufacturing case. This procedure is called forward allocation (Fig. 3.8) because environmental load is carried from left to right, that is in the same direction as the material flow in the SC.

Following the LCA philosophy, the environmental module also considers a backward allocation (Fig. 3.9) that is in the opposite direction to the SC material flow. For instance, the manufacturer is also responsible for the environmental impact generated by this product after manufacturing, that is, along other processes in which it participates, during its use, and finally, during the waste management and treatment of the generated residues.

Recycle processes and streams are treated by considering that the associated environmental load is included into the supplier LCA assessment. The model (Fig. 3.7) cuts the P_w stream, thus P_r and w_r now include the inputs and the emissions for the

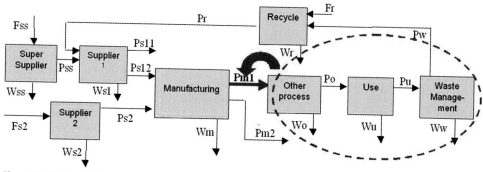

Figure 3.9 Backward allocation

recycling process plus the inputs and emissions for the supplier 1, respectively. The environmental load associated to stream P_w is considered to be zero.

3.4.1
Implementation Considerations

The user asks the system for an eco-label (ecological card). This eco-label can be expressed as a set of environmental loads as well as a set of environmental impacts. The system offers a table to enter data. These data belong to the following categories: inputs, emissions, products and functional unit, all referred to the main production process. This table would correspond to the manufacturing block in Fig. 3.7. Moreover, the system offers another table to introduce inputs and emissions associated to other processes such as those described as backward allocation.

According to the data entered, the system generates additional input data tables to be filled-up by the user with inputs, emissions, products and calculation basis for each new table. These new tables would correspond to the blocks on the left to the manufacturing block in Fig. 3.7.

Next, according to the kind of data entered in the tables, there are two types of calculation procedures. The tables whose inputs are all elementary flows must be calculated first. Otherwise, tables not having some elemental input have to be calculated afterwards. It is important to maintain the correct precedence in such a way that calculations follow the flow sheet from left to right.

Finally, a table that contains all the information entered to the system is built. With the final table, the life cycle assessment calculations are made using data saved in an impact category table and a coefficients impact table. The Unified Modeling Language (UML) representation [50] has been used to build the environmental module. The use case diagram shown in Fig. 3.10 summarizes the functionality of the module.

3.4.2
Industrial Testing

The environmental module has been tested in a real SC associated to automotive parts manufacturing. Specifically, the environmental impact associated to a certain component was evaluated and improvements were proposed to obtain the eco-labeling of the product.

Once selected the functional unit for the product chosen, the inventory analysis was carried out from the information collected on raw material consumption, emissions and product(s). In the impact assessment phase, with the inventory analysis results some impact indexes were calculated for the following categories:

- global warming
- stratospheric ozone depletion

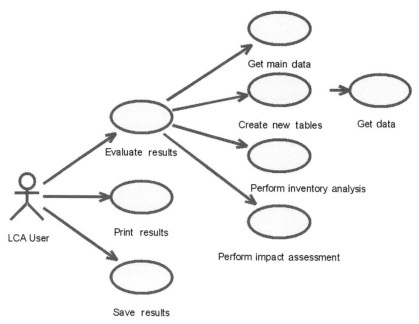

Figure 3.10 Use case diagram showing the functionality of the environmental module

- eco-toxicological impact
- photochemical oxidant formation
- acidification
- eutrophication.

Then allocation of environmental load was carried out satisfactorily backwards and forwards the supply chain. This permitted the maintenance of registers (eco-labeling) for each product. This register may give the manufacturer a substantial increase of penetration in the market. Moreover, the manufacturer can reduce the environmental impact of its products by selecting the "most ecological" supplier and/or modifying the process to reduce emissions. This could be done by incorporating other modules (planning, financial and optimization) in the final decision making.

3.5
Financial Module

The purpose of the financial module (FM) is to bridge the existing gap between supply chain financial decisions and production management by providing a common framework for integrated decision making that permits an optimum cash management thus avoiding "blind" financial/production disaggregated decision making occurring in industrial practice.

The methodologies currently used in production planning/scheduling try to optimize some performance measure without consideration of cash availability. Then, the output solution of the scheduling-planning model tries to fit the finances in an iterative trial and error procedure [51]. This procedure (called sequential procedure) usually incurs in substantial debt and has to pledge receivables (financial transaction of high cost to the manager). Moreover, the lack of synchronization between cash inflows and outflows results in cash balance fluctuations, which can be alleviated by considering the cash flows simultaneously with production decisions.

To achieve integration, the budgeting variables of liabilities and exogenous cash are calculated as a function of production planning and scheduling variables. Namely liabilities at a specific time period are a function of the cost of purchasing raw materials, the cost of materials processing and the cost of having to purchase part of the final product from another supplier or another plant. Exogenous cash flow incurred in every time period is due to the sale of products. The detailed formulation can be found in Refs. [51, 52].

In summary the proposed methodology contemplates:

- production expenses during the week consider an initial stock of raw materials and products,
- an initial working capital is considered,
- a short-tem financing source is represented by a constrained open line of credit,
- production liabilities incurred in every week period due to buy of raw materials,
- exogenous cash-flows due to sale of products,
- a portfolio of marketable securities is also considered.

3.5.1
Financial Module Interaction with the Multiagent System

The financial module constitutes the supporting tool for financial decisions within the multiagent system framework in a coordinated way with other decisions affecting the whole supply chain. This module permits coordination and integration of financial and operational decisions by exploiting the advantages offered by the multiagent system described previously.

The structure proposed for the integration of the financial module with MASC is shown in Fig. 3.11. The FM will be used by the SC central agent to identify the best opportunities for investment and financing the SC, as well as to evaluate the impact of operational decisions in the manufacturer's economy.

The real supply chain is modeled and represented by the multi agent system. The central agent interacts with the FM in order to maximize the net profit for a given budget provided by the budgeting model that uses the specific information given by the scheduling and planning model module for two sets of time periods. The first time period set corresponds to the scheduling and planning period, while the second set goes beyond the end of the planning horizon up to one year budgeting (see Fig. 3.12). It is important to note that the model incorporates a number of subjective constraints to allow for different profiles in financial risk management.

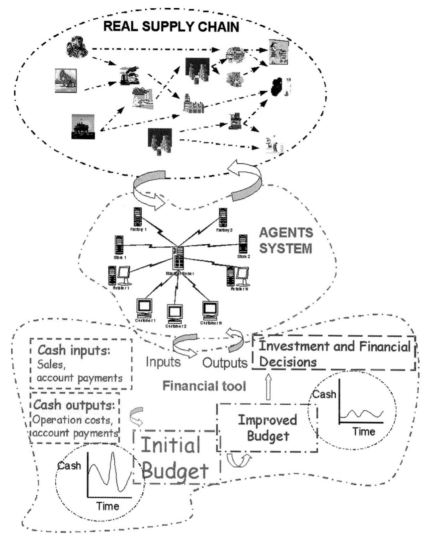

Figure 3.11 Integration of the financial module within the multiagent system

3.5.2
Testing Results in Industrial Scenarios

The benefits obtained by incorporating the financial module in the SC decision making have been assessed at different levels.

First, the use of an integrated model coordinating financial and operational decisions has been tested in a single manufacturer, specifically, a plant producing five different products and two different raw materials. Product switch-over basically

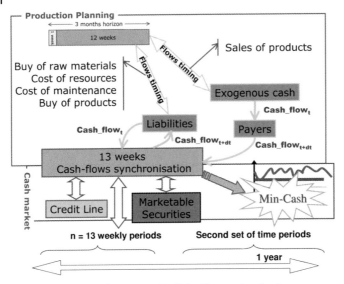

Figure 3.12 Budgeting horizons and its kink with operative planning model

depends on the nature of both substances involved in the precedent and following batch (precedence constraints). Cleaning times are constrained by the product sequence.

Comparative results obtained from the application of a sequential approach and those achieved with the FM are shown in Fig. 3.13. It can be seen that the integrated solution incurs less debt and avoids having to pledge receivables which means a 20 % savings for the firm.

The second case study deals with the SC of a large fruit cooperative made up of raw materials suppliers, manufacturing (fruit selection, clearing, and packaging), warehouses, distribution centers and clients. Here a deeper level of integration in the SC management was contemplated. Since this supply chain is driven by the raw materials arrival rather than by the customer demand, a main objective was to use the forecasting module (FOREST) to estimate the raw material arrival time thus reducing operational uncertainty. A second important objective was to integrate financial and production aspects linked to cash flow management across the supply chain. Both modules were used on-line through the Web, thus achieving high visibility (customer on-line information) and improved service (increased customer satisfaction). The net results were an increase of sales about 15 % and diminution of stocks about 20 % (saving of two million euro per campaign).

a) Sequential approach

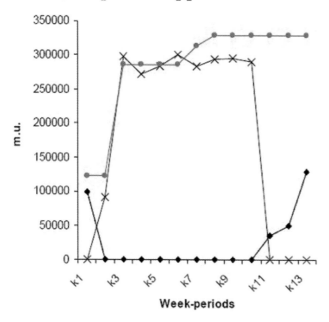

b) Integrated approach

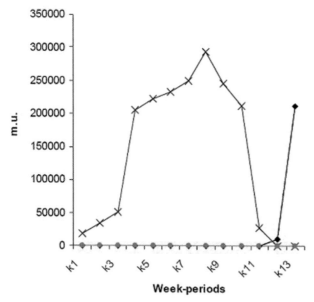

Figure 3.13 Comparative results between sequential and integrated approach: debt incurred, marketable securities and amount accumulated of pledged receivables at every week period (legend: ——×—— Marketable securities; —●— Debt; —◆— Receivables pledged)

3.5.3
Negotiation Module

When adopting an agent-oriented view of computation, it is readily apparent that most problems require or involve multiple agents as indicated before. Moreover, these agents will need to interact with one another, either to achieve their individual objectives or to manage the dependencies that follow from being situated in a common environment. These interactions can vary from simple information interchanges, to requests for particular actions to be performed and on to cooperation (working together to achieve a common objective) and coordination (arranging for related activities to be performed in a coherent manner). However, perhaps the most fundamental and powerful mechanism for managing interagent dependencies at run-time is negotiation – the process by which a group of agents come to a mutually acceptable agreement.

Automated negotiation among autonomous agents is needed when agents have conflicting objectives and desire to cooperate. This typically occurs when agents have competitive claims on scarce resources, not all of which can be simultaneously satisfied. These resources can be commodities, services, time, money, etc. Specifically, the main objective of the negotiation module developed is to enhance profitable partnerships in SCs. This goal is divided into the following steps:

- Identify the most profitable relationships and enhance them.
- Integrate supply contract negotiations into supply chain management.
- Evaluate different negotiation tactics according to the SC performance and partners behaviors.
- Develop learning techniques.

The proposed approach [53] takes into account the tradeoff between the quality of the offers made to customers, i.e., the level of satisfaction perceived by the client, and the expected profit to be achieved in the short term operation of the SC. Therefore, a two-stage stochastic formulation is derived that considers the uncertainty associated with reactions to future demand, in order to compute a set of Pareto optimal solutions to the proposed problem. Each of these solutions comprises an SC schedule and a set of values for the parameters of the offers. Through comparison of the Pareto curve and the solution that would be obtained without negotiation, a set of offers representing contracts that are desirable from the supplier's perspective is obtained. This set of values may be offered by the supplier in order to reach an agreement with the customer during the negotiation procedure. This approach facilitates a rational negotiation, in the sense that it enables the negotiator to simultaneously process much more data related to production and transport plans and customer preferences, thus avoiding having to rely exclusively on the negotiator's beliefs and interests.

3.5.4
Motivating Example

Inspired in a real industrial case, a relatively simple linear SC is considered to illustrate the performance and results of the negotiation module. It entails a batch plant which produces three different products. The manufacturer has a warehouse (W) at which the products are stored when they leave the plant. As the plant has a limited capacity and is imagined to be next to the factory, no transport is necessary. There is also a distribution center (DC) from which customers are served. Three occasional orders are met if the DC has some amount of the product requested in stock, provided that it is not planning to use that stock to satisfy contractual requirements. If an order has only been partially met, there is no penalization but that particular sale will not be able to be carried out at a later point, when the merchandise reaches the DC. Moreover, the possibility of signing a contract with a customer is considered.

The main objective of the proposed example is to observe how most of the activities executed by a SC can be integrated using the proposed negotiation model, rather than to study very complex structures involving many entities. The proposed approach provides a set of Pareto solutions to be used by the decision maker during the negotiation procedure (Fig. 3.14). There is a tradeoff between the expected profit and the quality of an offer sent to a customer, in this case the level of consumer satisfaction (CSat) attained by an offer, as well as the connection between customer relationship management (customer satisfaction) and production activities (schedules). For instance, for the stochastic Pareto solution the difference between the best-case and the worst-case values for CSat = 80 % is approximately 450 m.u. (20%), while the deterministic solution, shows, this difference equal to 1140 m.u. (55 %). Therefore, the stochastic treatment of the negotiation problem minimizes the impact of the uncertain environment by both increasing the expected profit and reducing the variability of the solution compared with the deterministic solution, which makes it very attractive from the perspective of the decision maker.

Indeed more complex storage policies, plant flexibility, number of entities in the SC network and so on can be addressed using the same approach.

In addition, since future predictions related to market behavior cannot be perfectly forecasted, a number of the parameters in the associated scheduling problem, such as product demands and prices, were considered to be uncertain parameters. The two-stage stochastic formula developed has allowed this situation, commonly found in practice, to be handled properly, which thus reduces the impact of the uncertainty on what profit is achieved in short-term planning. The usefulness of signing contracts as a way of reducing uncertainty has also been shown by means of the aforementioned stochastic formulation.

The proposed strategy represents a method for facilitating rational negotiation, in the sense that it enables the negotiator to process a far greater amount of production, transport planning and customer preference data simultaneously and thus prevents beliefs and interests from being relied on exclusively.

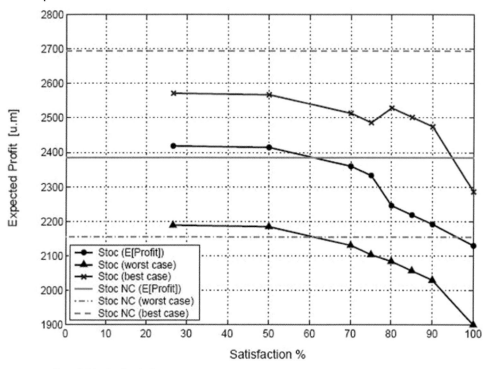

Figure 3.14 Stochastic Pareto curve

3.6
Multiagent Architecture Implementation and Demonstration

The multiagent system framework has been implemented as a Web service. Each agent receives and transmits relevant information for the optimization of the SC. All agents depend on a central agent that coordinates information handling and that may modify a particular agent's decision should it be necessary for the whole SC negotiation optimization. Web services (agents) are programmed in C# using the tools of Visual Studio.NET by Microsoft, while XML under the SOAP protocol is used for communication between them. Each agent may use distributed modules (forecasting, planning and scheduling, optimization, environmental, financial and diagnosis) to support his activities.

3.6.1
Manager Agent System

The manager agent system has a central agent that coordinates relevant information flow from/to the other agents in the network, and will eventually take decisions for the whole SC optimization. This way, the SC behavior can be accommodated to a full range of scenarios from a decentralized operation to a fully centralized management where the director (central) agent has the control on every individual SC activity. In general, the central agent will perform the following functionalities:

- decision support,
- real-time information on SC activities,
- SC performance optimization,
- simulation and performance indicators calculation,
- client/supplier selection,
- graphic representation (Pert, Gantt).

These and other functionalities (forecasting, environmental assessment, financial assessment, SC retrofit, etc.) will require the use of the appropriate module. Figure 3.15 shows the manager agent utility system.

Main performance indicators and expected ratio are indicated on Table 3.2. Obviously, the expected ratio will depend on the SC original situation and the type of SC. The table should be continuously updated once intermediate objectives are achieved.

A key component in the system architecture is the database that must be made available to each agent locally. For instance, the basic design of the database of the central agent is shown in Fig. 3.16 for the specific scenario contemplating retailers of four different items (computers, bakery, books, and furniture).

3.6.2
Graphic Interface

The graphic interface has a double objective. On the one hand a graphic user interface (GUI) is needed for the real client that requires interaction with the multiagent system (specific demand implementation, request of information). Otherwise, a graphic interface is needed for the SC manager. In this regard a client GUI is provided to perform the SC simulation.

The SC client makes the command through the Web application shown in Fig. 3.17, which enables him to communicate with the central agent who consequently

Table 3.2 Indicators range dashboard

Elementary indicator	Ratio expected (min. to max. %)
I1 Modeling time reduction	60–80 %
I2 Forecast accuracy	15–65 %
I3 Inventory reduction	20–85 %
I4 Means of production capacity utilization increase	5–60 %
I5 Cycle process time improvement	15–70 %
I6 Supply chain costs reduction	10–30 %
I7 Delivery performance improvement	15–45 %

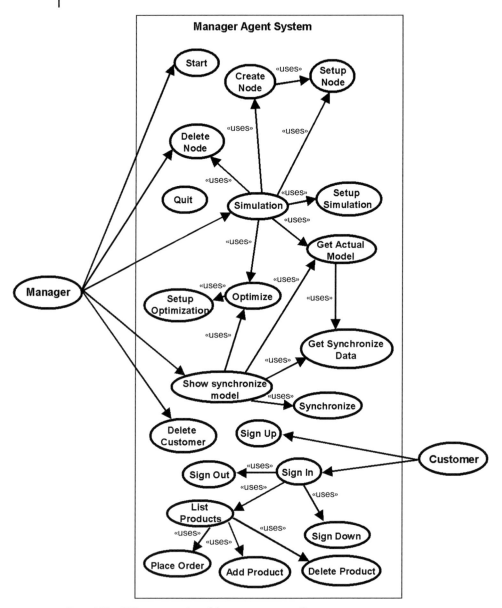

Figure 3.15 UML representation of the manager agent utility system

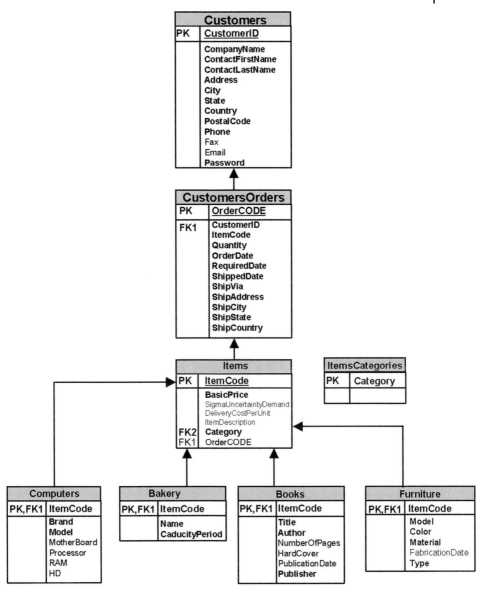

Figure 3.16 Basic design of the database of the central agent for the specific scenario contemplating retailers of four different items (computers, bakery, books, and furniture)

will update the database. Once the client signs up in the Web service and provides the information requested (Fig. 3.18) he has access to the services and information offered (Fig. 3.19).

Figure 3.17   System access Web page

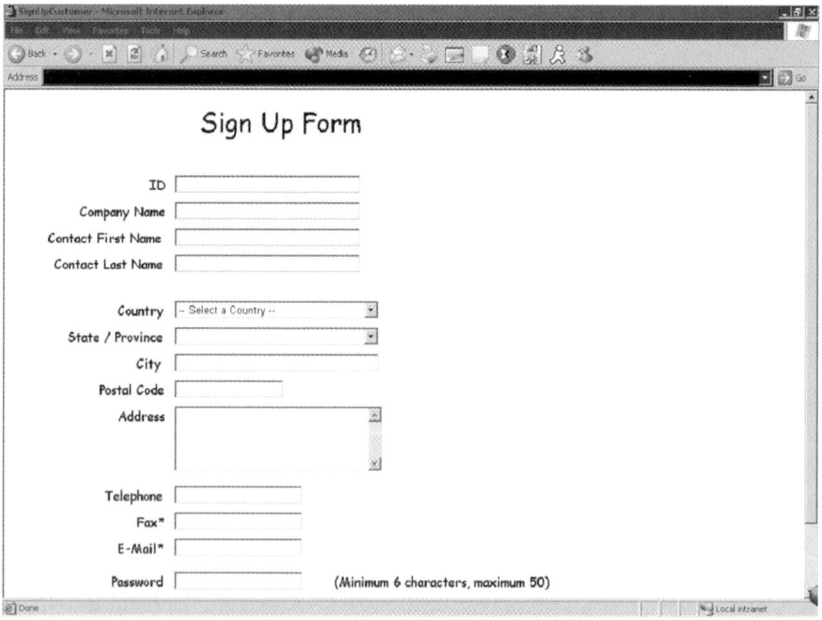

Figure 3.18   Client sign-up form

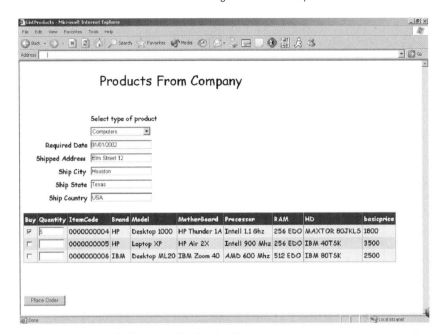

Figure 3.19 Example of information offered to the client

The client agent may be interested to know the real SC system behavior in front of new demand. In this case, a simulation of the real client is realized that analyzes alternative possible scenarios. Therefore, the client agent is provided with:

- *demand generator* supplied with different patterns (stochastic, probability distribution functions),
- *connectivity* with the central agent to receive/transmit messages,
- *graphic user interface*.

The interfaces created for the rest of the agents can be seen in the following figures. The central agent interface is shown in Fig. 3.20. It collects all information from the database (on products, transport, warehouses, factories, environmental impact, financial, etc.) and permits simulation (at the left panel) and optimization (at the right panel) of the SC: Multiobjective optimization is carried out using any of the solvers offered in the center panel. The forecasting module interface is shown in Figure 3.21 giving the demand forecast in terms of dates and amounts of specific products. The environmental module interface appears in Fig. 3.22. Here the left panel permits the input of raw materials needed for each specific product and the right panel has the commands to perform the life cycle analysis of the entire SC. Figure 3.23 shows the financial module interface. At the left initial assets, liability and equity can be introduced, which appear optimized at the right after optimization under the selected constraints at the center (minimum cash, debt, interest rate, etc). Finally, the negotiation agent interface can be observed in Fig. 3.24 for a certain product. It shows the evolution towards satisfaction for both, customer

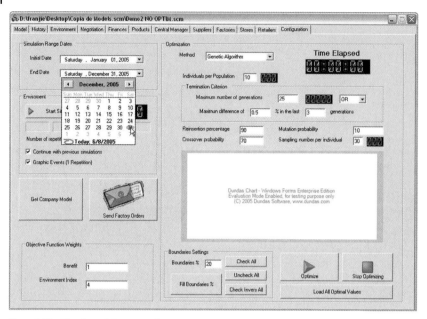

Figure 3.20 The Central Agent interface

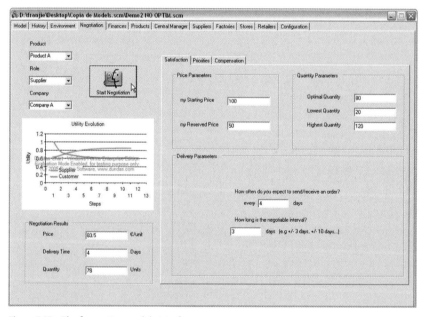

Figure 3.21 The forecasting module interface

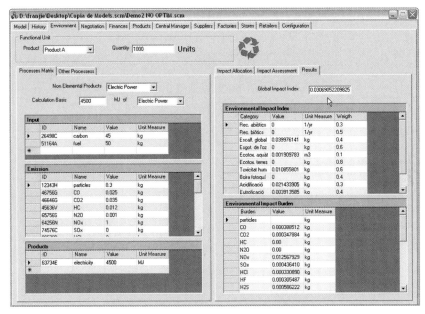

Figure 3.22   The environmental module interface

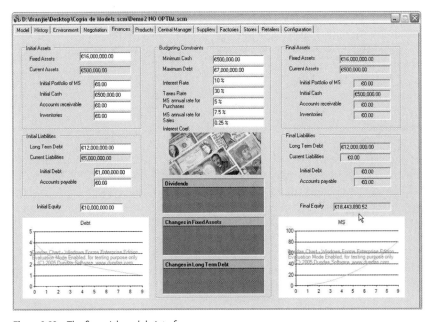

Figure 3.23   The financial module interface

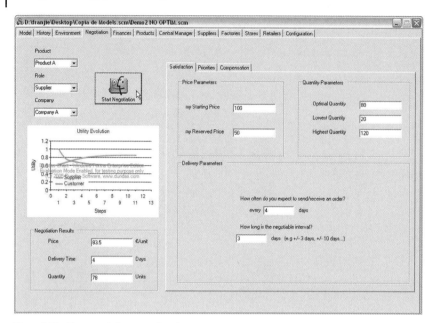

Figure 3.24 The negotiation agent interface

and supplier once the negotiation is initiated in terms of quantity, price and delivery time.

The planning and scheduling module is not shown, since it is fully described elsewhere [54, 55]. The same can be said regarding the real-time monitoring and diagnosis module [56].

3.6.3
Demonstration

The multiagent system described has been tested in industrial scenarios. Some of the scenarios have been partially presented in previous sections of this chapter to show relevant components of the SC system. A global on-line through the Web demonstration took place recently at the CHEM Users Committee held in Lille, France [57]. Here operational tactical and strategic activities were shown to successfully cooperate in the SC, from demand forecasting to diagnosis, control and retrofit considerations.

The demonstration contemplates the whole SC of a cosmetics manufacturing group of enterprises with the following characteristics:

- multiproduct manufacturing plants located in Europe (Oviedo, Tarragona in Spain and one in Italy), United States (California and Florida) and Mexico (Queretaro),
- warehouses for final products,
- distribution centers from which customers are served,
- transport system for distribution to retailers and clients.

The demonstration is initiated by forecasting from historical data the procurement needs of specific products at one of the warehouses in France. A robust demand is obtained using the forecasting module. Then, the following real-time sequence of decisions and activities is carried out by the multiagent system:

- The negotiation module is used to select and agree the most satisfactory provider (the factory situated in Tarragona) for the specific products and in the amounts, prices and due dates envisaged.
- The environmental impact is assessed for the complete life cycle of the products contemplated by means of the environmental module described before.
- Financial evaluation is carried out, taking into account budgeting, cash flow and additional considerations provided by the financial module.
- The central agent collects all the preceding information and requests additional manufacturing data from Tarragona plant. Simulation of the whole SC is then carried out checking for feasibility. Finally, multiobjective optimization (profit, cash flow, debt, environmental impact, due dates) is realized. Optimum values are transmitted over the Web to the plant in Tarragona for manufacturing.
- The production planning and scheduling module calculates the optimal production schedule for the forecasted demand, which is automatically performed in the plant.
- An incidence occurs during plant operation (the reactor heating system breaks down). The monitoring and diagnosis module detects and isolates the fault. An alarm is issued and diagnosed. As a consequence, a rescheduling procedure takes place to find an alternative production route which is implemented again in real-time.
- The operator repairs the reactor which comes back to operation. The monitoring system reacts sending a new plan, which coincides with the original since it was optimum, resuming the plant operation using the repaired reactor.

3.7
Concluding Remarks

The supply chain of a manufacturing enterprise is, nowadays, a world-wide network of suppliers, factories, warehouses, distribution centers and retailers through which raw materials are acquired, transformed and delivered to customers. In this sense, the whole supply chain can be considered as a dynamic virtual enterprise: by the adequate management of its supply chain, the manager can easily find adequate solutions to cope with the dynamics of its production scenario, which includes drastic and unexpected changes in materials or production resources availability, in the market conditions, or even in politics.

This chapter has presented recent advancement on integrated solutions for on-line supply chain management. Specifically, an environment is presented that encompasses the SCM characteristics identified in a preliminary review of the state of the art. It reported the integration of negotiation, environmental, forecasting and financial decisions in a reactive mode as an example of a new technology that may lead to better, fully integrated, easier to use and more comprehensive tool for SCM. A brief description of the architecture and functionalities of the solution implemented has also been presented.

Acknowledgements

The authors wish to acknowledge support of this research work from the European Community (Contract No GIRD-CT-2001-00466), the CICyT-MEC (project No DPI2003-0856), and the CIRIT-Generalitat de Catalunya (project No I-353). Contribution from Fernando Mele, Gonzalo Guillen and Francisco Urbano, predoctoral students of the research group CEPIMA is also much appreciated (Chemical Engineering Department, Universitat Politècnica de Catalunya).

References

1 *Vidal C. J. Goetshalckx M.* Strategic Production-Distribution Models: A Critical Review with Emphasis on Supply Chain Models. Engineering Journal Operational Research 98 (1997) p. 1–18

2 *Applequist G. E. Pekny J. F. Reklaitis G. V.* Economic Risk Management for Design and Planning of Chemical Manufacturing Supply Chains. Computers and Chemical Engineering 24 (2000) p. 2211–2222

3 *Badell M. Romero J. Huertas R. Puigjaner L.* Planning, Scheduling and Budgeting Value-Added Chains. Computers and Chemical Engineering 28 (2004) p. 45–61

4 *Wu J. Cobzaru M. Ulieru M. Norrie D.* SC-Web-CS: Supply Chain Web-Centric Systems. Proceedings of the International Conference on Artificial Intelligence and Soft-Computing, Bant, Canada (2000) pp. 501–507

5 *Grossmann I. E. McDonald C. M.* Foundations of Computer-Aided Process Operations: A View to the Future Integration or R&D Manufacturing and the Global Supply Chain, CACHE Corp., Austin, Texas 2003

6 *Wu J. Ulieru M. Cobzaru M. Norric D.* Agent-Based Supply Chain Management System: State-of-the-Art and Implementation Issues, in Proceedings of European Symposium of Computer-Aided Process Engineering-14 (ESCAPE-14), Lisbon, Portugal, Elseiver, Amsterdam 2004

7 *Lee H. L. Billington C.* Managing Supply Chain Inventory: Pitfalls and Opportunities. Sloan Management Review, Spring (1992) p. 65–73

8 *Geoffrion A. M. Powers R. F.* Facility Location Analysis is Just the Beginning. Interfaces 10 (1980) p. 22–30

9 *Erengunc S. S. Simpson N. C. Vakharia A. J.* Integrated Production/Distribution Planning in Supply Chain: An Invited Review. European Journal of Operation Research 115 (1999) p. 219–236

10 *Williams J. F.* Heuristic Techniques for Simultaneous Scheduling of Production and Distribution in Multi-Echelon Structures: Theory and Empirical Comparisons. Management Science 27 (1981) p. 336–352

11 *Williams J. F.* A Hybrid Algorithm for Simultaneous Scheduling of Production and Distribution in Multi-Echelon Structures. Management Science 29 (1983) p. 77–92

12 *Ischii K. K. Takahashi Muramatsu R.* Integrated Production Inventory and Distribution Systems. International Journal of Production Research 26 (1988) p. 473–482

13 *Cohen M. A. Lee H. L.* Resource Deployment Analysis of Global Manufacturing and Distribution Networks. Journal of Manufacturing and Operations Management 2 (1989) p. 81–104

14 *Cohen M. A. Moon S.* Impact of Production Scale Economizes Manufacturing Complexity and Transportation Costs on Supply Chain Facility Networks. Journal of Manufacturing and Operations Management 3 (1990) p. 35–46

15 *Newhart D. D. Stott K. L. Vasko F. J.* Consolidating Product Sizes to Minimize Inventory Levels for a Multi-Stage Production and Distribution Systems, Journal of the Operational Research Society 44(7) (1993) p. 637–644

16 *Arntzen B. C. Brown G. G. Harrison T. P. Trafton L. L.* Global Supply Chain Management at Digital Equipment Corporation. Interfaces 25 (1995) p. 69–93

17 *Voudouris V. T.* Mathematical Programming Techniques to De-bottleneck the Supply Chain of Fine Chemical Industries. Computers and Chemical Engineering 20(Suppl.) (1996) p. S1269–1275

18 *Camm J. D. Charman F. A. Evans J. R. Sweeney D. J. Wegryn G. W.* Blending ORIMS Judgement and GIS: Restructuring P&G's Supply Chain. Research 27 (1997) p. 120–142

19 *Papageorgiu L. Rotstein G. Shah N.* Strategic Supply Chain Optimization for the Pharmaceutical Industries. Industrial and Engineering Chemistry Research 40 (2001) p. 275

20 *Davis T.* Effective Supply Chain Management. Sloan Management Review Summer (1993) p. 35–46

21 *Cohen M. A. Lee H. L.* Integrated Analysis of Global Manufacturing and Distribution Systems: Models and Methods. Operations Research 36 (1988) p. 216–228

22 *Svoronos A. Zipkin P.* Evaluation of One-for-One Replenishment Policies for Multi-Echelon Inventory Systems. Management Sciences 37 (1991) p. 68–83

23 *Pyke D. F. Cohen M. A.* Performance Characteristics of Stochastic Integrated Production: Distribution Systems. European Journal of Operational Research 68 (1993) p. 23–48

24 *Pyke D. F. Cohen M. A.* Multi-Product Integrated Production-Distribution Systems. European Journal of Operational Research 74(1) (1994) p. 18–49

25 *Lee H. L. Padmanabhan V. Whang S.* Information Distortion in a Supply Chain: The Bullwhip Effect. Management Science 43 (1997) p. 546–558

26 *Owen S. H. Daskin M. S.* Strategic Facility Location: A Review. European Journal of Operation and Research 111 (1998) p. 423–447

27 *Mobasheri F. Orren L. H. Sioshansi F. P.* Scenario Planning at Southern California Edison. Interfaces 19 (1984) p. 31–44

28 *Mulvey J. M.* Generation Scenarios for the Towers: Instrument System. Interfaces 26 (1996) p. 1–15

29 *Jenkins L.* Selecting Scenarios for Environmental Disaster Planning. European Journal of Operation and Research 121 (1999) p. 275–286

30 *Gupta A. Maranas C. D. McDonald C. M.* Mid-Term Supply Chain Planning Under Demand Uncertainty: Customer Demand Satisfaction and Inventory Management. Computers and Chemical Engineering 24 (2000) p. 2613–2621

31 *Tsiakis P. Shah N. Pantelides C. C.* Design of Multi-Echelon Supply Chain Networks Under Demand Uncertainty. Industrial and Engineering Chemistry Research 40 (2001) p. 3585–3604

32 *Lababidi H. M. S. Ahmed M. A. Alatigi I. M. EL-Enzi A. F.* Optimizing the Supply Chain of a Petrochemical Company Under Uncertain Operating and Economic Conditions. Industrial and Engineering Chemistry Research 43 (2004) p. 63–73

33 *Guillén G. Bonfill A. Espuña A. Puigjaner L.* Integrating Production and Transport Scheduling for Supply Chain Management Under Market Uncertainty, in Proceedings of European Symposium of Computer-Aided Process Engineering-14 (ESCAPE-14), Lisbon, Portugal, Elseiver, Amsterdam 2004

34 *Baumol W. J.* The Transactions Demand for Cash: An Inventory Theoretic Approach. The Quarterly Journal of Economics 66 (1952) p. 545–556

35 *Miller M. H. Orr R. A.* A Model of the Demand for Money by Firms. The Quarterly Journal of Economics 80 (1966) p. 413–435

36 *Christy D. P. Grout J. R.* Safeguarding Supply Chain Relationships. International Journal of Production Economics 36 (1994) p. 233–242

37 *Romero J. Badell M. Bagajewicz M. Puigjaner L.* Integrating Budgeting Models into Scheduling and Planning Models for the

Chemical Batch Industry. Industrial and Engineering Chemistry Research 42 (2003) p. 6125–6134

38 *Badell M. Romero J. Puigjaner L.* Joint Financial and Operating Scheduling/Planning in Industry, in Proceedings of European Symposium of Computer-Aided Process Engineering-14 (ESCAPE-14), Lisbon, Portugal, Elseiver, Amsterdam 2004

39 *Forrester J. W.* Industrial Dynamics. MIT Press, Cambridge, MA 1961

40 *Towhill D. R.* Industrial Dynamics Modeling of Supply Chains. Logistics Information Managment 9(4) (1996) p. 43–56

41 *Backs T. Bosgra D. Marquardt W.* Towards Intentional Dynamics in Supply Chains Conscious Process Operations in Pekny, J.F. and Blau G.E. (Eds.) Proceedings of Third International Conference on Foundations of Computer-Aided Process Operations, AIChE, New York, (1998) p. 5

42 *Perea-Lopez E. Grossmann I. Ydstie B. E. Tcihmassebi T.* Industrial and Engineering Chemistry Research 40 (2001) p. 3369–3383

43 *Brown M. W. Rivera D. E. Carlyle W. M. et al.* A Model Predictive Control Framework for Robust Management of Multi-Product, Multi-Echelon Demand Networks. In Proceedings of 15th IFAC world Congress, Barcelona, Spain, 2002

44 *Mele F. D. Forquera F. Rosso E. Basualdo M. Puigjaner L.* A Comparison Between Chemical and Model Predictive Control Over Supply Chain Dynamic Model, in Proceedings of 9th Mediterranean Congress of Chemical Engineering, Expoquimia, Barcelona, Spain 2002

45 *Mele F. D., Espuña A., Puigjaner L.* Supply chain management, through combined Simulation-Optimisation approach, in Proceedings ESCAPE-15 (L. Puigjaner, A. Espuña, Eds.), Elsevier, Amsterdam, (2005) p. 1405–1410.

46 *Garcia-Beltrán C. and Feritil S.* Multi-Agent-Based Decision System for Process Reconfiguration, in Latino-American Control Conference, Guadalajara, Mexico, 2002

47 *Lin J. You J.* Smart Shopper: An Agent-Based Web Approach to Internet Shopping. IEEE Trans Fuzzy Systems 11 (2003) p. 226–237

48 *Report D2-2* Final Specifications of the Supply Chain Multi-Agent Architecture. Project I-303 (GICASA-D) 2003

49 *Report D19* Environmental Impact Considerations Based on Life Cycle Analysis. Project G1RD-CT-2000-00318 2003

50 *Muller P. A.* Modelado de Objetos con UML. Ediciones Gestión 2000 S.A. Barcelona 1997

51 *Romero J. Badell M. Bagajewicz M. Puigjaner L.* Integrating Budgeting Models into Scheduling and Planning Models for the Chemical Industry. Industrial and Engineering Chemistry Research 42 (2003) p. 6125–6134

52 *Badell M. Romero J. Huertas R. Puigjaner L.* Planning, Scheduling and Budgeting value-aided chains, Computers Chemical Engineering 28 (2004) p. 45–61

53 *Guillén G. Pina C. Espuña A. Puigjaner L.* Optimal Offer Proposal Policy in an Integrated Supply Chain Management Environment. Industrial and Engineering Chemistry Research 44 (2005) p. 7405–7419

54 *Puigjaner L.* Handling the Increasing Complexity of Detailed Batch Process Simulation and Optimization. Computers and Chemical Engineering 23(Suppl.) (1999) p. S929–S943

55 *Arbiza M. J. Cantón Espuña J. A. Puigjaner L.* Objective-Based Schedule Selector: a Rescheduling Tool for Short-term Plan Updating, in Barbosa-Póvoa A. (Ed.) European Symposium on Computer-Aided Process Engineering-14, Lisbon, Portugal CD-ROM 2004

56 *Ruiz D. Benqlilou C. Nougués J. M. Puigjaner L.* Proposal to Speed Up the Implementation of an Abnormal Situation Management in the Chemical Process Industry. Industrial and Engineering Chemistry Research 41 (2002) p. 817–824

57 *Puigjaner L.* Real-Time, Optimization of Process Operations: An Integrated Solution Perspective. CHEM User's Committee Seminar, Lille, France, 2004

4
Databases in the Field of Thermophysical Properties in Chemical Engineering

Richard Sass

4.1
Introduction

Process synthesis, design, and optimization, and also detail engineering for chemical plants and equipment depend heavily on availability and reliability of thermophysical property data of pure components and mixtures involved. To illustrate this fact we can analyze the needs for one of the essential process engineering processes, the separation of fluid mixtures. For the design of such a typical separation process, e.g., distillation, we require thermodynamic properties of mixture, in particular for a system that has two or more phases at a certain temperature or pressure. We require the equilibrium constants of all components in all phases.

The quality of data inside the data calculation modules is essential and can have extensive effects. Inaccurate data may lead to very expensive misjudgements whether it is to proceed with a new process or modification of it or not to go ahead. Inadequate or unavailable data may cause a promising and profitable process to be delayed or in the worst case be rejected, only for the reason that it was not properly modelled in a simulation. Another potential danger, partially generated by the marketing statements of simulation software producers, is that the credibility of the results of a thermophysical model calculation generated by computer software is very high, even if the result is wrong. So the expert has the duty to prove that the most sophisticated software will not lead automatically to the most cost-effective solution in order to save effectively energy, if there is not a background with an accurate database of physical and thermodynamic data.

Computer Aided Process and Product Engineering. Edited by Luis Puigjaner and Georges Heyen
Copyright © 2006 WILEY-VCH Verlag GmbH & Co. KGaA, Weinheim
ISBN: 3-527-30804-0

4.2
Overview of the Thermophysical Properties Needed for CAPE Calculations

Without access to a numerical database and if the available literature and notes do not contain a value, the only possibility is to measure properties or to calculate them with a group contribution method or another estimation routine. The first alternative is expensive and time-consuming; the second one will produce data with unknown reliability in most cases, especially when molecules with two or more nonhydrocarbon functional groups in near proximity are involved. With the help of thermophysical databases with experimental values of pure component and mixtures, this problem can be solved. A description of what data are needed and which types of data are available follows. The properties required for the design of a thermal or chemical process depends upon the specific case and the temperature, pressure and concentration range. A short overview of the data needed in the simulation and design of processes is given in Table 4.1.

Table 4.1 Important categories of property data

Property type	Specific properties
Phase equilibria	Boiling and melting points, vapor pressure, fugacity and activity coefficients, solubility (Henry's constants, Ostwald or Bunsen coefficients)
PVT behavior	Density, volume, compressibility, critical constants
Caloric properties	Specific heat, enthalpy, entropy
Transport properties	Specific heat, latent heat, enthalpy, entropy, viscosity, thermal conductivity, ionic conductivity, diffusion coefficients
Boundary properties	Surface tension
Chemical equilibrium	Equilibrium constants, association/dissociation constants, enthalpies of formation, heat of reaction, Gibbs energy of formation, reaction rates
Acoustic	Velocity
Optical	Refractive index, polarization
Safety characteristics	Flash point, explosion limits, autoignition temperature, minimum ignition energy, toxicity, maximum working place concentration
Molecular properties	Virial coefficients, binary interaction parameters, ion radius and volume

4.3
Sources of Thermophysical Data

For years, the most popular way to find thermophysical property data was to take a look inside favorite book collections, starting with the Handbook of Chemistry and Physics up to data collection handbooks issued by data producers as DIPPR [1] and TRC [2], the Landolt-Börnstein [3] and the DECHEMA Chemistry Data Series [4]. Despite the inconvenience in using handbooks for data searches, a lot of users still appreciate the fast access to the data and, in comparison to the databases, relatively moderate price. An overview by Hochschule Merseburg University of Applied Science gives a list of available books and publications on thermophysical property data (www.fh-merseburg.de/PhysChem).

Nowadays, mainly under the pressure of having data available in a short time for calculations and due to the full-time access to networks, the easiest way to find data is with access to databases, accessible or as in-house versions or on-line via hosts or via the World-Wide Web.

Two types of collections and/or databases can be distinguished: bibliographical ones and numerical ones. A bibliographical collection or database is a literature source containing only references. Knowing the chemical species one needs data for, one can find literature references containing that data. Afterwards one has to go to the library and look up the different references to get the data. A numerical database or collection typically contains the literature references as well as measurement data. The numerical data could be accessed and used directly. In some cases this approach is combined with a critical review and selection of the available data, so that only thermodynamically consistent and proven data are contained in the collection. The approach could also be combined with model parameter fitting and recommendation, so that end-users only have to transfer the recommended parameters into their applications to implement a tested model with a defined reliability over all known measurement data points. In the following, a survey on the existing and still maintained collections and databases on solutions is given.

4.4
Examples of Databases for Thermophysical Properties

Due to the fact that dozens of sources for thermodynamic data are now available on the Web, only a few major providers will be mentioned in this chapter. To have access to a larger overview about what is available on the Web, a look at the pages, e.g., of the University of Illinois should be considered (http://tigger.uic.edu/~mansoori/Thermodynamic.Data.and.Property-html).

A few examples for the largest and most famous databases are shown in Table 4.2. Three examples of databases are described as follows.

Table 4.2 Provider list for thermophysical data

Producer	Database name	URL
DECHEMA	DETHERM	www.dechema.de/detherm-lang-en.html
DDBST	DDB	www.ddbst.de/new/Default.htm
NIST	Properties of fluids	http://properties.nist.gov/
NIST	Chemistry WebBook	http://webbook.nist.gov/chemistry/
IUPAC-NIST	Solubility Database	http://srdata.nist.gov/solubility/
K&K Associates	Thermal Resource Center	www.tak2000.com/
FIZ Chemie	INFOTHERM	www.fiz-chemie.de
CERAM Research limited	Thermophysical Properties Database	www.ceram.co.uk/thermet.html
API	Technical database	www.dnv.com/software/all/api/index.asp
MDL	CrossFire Beilstein	www.mdl.com/products/knowledge/crossfire-beilstein/
TPC, Academy of Science Russia	THERMAL	www.chem.ac.ru/Chemistry/Databases/THERMAL.en.html
AIChE	DIPPR	http://dippr.byu.edu/
G&P Engineering Software	MIXPROPS	www.gpengineeringsoft.com/pages/pdtmixprops.html
G&P Engineering Software	PHYPROPS	www.gpengineeringsoft.com/pages/pdtphysprops.html
Ecole Polytechnique de Montreal	FACT	www.crct.polymtl.ca/fact/index.php
S. Ohe	Fundamental Physical Properties	http://data-books.com/bussei-e/bs-index.html
Prode	Prode Properties	www.prode.com/en/ppp.htm
NEL	PPDS	www.ppds.co.uk/Products/
THERMODATA	THERMODATA	http://thermodata.online.fr
Chinese Academy of Science	Engineering Chemistry Database	http://chinweb.ipe.ac.cn/

4.4.1
NIST Chemistry WebBook

The NIST Chemistry WebBook provides access to data compiled and distributed by NIST under the Standard Reference Data Program [5].

The NIST Chemistry WebBook [6] contains:

- thermochemical data for over 7000 organic and small inorganic compounds:
  - enthalpy of formation
  - enthalpy of combustion
  - heat capacity
  - entropy
  - phase transition enthalpies and temperatures
  - vapor pressure
- reaction thermochemistry data for over 8000 reactions:
  - enthalpy of reaction
  - free energy of reaction
- IR spectra for over 16,000 compounds;
- mass spectra for over 15,000 compounds;
- UV/V is spectra for over 1600 compounds;
- electronic and vibrational spectra for over 4500 compounds;
- constants of diatomic molecules (spectroscopic data) for over 600 compounds;
- ion energetics data for over 16,000 compounds:
  - ionization energy
  - appearance energy
  - electron affinity
  - proton affinity
  - gas basicity
  - cluster ion binding energies
- thermophysical property data for 34 fluids:
  - density, specific volume
  - heat capacity at constant pressure (C_p)
  - heat capacity at constant volume (C_v)
  - enthalpy
  - internal energy
  - entropy
  - viscosity
  - thermal conductivity
  - Joule-Thomson coefficient
  - surface tension (saturation curve only)
  - sound speed.

Data on specific compounds in the Chemistry WebBook based on name, chemical formula, CAS registry number, molecular weight, chemical structure, or selected ion energetics and spectral properties can be searched for.

4.4.2
DETHERM

The DETHERM [7] database provides thermophysical property data for about 24,000 pure compounds and 146,000 mixtures. DETHERM contains literature values, together with bibliographical information, descriptors and abstracts. At the time 5.2 million data sets are stored. DETHERM is a collection of data packages produced by well known providers of thermophysical packages, unified under a common graphical user interface. The database files in Table 4.3 are part of DETHERM.

An example for the actual possibilities for presentation of the results is seen in Fig. 4.1.

Table 4.3 Content of DETHERM

Dortmunder Datenbank DDB	Phase equilibrium data
(Prof. Gmehling, University of Oldenburg)	• Vapor-liquid equilibria • Liquid-liquid equilibria • Vapor-liquid equilibria of low boiling substances • Activity coefficients at infinite dilution • Gas solubilities • Solid-liquid equilibria • Azeotropic data Excess properties • Excess enthalpies • Excess heat capacities • Excess volume Pure component properties • Transport properties • Vapor pressures • Critical cata • Melting points • Densities • Caloric properties • Others
Electrolyte data collection ELDAR *(Prof. Barthel, University of Regensburg, LS Chemie IV)*	Caloric data Electrochemical properties Phase equilibrium data PVT properties Transport properties
Thermophysical database INFOTHERM *(FIZ CHEMIE)*	PVT data Transport properties Surface properties Caloric properties Phase equilibrium data • Vapor-liquid equilibria • Gas-liquid equilibria • Liquid-liquid equilibria • Solid-liquid equilibria Pure component basic data

Table 4.3 Content of DETHERM (Fortsetzung)

Dortmunder Datenbank DDB	Phase equilibrium data
Thermophysical Parameter Database	Phase equilibria
COMDOR (Leuna GmbH in Cooperation with FIZ Chemie)	Excess enthalpies Transport and surface properties Caloric and acoustic data
Data Collection C-DATA (Institut for Chemical Technic, Prag)	Twenty physicochemical properties for 593 pure components
Basic Database Böhlen BDBB (Sächsische Olefinwerke AG Böhlen, now DOW Chemical)	Pure component database of the Sächsische Olefinwerke with chemical and physical basic data for 1126 pure substances (mainly for the fields of petroleum and coal chemistry)
Additional (DECHEMA e.V.)	Vapor pressures Transport properties • Thermal conductivities • Viscosities Caloric properties PVT data • PVT data • Critical data Eutectic data Solubilities Diffusion coefficients

4.4.3
DIPPR Database [1]

The major content of the database of the Design Institute for Physical Property Data (DIPPR), a subsidiary of the American Institute of Chemical Engineers (AIChE) [8], are mainly data collections of pure component properties but also data for selected properties of mixtures and the results of a project related to environmental, safety and health data. In total data of 1700 compounds in the database cover mainly the components of primary interest to the process industries. The special focus of the DIPPR database is to provide reliable data of thermophysical properties, including the temperature dependency of the properties, which are approved by technical committees, where industrial experts are involved in the design of the database and in the evaluation of the data.

Table 4.4 gives an overview of the content of the DIPPR database.

The use of these databases is meanwhile a standard option in the preparation of the process design. A bigger difficulty is the absence of thermophysical data for newer processes involving electrolytes and of solutions containing biomaterial. This specific topic will be explained in the next chapter.

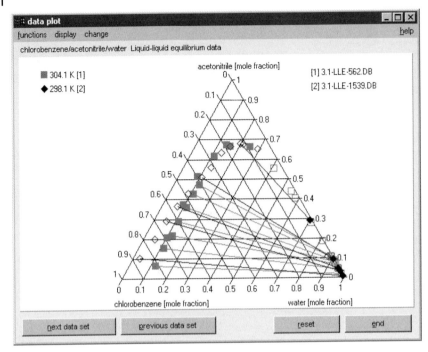

Figure 4.1 Joint graphical display of two different LLE data sets in DETHERM for the system chlorobenzene/acetonitrile/water

Table 4.4 Properties in the DIPPR 801 Database [9]

Constant properties: property	Units
Acentric factor	-
Auto ignition temperature	K
Dipole moment	C m
Absolute entropy of ideal gas at 298.15 K and 1 bar	J (kmol K)$^{-1}$
Lower flammability limit temperature	K
Upper flammability limit temperature	K
Lower flammability limit percent	Vol % in air
Upper flammability limit percent	Vol % in air
Flash point	K
Gibbs energy of formation for ideal gas at 298.15 K and 1 bar	J kmol^{-1}
Standard state Gibbs energy of formation at 298.15 K and 1 bar	J kmol^{-1}

Table 4.4 Properties in the DIPPR 801 Database [9] (Fortsetzung)

Constant properties: property	Units
Net standard state enthalpy of combustion at 298.15 K	J kmol^{-1}
Enthalpy of formation for ideal gas at 298.15 K	J kmol^{-1}
Enthalpy of fusion at melting point	J kmol^{-1}
Standard state enthalpy of formation at 298.15 K and 1 bar	J kmol^{-1}
Heat of sublimation	J kmol^{-1}
Liquid molar volume at 298.15 K	m^3 kmol^{-1}
Melting point at 1 atm	K
Molecular weight	kg kmol^{-1}
Normal boiling point	K
Parachor	–
Critical pressure	Pa
Radius of gyration	m
Refractive index	–
Solubility parameter at 298.15 K	(J m^{-3})$^{1/2}$
Standard state absolute entropy at 298.15 K and 1 bar	J (kmol K)$^{-1}$
Critical temperature	K
Triple point pressure	Pa
Triple point temperature	K
Critical volume	m^3 kmol^{-1}
van der Waals area	m^2 kmol^{-1}
van der Waals reduced volume	m^3 kmol^{-1}
Critical compressibility factor	–

Temperature-dependent properties: property	Units
Heat capacity of ideal gas	J (kmol K)$^{-1}$
Heat capacity of liquid	J (kmol K)$^{-1}$
Heat capacity of solid	J (kmol K)$^{-1}$
Heat of vaporization	J kmol^{-1}

Table 4.4 Properties in the DIPPR 801 Database [9] (Fortsetzung)

Temperature-dependent properties: property	Units
Liquid density	$kmol\ m^{-3}$
Second virial coefficient	$m^3\ kmol^{-1}$
Solid density	$kmol\ m^{-3}$
Surface tension	$N\ m^{-1}$
Thermal conductivity of liquid	$W\ (m\ K)^{-1}$
Thermal conductivity of solid	$W\ (m\ K)^{-1}$
Thermal conductivity of vapor	$W\ (m\ K)^{-1}$
Vapor pressure of liquid	Pa
Vapor pressure of solid or sublimation pressure	Pa
Viscosity of liquid	Pa s
Viscosity of vapor	Pa s

4.5
Special Case and New Challenge: Data of Electrolyte Solutions

A much bigger challenge than these normal solutions is the modeling of electrolyte containing solutions. The modeling of electrolyte solutions or, more generally speaking, liquids containing fractions of electrolytes is nowadays still an exhausting task. Chemical and process engineers for example are nowadays able to model or even predict a vapor-liquid equilibrium, the density or viscosity of a multicomponent mixture containing numerous different species with sufficient reliability. But if only traces of salt are contained in the mixture, nearly all models tend to fail. The modeling results however have a great impact on the design and construction of single chemical apparatus as well as whole plants or production lines. Proper functioning could only be guaranteed based on reliable results.

Another area influenced greatly by electrolyte modeling is biochemical engineering. For example nobody knows how to predict quantitatively salting-out effect of proteins, crystallization processes of biomolecules, the influence of ions on nanoparticle formation, their size morphology and crystal structure, zeolite synthesis and so on. But the development of new production processes in this intensively growing area requires accurate macroscopic physical property models capturing accurately the underlying physics. In some cases there is a limited understanding of these mechanisms, but no real predictability.

The chemical engineer developing new production processes as well as the physical chemist developing models are therefore having a pressing need to access reliable thermophysical property data. Process as well as model development, either predicting or even only interpolating, requires multitudinous amounts of reliable thermophysical property data for electrolytes and electrolyte solutions. Among the most important property types are:

- vapor-liquid equilibrium data
- activity coefficients
- osmotic coefficients
- electrolyte and ionic conductivities
- transference numbers
- viscosities
- densities
- frequency dependent permittivity data.

How does one find such data?

4.5.1
Reliable Data Sources

Thermophysical property data for electrolytes and electrolyte solutions are measured from numerous researchers and scientists and are published typically in a large number of journals and publications. But people requiring such data will not search the primary publications, because this is too time-consuming. And in most cases it is even impossible, because industrial users do not have access to all the required literature immediately. Instead the preferred way will be to check either a printed data collection or to search within an electronic database for the components, mixtures and properties one needs.

Such printed data collections or databases are typically compiled and/or maintained by individuals or groups having a well known reputation in that field. Therefore they have an overview of the primary literature publishing physical property data and are able to continuously add new data to their collections. In most cases these groups also use their own collections for model development. In the following pages, a survey of maintained databases for electrolyte properties is given.

4.6
Examples of Databases with Properties of Electrolyte Solutions

4.6.1
The ELDAR Database [10]

The Electrolyte Database Regensburg ELDAR is a numerical property database for electrolytes and electrolyte solutions. It contains data on pure substances and aqueous as well as organic solutions. The data collection for ELDAR started in 1976 within the framework of the DECHEMA study [11] *Research and Development for Sav-*

ing the Raw Material Supply which was supported by the German Ministry for Research and Technology (BMFT). The work of this study 1981 led to development of ELDAR. From the beginning up to now the ELDAR database development was headed by the Institute of Physical and Theoretical Chemistry of the University of Regensburg. The database was designed as a literature reference, numerical data and also model database for fundamental electrochemical research, applied research and also the design of production processes.

The database is still maintained and has roughly doubled its size since beginning. It contains data of more than 2000 electrolytes in more than 750 different solvents. Nowadays ELDAR contains approximately:

- 7400 literature references
- 45,400 data tables
- 595,000 data points.

ELDAR contains data on physical properties like densities, dielectricity coefficients, thermal expansion, compressibility, PVT data, state diagrams, critical data, thermodynamic properties like solvation and dilution heats, phase transition values (enthalpies, entropies, Gibbs free energies), phase equilibrium data, solubility, vapor pressures, solvation data, standard and reference values, activities and activity coefficients , excess values, osmotic coefficients, specific heats, partial molar values, apparent partial molar values and transport properties like electrical conductivities, transference numbers, single ion conductivities, viscosities, thermal conductivities and diffusion coefficients.

ELDAR is distributed as part of DECHEMA's numerical database for thermophysical property data, which is called DETHERM. To access ELDAR one could therefore use several options:

- in-house client-server installation as part of the DETHERM database [7];
- Internet access using DETHERM ... on the WEB [7];
- on-line access using host STN International [12].

To get an overview of the data available, the Internet access option could be recommended, because existence of data for a specific problem could be checked free of charge and even without registration.

4.6.2
The Electrolyte Data Collection

The Electrolyte Data Collection is a printed publication which is part of DECHEMA's Chemistry Data Series. The Electrolyte Data Collection is published by Barthel and his coworkers from the University of Regensburg. The printed collection and the database ELDAR have complementary functions. The data books give a clear arrangement of selected recommended data for each property of an electrolyte solution. The electrolyte solutions are classified according to their solvents and solvent mixtures. All solution properties have been recalculated from the original measured data with the help of compatible property equations. A typical page of the books contains the following for the described system:

- general solute and solvent parameters
- fitted model parameter values
- measured data together with deviations against the fit
- a plot
- literature references.

The Electrolyte Data Collection has nowadays 18 volumes and consists of 9500 printed pages. Covered properties are:

- conductivities
- transference numbers
- limiting ionic conductivities
- dielectric properties of water, aqueous and nonaqueous electrolyte solutions
- viscosities of aqueous and nonaqueous electrolyte solutions.

4.6.3
ICV-SEP Data Bank for Electrolyte Solutions

The Engineering Research Center Phase Equilibria and Separation Processes (ICV-SEP) of the Technical University of Denmark (DTU) is operating a data bank for electrolyte solutions [13]. It is a collection of scientific papers containing experimental data for aqueous solutions of electrolytes and/or nonelectrolytes and also theoretical papers related to electrolyte solutions. The database is a mixture between a literature reference database and a numerical database. Currently references to more than 4000 papers are stored in the database. In addition experimental data from around 2000 of these papers are stored electronically as well. Most of the experimental data concern aqueous solutions. The access to the literature reference database is free of charge, but requires a registration. The access to the numerical database is restricted to members of an industrial consortium supporting the work of ICV-SEP.

4.6.4
The Dortmund Database DDB [14]

The Dortmund Database Software and Separation Technology from the University of Oldenburg is well known for its data collections in the areas of vapor-liquid equilibria and related properties. While the major part of the data collections is dealing with nonelectrolyte systems, two collections contain exclusively electrolyte data. They are focused on:

- vapor-liquid equilibria
- gas solubilities.

The two collections together currently contain 3250 data sets. Access to these collections is possible either on-line using the DETHERM on the Web or in-house using special software from DDBST or DECHEMA.

4.6.5
Closed Collections

In addition to the above-described publicly available and still maintained collections, do other old electrolyte data collections exist? Among them is for example the ELYS database, which was compiled by Lobo, Department of Chemistry, University of Coimbra, Portugal, or the DIPPR 861 Electrolyte Database Project. But these closed collections are typically not maintained any more and also not publicly available. It is likely the references and/or data published in these collections could also be found inside the aforementioned living collections.

4.7
A Glance at the Future of the Properties Databases

Most of the engineers in chemical companies trust in the power of their evaluation of the equations of state for the calculation of the optimal point of work. Nevertheless the opinion that databases have less importance these days is growing, mainly when budgetary elements come into consideration. The knowledge of the importance of a correct process design is run over by considerations that a saving of one euro per kg for a product which costs 50 euros per kg is not very relevant. That is not the case for basic chemicals, where saving of the same order of magnitude represents 25 % of the total costs and 10 % of the used energy. Unfortunately the production of these chemicals is today mainly transferred into low cost countries, i.e., not very relevant for research purposes. A lot of companies made the outsourcing of their measurements, so that only a limited amount of experts in companies maintain the knowledge for these activities.

When we look at the constraints to find new methods for the design of biologic or polymer solutions, we must be sceptical to find enough people to manage future visions for models with the knowledge what was in the past. In the time of a rise in steel consumption in the Chinese industry, where in an unexpected way a demand of coal energy started again, it may be that the properties will rise in interest. From a governmental funding point of view, it is a good sign that new projects are coming up in order to find a new approach to build evaluated databases.

References

1 Design Institute for Physical Properties (DIPPR): *www.aiche.org/dippr/*, **2006**

2 TRC Thermodynamic Tables: *www.trc.nist.gov/tables/trctables.htm*, **2006**

3 *Landolt-Börnstein: www.springeronline.com/ sgw/cda/frontpage/0%2C11855%2C1-10113-2-95856-0%2C00.html*, **2006**

4 DECHEMA Chemistry Data Series: *www.dechema.de/CDS-lang-en.html*, **2006**

5 NIST Standard Reference Data Program: *www.nist.gov/srd/*, **2006**

6 NIST Chemistry WebBook: *http://webbook.nist.gov*, **2006**

7 DECHEMA DETHERM database: *www.dechema.de/detherm-lang-en.html*, **2006**

8 AIChE: *www.aiche.org*, **2006**

9 DIPPR Project 801: *http://dippr.byu.edu*, **2006**

10 University of Regensburg: *www.uni-regensburg.de/Fakultaeten/nat-Fak-IV/Physika-lische-Chemie/Kunz/*, **2006**

11 DECHEMA, Forschung und Entwicklung zur Sicherung der Rohstoffversorgung. Programmstudie Chemische Technik: Rohstoffe, Prozesse, Produkte, Vol. 6, DECHEMA Deutsche Gesellschaft für Chemisches Apparatewesen e.V., Frankfurt am Main, **1976**

12 STN: *www.stn-international.de/*, **2006**

13 ICV-SEP Data bank for electrolyte solutions: *www.ivc-sep.kt.dtu.dk/databank/databank.asp*, **2006**

14 Dortmund Database DDB: *www.ddbst.de*, **2006**

5
Emergent Standards

Jean-Pierre Belaud and Bertrand Braunschweig

5.1
Introduction

Software standards in computer-aided process and product engineering are needed in order to facilitate application and software components interoperability. In the past, end-user organizations, software companies, governmental organizations and universities have spent hundreds of thousands, if not millions of euros, dollars and yens to develop bridges between software systems such as for transferring simulation data to an engineering database in order to provide the values for basic design; for integrating real time data coming from several process control systems into a common information network for the operators; for allowing a process simulation tool to use pure component data from a physical properties data bank; for using a specialized unit operation simulation model within a commercial process simulation environment, etc.

This question has been a subject of concern for years, as a source of unnecessary costs, delays and moreover of inconsistencies between data produced and consumed by different nonintegrated systems using different bases, different calculation principles, different units of measurements, running on different computers under different operating systems and written in different languages. This need in the domain of computer-aided process engineering has been described elsewhere; see, for example, Braunschweig and Gani (2002).

Software standards remove this problem by providing the desired interoperability between software tools, platforms and databases. With appropriate machine-to-machine interface standards, using the best available tools together becomes a matter of plug-and-play, supposedly as easy as connecting USB devices or hi-fi systems[1]. Moreover, not only do these standards enable several software pieces available on your local PC to be put together, but they allow, thanks to the use of *middleware*, heterogeneous software modules available on your organizations' intranet, or on the

---

[1] Assuming that there is one commonly agreed standard and not several, e.g., see the problems of the multiple standards for writable DVDs and the lack of interoperability that this multiplicity generates.

Computer Aided Process and Product Engineering. Edited by Luis Puigjaner and Georges Heyen
Copyright © 2006 WILEY-VCH Verlag GmbH & Co. KGaA, Weinheim
ISBN: 3-527-30804-0

internet to interoperate, e.g., thanks to *Web services* technologies. Of course, such a facility has significant organizational, economic and technical consequences. We will briefly examine these consequences at the end of this chapter.

However, our main focus will be on technologies, starting with a discussion on the concepts of openness and of open standards development. Then, we will examine some of the most significant operational standards in the domain of computer-aided process and product engineering, namely the CAPE-OPEN standard for process modeling tools, the OPC standard for process control systems. Following this, we will look at some of the current software interoperability technologies that we think will power future systems, i.e., XML and Web services technologies, leading to what is now called service-oriented architectures. Further on, we will shortly address standards for multiagent systems and the emerging Semantic Web standards, which should play a major role in the longer term, moving from *syntactic* to *semantic* interoperability of CAPE systems and services. We will conclude with a brief look at the organizational and economic consequences of the trend towards interoperability and standards.

This chapter deals essentially with software-oriented standards, i.e., standards related to the use of one piece of software from within another piece of software. *Data-oriented standards* allowing to exchange data (from databases, files, etc.) between many software applications are only marginally addressed, e.g., in the POSC section.

5.1.1
Open Concepts

There is a clear fact that the emergence of the World-Wide Web was done with concepts of common development and usage. These concepts called here open concepts commonly encompass open standards, open computing, standardization processes and open software. In the first years of e-business, (open) standards were essential to the development of the Web, to e-commerce and to inter/intra-organizational integration. Standardized information technologies such as TCP/IP, HTTP, HTML, XML, CORBA-IIOP, Web services-SOAP, etc., achieve interaction and information exchange with external or internal, homogeneous or heterogeneous, and remote or nearby applications. These technologies are now core technologies of our networked environment. For the next generation of information systems and of computer technologies, open concepts should again play a key role for emergent information technologies (IT) standards introduced in Section 5.3. Heintzman (2003) gives a good introduction to open concepts for the domain of IT, through formal definitions, a brief history from the 1970s to the modern day battle of openness, and addresses commercial challenges of open projects from an IBM perspective. There is no reason why process engineering would escape from this trend, even if this field is a niche business and therefore more restricted and less global. Section 5.2 illustrates concrete technologies using open concepts in the field of CAPE. For example, CAPE-OPEN (CO) is a significant technology for interoperability and integration of process engineering software components allowing engineering based on *off-the-shelves components*.

5.1.2
Open Standards and Standardization Process

In order to develop modern software applications and systems, technology selection involves many criteria. One main issue is to know if the technology is an (open) standard technology or a proprietary technology. Open standard technologies are freely distributed data models or software interfaces. They provide a basis for communication, common approaches and enable consistency (Fay 2003), resulting in improvements of developments, investments and maintenance. Clearly the common effort to develop an IT or a CAPE standard and its world-wide adoption by the community can be a source of cost reduction, because not only is the development cost shared but also the investment is expected to be more future-proof.

Open standards are developed by software and/or business partners who collaborate within neutral organizations (such as W3C, OASIS, OMG, etc., for IT and CO-LaN, POSC, etc., for process engineering) in accordance with a standardization process. Such organizations represent a new kind of actor additional to more traditional actors, i.e., academics, software/hardware services suppliers and end-user companies. In the information and communication industry Warner (2003) calls this standardization process *block-alliance* in committee-based standard setting and examines it with block-alliance in market-based standard battle. The latter, which is beyond our scope, leads to *de facto* standards if the resulting technology successfully matches the market. However, both approaches are not so distinct since a standardization process can be a means in a business strategy. For example the Java platform and UML mix

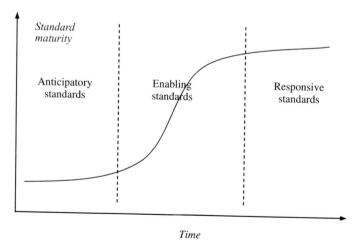

Figure 5.1   Timing of standard

committee-based and market-based processes. If we consider the S-curve lifecycle of a simple technology, Sherif (2003) classifies the technological innovations in terms of market innovation and of technological competencies with radical, platform, incremental and *architectural innovations*. Weiss and Cargill (1992) show the ideal relationship between these types of innovation and the standardization process timing with the type of standards needed at each phase (Fig. 5.1). As an illustration we would say that the CO standard is in the second phase: initial products are commercialized; CO technology is now well disseminated; there is a well-established organization releasing formal specifications; development tools, labeling process and promotion actions support the CO standard.

5.1.3
Open Computing, Open Systems, and Open/Free Software

By extension of the open standards paradigm, building modern software solutions can be based on an open computing paradigm. Open computing means that there is a standardization of information exchange. Then the resulting open system is a system whose characteristics comply with standards made available throughout the industry and therefore that can be connected to other systems complying with the same standards (IBM Glossary 2004). Open computing promises many benefits: flexibility/agility, integration capability, software editor independence, development cost and adoption of technological innovation. While always giving priority to the quality of business models available in a specific CAPE tool, process engineers can now privilege open CAPE systems, ensuring the exchange of information between CAPE solutions of distinct editors thus making it possible to benefit from various fields of expertise. This communication can be done statically with *data models* or dynamically with application programming interface (API). Open computing in CAPE is illustrated in Section 5.2.

The tools for application engineering or for software development can be open source software tools or commercial software tools. Heintzman (2003) identifies several types of projects for the development and management of open source software: academic projects (especially viewed as a new media for collaboration, innovation promotion and dissemination), foundation projects (for base software such as Linux, Apache, Eclipse, Mozilla, etc.), middleware projects (advanced software such as JBoss, MySQL, etc.), niche projects (very specific software available on the Internet[2]). Open source software projects in the CAPE field are not significant at present but they could occur in academic or niche projects, the only known example at the time of writing being SIM42 project (Sim42 Foundation 2004), which develops an open source chemical engineering simulator.

---

2 For example SourceForge.net is the largest repository of open source software projects with more than 118 000 projects at the beginning of 2006.

5.2
Current CAPE Standards

For several decades, experts and process engineers concentrated on the creation, evolution and improvement of models of thermodynamic and physical properties, unit operation, numerical methods, etc. Thus many CAPE software solutions allowing a more or less rigorous representation were developed. Each one is unique and dependent on the know-how of its author or editor. Particularly, in addition to the specific modeling activity, each one is characterized by selected computing technologies, i.e., supporting environment, implementation languages, persistence system, logical architecture, etc. This results in heterogeneity of available solutions and an impossibility of exchanging information between the different tools. Dual bridges between certain tools exist but this option remains proprietary and only operational for a limited number of associations of tools. Now the demand of users of CAPE tools turn to open systems, ensuring process, model and data exchange with third-party tools. In the same way, process engineers wish to be able to integrate their know-how easily and thus to deploy a final solution specific to their needs from best-in-class software components. Open computing and its related IT and CAPE standards allow to build a user-centered modeling and simulation environment from enterprise internal components and selected off-the-shelf components. Several initiatives that promote a standard for process information exchange can be identified, according to two types of techniques[3], data models and API:

- data models such as pdXML, energy eStandards from POSC and Physical Property Data eXchange from DECHEMA;
- APIs such as OPC from OPC Foundation, Physical Properties Package from IK-CAPE and CAPE-OPEN from the CO-LaN.

Open software architectures can now be exploited by the new generations of CAPE software solutions in order to provide better enterprise process applications integration. As an illustration of interest, Fieg et al. (1995), Mahalec (1998), Braunschweig et al. (2000), White (2000), Braunschweig and Gani (2002) and Belaud et al. (2002) discuss open computing, its resulting and its expected benefits. The next sections introduce CAPE-OPEN, OPC and energy eStandards.

5.2.1
CAPE-OPEN Standard for Modeling and Simulation

To solve problems, process engineers typically use a collection of in-house, commercial and/or academic software. Each user requires a broader access to available information and models to fit with the demand on the one hand, and has the constraint to match easily the old and the new, on the other hand. Information technologies play a predominant role to improve CAPE tools in supporting process engineers who

---

3 In some cases this distinction is not so obvious as some work both ways. Moreover, the XML technology adopted by some standards does not really comply with this classification.

face these new challenges of interoperability. It is quite obvious that work is needed to develop and establish open systems for CAPE related software. Development of open systems requires the establishment of open standards. The CAPE-OPEN standard, through which a host tool and any external tool can communicate, is the answer to this question, as it provides an *open communication system for process simulation,* allowing the final users to employ various elements within any other. Specifically, since 1995, an international group of operating companies, software suppliers and academics, developed, through the CAPE-OPEN initiative, an open communication system for key simulation elements, and demonstrated its effectiveness on numerous examples. Through this it also promoted the adoption of the open system by the major providers and users of process simulation.

The CAPE-OPEN standard (Belaud and Pons 2002, present version 1.0) consists in a technical architecture, interface specifications and implementation specifications. The technical architecture relies on modern development tools and up-to-date information technologies such as object-oriented paradigm, component-based approach, Web-enabled distributed architecture, middleware technology and uses the Unified Modeling Language (UML) notation. The interface specifications identify a conceptual model and the implementation specifications give the corresponding platform specific model for COM and CORBA. The specifications cover major application areas, e.g., unit operation, thermodynamic and physical properties, numerical solvers, optimization, planning and scheduling, chemical reactions systems, etc. CAPE-

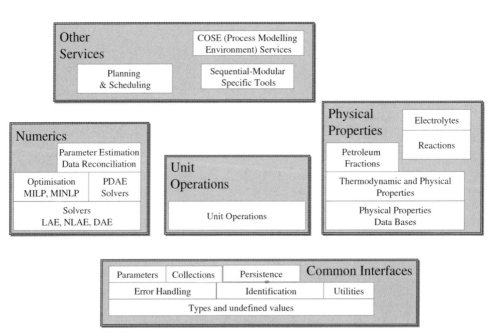

Figure 5.2 CAPE-OPEN version 1.0 specifications

OPEN compliant software environments and components are now available on the market. Belaud et al. (2003) deal with the unit operation interface and show an example for a fixed bed reactor for butane isomerization. The CAPE-OPEN standard is free of charge and is managed by the CO-LaN consortium (www.colan.org, and Pons et al. 2003), which gathers operating companies, software suppliers and academic institutes.

In addition to publishing the standard specifications, CO-LaN provides tools for supporting the transition to CAPE-OPEN technology:

- migration tools, that is, software that automate the migration of existing components to CAPE-OPEN compliance,
- code examples for re-use,
- software testers that check compliance with the standard,
- guidelines and other helpful documents.

Recent announcements from software suppliers, end-users and research institutions demonstrate that CAPE-OPEN is increasingly accepted by the CAPE community. Its main technological benefits are:

- for suppliers: increased usage of CAPE tools and reduced development and integration costs,
- for users: "develop your expertise once, plug and run everywhere" and access to best-in-class solutions,
- for academics: improved dissemination of research results and better matching with industrial needs.

Organizations who adopt the CAPE-OPEN standard, and possibly become members of the CO-LaN, will be the first ones to harvest the benefits of open standard interfaces in process modeling and simulation.

5.2.2
Extensions to the CAPE-OPEN Standard

The 1.0 version of the CAPE-OPEN standard offers the following interface specifications as shown in Fig. 5.2. Details on these specifications are available elsewhere and on CO-LaN's Web site. Although addressing a broad range of applications of CAPE modeling and simulation, the specifications are subject to improvements and extensions. At the time of writing this chapter, two such projects are active:

- Improvement and refactoring of the *thermodynamic and physical properties* specifications. This work will eventually deliver version 1.1 of the specification which should be restructured in a more logical way, better documented, and therefore easier to use.
- Extension of the *unit operation* (UO) specification. The UO CAPE-OPEN standard, in version 1.0, does only address steady state simulation; although several tests have shown that CAPE-OPEN unit operations could be used, with limitations, in

dynamic simulation, work is going on to provide a specification fully compliant with all possible uses in dynamic simulation. A new version of the UO standard will be released after sufficient testing in a number of dynamic process modeling environments.

The decision to launch a new improvement/extension project is taken by CO-LaN's board of directors following proposals presented by special interest groups or by CO-LaN members.

5.2.3
OPC for Process Control and Automation

Since 1996 the OPC Foundation (OPC Foundation 1998) has been a nonprofit organization which ensures the definition and the use of interfaces for applications in control and automation of processes. It is dedicated to ensuring interoperability in automation by creating and maintaining open specifications that standardize the communication of acquired process data, alarm and event records, historical data, and batch data to multi-supplier enterprise systems and between production devices. The vision of OPC is to be the foundation for interoperability for moving information vertically from the factory floor through the enterprise of multi-vendor systems, as well as providing interoperability between devices on different industrial networks from different vendors. The foundation gathers more than 300 members, suppliers and users of control systems, instrumentation, and process control systems. It is worth noting that Microsoft is a member and acts as a technology advisor.

The OPC-OLE for process control standard (Iwanitz and Lange 2002) is based on Microsoft OLE-ActiveX/(D)COM technology and standardizes the communication of OPC compliant data sources[4] and OPC compliant applications[5] through different connections (radio, serial, Ethernet and others) on different operating systems (Windows, Unix, VMS, DOS and others). Many specifications are available:

- OPC Data Access provides access to real-time process data,
- OPC Historical Data access is used to retrieve process data for analysis,
- OPC Alarms and Events is used to exchange and acknowledge process alarms and events,
- OPC Data eXchange defines how OPC servers exchange data with other OPC servers;
- OPC XML encapsulates process control data making it available across all operating systems.

As for CAPE-OPEN, the OPC foundation provides several tools and technologies supporting application and migration to the OPC standard, including self-testing software.

---

4 programmable logic controllers, distributed control systems, databases and other devices

5 human machine interface, trending subsystems, alarm subsystems, spreadsheet, historians, enterprise resource planning, etc.

5.2.4
Energy eStandards for Oil and Gas Processes

POSC is an international not-for-profit membership corporation. It unites industry people, issues and ideas to facilitate *exploration and production information sharing and business process integration* in the petroleum industry. Since 1990, membership has grown to over 100 companies. The membership includes world-wide representation of major and national oil companies, suppliers of petroleum exploration and production software and services, government agencies, computing and consulting companies, and research and academic institutions.

POSC provides open specifications for information modeling, information management, and data and application integration over the life cycle. These specifications are gathered in the energy eStandards project that relies principally on XML technologies (DTD, XML, Schema, etc.) for leveraging Internet technologies in the integration of oil and gas business processes. The set of standards are classified according to POSC areas: internet data exchange standards, practical exploration and production standards, data management standards, standards usability and application interoperability standards. For example, in the data management standards area the Epicentre standard provides a logical data model for upstream information. Also in the Internet data exchange standards area, ChemicalUsageML is a specification for the transfer of information about potential chemical hazards, and WellLogML is an XML DTD and a XML schema for well log data representation.

These standards are not directly related to CAPE applications. However, the scope of POSC encompasses both underground applications (geology, geophysics, reservoir, drilling) and offshore applications (production, transportation). The second application area has many similarities with downstream areas such as petroleum refining, as it essentially involves the design, operation and monitoring of continuous processes. Some of the POSC projects such as POSC-CAESAR delivered technologies applicable to CAPE in general. Since these are data-oriented standards we do not address them in this chapter. Commonalities can also be found with a number of data modeling projects undertaken by the chemical engineering community such as PI-STEP, PDXI or pdXML (Teague 2002 and Teague 2002b).

5.3
Emergent Information Technology Standards

Although not yet fully exploited by the CAPE community, a number of emergent IT standards will become important for our applications in the near future. Complementing some of the technologies presented in the previous section, these new IT standards support Internet-based computing and take advantage of Web technologies. We will first look at Web services together with their newly developed business standards, leading to service-oriented architectures; then we will go a step further and introduce IT standards for multi-agents architectures and the recently pub-lished[6] Semantic Web standards.

5.3.1
Web Services and Business Standards

Web technologies are being used more and more for application to application communication. Before the twenty-first century, software suppliers and IT experts promised this interconnected world thanks to the technology of Web services. Web services propose a *new paradigm for distributed* computing (Bloomberg 2001) and are one of today's most advanced application integration solutions (Linthicum 2003). They help business applications to contact a service broker, to find and to integrate the service from the selected service provider.

For example, during a simulation, the simulation environment, in need of an external thermodynamic service, contacts a UDDI directory in order to take advantage of a particular thermodynamic model (yellow page function). Once the producer of such services (a company) is selected, the simulation environment recovers the signatures of all available services using the associated WSDL descriptions[7]. These phases of discovery and description can be carried out dynamically or statically during the development process. Then the simulation environment connects to the specific thermodynamic service and uses it with SOAP[8] communication protocol. This scenario can take place on the Internet or on company intranets or extranets; it uses a set of technologies: UDDI, WSDL and SOAP, proposed by the Web services community to ensure interworking and integration of Web services.

However, even if the idea of Web services has generated too many promises[9], Web services should be viewed for now as a part of a global enterprise software solution and not as a global technical solution. In a project, Web services can be used within a general architecture relying on Java EJB or on Microsoft's .NET framework. Many projects already utilize Web services, sometimes with nonstandard technologies, particularly for noncritical intranet applications. Even if Web services miss advanced functionalities, many advantages like lower integration costs, the re-use of legacy applications, the associated standardization processes and Web connectivity can plead in favor of this new concept for software interoperability and integration (Manes 2003).

5.3.1.1
Definition
A Web service is a standardized concept of functions invocation relying on Web protocols, independent of any technological platform (operating system, application

---

6 at the time of writing this section (early 2004)

7 Web Service Description Language, somewhat equivalent to OMG's CORBA and to Microsoft's COM IDL

8 simple object access protocol, known as the "piping" between Web services

9 Early standards, security, orchestration, transaction, reliability, performance, ethic and economic models are the main concerns.

server, programming language, database, and component model). BearingPoint et al. (2003) focus on the evolution from software components to Web services and write: "a Web service is autonomous and modular application component, whose interfaces can be published, sought and called through Internet open standards." We see the introduction of Web services as a move from component architectures towards *internet awareness,* this context implying the use of associated technologies, i.e., HTTP and XML, and an e-business economic model. Current component technology based on EJB, .NET and CCM being not fully suitable, Web services provide a new middleware for providing functionality anywhere, anytime and to any device.

5.3.1.2
Key Principles
IBM and Microsoft's initial view of Web services, first published in 2000, identified three kinds of roles (Fig. 5.3):

- A service provider publishes the availability of its services and responds to requests to use its services.
- A service broker registers and categorizes published service providers and offers search capabilities.
- A service requester uses service brokers to find a needed service and then employs that service.

These three roles make use of proposed standard technologies: UDDI from the OASIS consortium, WSDL and SOAP from the World-Wide Web consortium (W3C). UDDI acts as a directory of available services and service providers; WSDL is an XML vocabulary to describe service interfaces. SOAP is an XML-based transfer protocol that allows you to send requests to services on through HTTP. Further domain-specific technologies related to Web services are being developed, e.g., the following proposed by the OASIS consortium, a consortium of companies interested in the development of e-business standards ebXML, supported by Sun Microsystems, is a global framework for e-business data exchange; BPEL (formerly called BPEL4WS), is a proposed standard for the management and execution of business processes based on Web services; SAML aims at exchanging authentication and authorization information; WS-Reliable Messaging is for ensuring reliable message delivery for Web services; WS-Security aims at forming the necessary technical foundation for higher-level security services, etc. A recent glossary of technologies related to Web services, each one defined by only a few lines of text, is 16 pages long (Cutter Consortium 2003).

Simply stated, the interface of a Web service is documented in a file written in WSDL and the data transmission is carried out through HTTP with SOAP. SOAP can also be used to query UDDI for services. The functions defined within the interface can be implemented with any programming language and be deployed on any platform. In fact any function can become a Web service if it can handle XML-based calls. The interoperability of Web services is similar to distributed architectures based on standard middleware such as CORBA, RMI or (D)COM but Web services offer a loose coupling, a nonintrusive link between the provider and the requester,

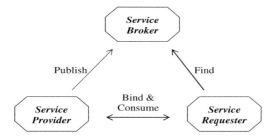

Figure 5.3 Key principles of Web services

due to the loosely-coupled SOAP middleware. Bloomberg (2001) compares these different architectures.

Oellermann (2002) discusses the creation of enterprise Web services with real business value. Basically he reminds that a Web service must provide the user with a service and needs to offer a business value. The technically faultless but closed .NET "my services" project from Microsoft demonstrates that this is always challenging to convince final users. With Google Web API beta (2004), software developers can query the Google search engine using Web services technology. Indeed Google search engine is available as a Web service since mid-2002. Search requests submit a query string and a set of parameters to the Google Web APIs service and receive in return a set of search results. A developer's kit provides documentation and example code (Java, C# and Visual Basic) for using this Web service from any platform that supports it.

5.3.1.3
SOAP: a Loosely-coupled Middleware Technology

HTML-HTTP act as loosely-coupled middleware technology between the Web client (navigator) and the business logic layer (Web server). Around the year 2000 Microsoft and IBM proposed to use the XML data format over the Internet protocols: HTTP as *transport layer* and XML as *encoding format* now constitute the key underlying technologies for Web services.

On top of these, SOAP (currently in version 1.2) was delivered in June 2003, as a lightweight protocol for exchange of information in a decentralized and distributed environment. SOAP can handle both the synchronous request/response pattern of RPC architectures and the asynchronous messages of messaging architectures. An example of SOAP request message in a synchronous manner can be found in Google Web APIs beta (2004). A SOAP request is sent as a HTTP POST. The XML content consists in three main parts:

- The envelope defines the namespaces used.
- The header is an optional element for handling supplementary information such as authentification, transactions, etc.
- The body performs the RPC call, detailing the method name, its arguments and service target.

Whereas OMG CORBA, Java RMI, Microsoft (D)COM and .NET Remoting try to adapt to the Web, SOAP middleware ensures a native connectivity with it since it builds on HTTP, SMTP and FTP and exploits the XML Web-friendly data format. The many reasons for the success of SOAP are its native Web architecture compliancy, its modular design, its simplicity and extensibility, its text-based model[10], its error handling mechanism, its ability for being the common messaging layer of Web services, its standardization process and its support from major software editors.

With so many advantages for integration and interoperability one could expect a massive adoption by software solutions architects. However the deployment of Web services still remains limited. In addition to technical issues, three main reasons can be noted:

- Web services are associated to SOAP, WSDL and UDDI. The UDDI directory of Web services launched in 2000 by IBM, Microsoft, Ariba, HP, Oracle, BEA and SAP, was operational at the end of 2001 with three functions (white, yellow and green pages). However due to technical and commercial reasons this world-wide repository that meets an initial need (to allow occasional, interactive and direct interoperability) founded on the euphoria of e-business years does not match the requirements of enterprise systems. Entrusted to OASIS in 2002, UDDI version 3.0 proposes improvements in particular for intranet applications.
- The simplicity and interoperability claimed by Web services are not so obvious. Different versions of SOAP and incompatibilities of editors' implementations are source of difficulties, to such a degree that editors created the WS-I consortium to check implementations of standards of Web services across platforms, applications, and programming languages.
- The concept was initially supported by a small group of editors (with Microsoft and IBM leading); now the "standards battle"[11] and the multiplication of proposed standards weaken the message of Web services (Koch 2003).

5.3.1.4
Service-oriented Architecture

In order to better integrate the concept of Web services in enterprise systems, IT editors now propose the service-oriented architecture (SOA) approach (Sprott and Wilkes 2004). Beyond the marketing hype, a consensus is established on the concept of service as an autonomous process, which communicates by message within an architecture that identifies applications as services. This design is based on coarse-grained, loosely-coupled services interconnected by asynchronous or synchronous communication and XML-based standards. The definition and elements of SOA are not well established yet. Sessions (2003) wonders whether a SOA is (1) a collection of components over the Internet, (2) the next release of CORBA or (3) an architecture for publishing and finding services.

An SOA is only an evolution of Web-distributed component-based architectures to get applications integration easier, faster, cheaper and more flexible, improving

---

[10] In contrast to binary and not self-describing CORBA, RMI, (D)COM, .NET protocols.

[11] with BEA, IBM and Microsoft on one side and Iona, Oracle and Sun from the other side

return on investment. In fact the main innovations are in the massive adoption of Web services[12] by the industry and in the use of the XML language to describe services, processes, security and exchanges of messages. This promises more future-proof IT projects than in the past.

Despite limitations of Web services, the technology now appears to be complementary to solutions based on classic middleware bus, as well as to enterprise application integration solutions. Its loose coupling brings increased flexibility and facilitates the re-use of legacy systems. Moreover Web services can be used like low-cost connectors between distinct technological platforms like COM, .NET and J2EE. The next release of Microsoft's Windows Vista operating system will include Indigo, a new interoperability technology based on Web services, for unifying Microsoft's proprietary communication mode; Abitboul, research director at INRIA, estimates that Web services will represent, in the long run, the natural protocol for accessing information systems. Thus it seems that we are only at the start of Web services and SOA. Andrews (2004) predicts dramatic changes in the Web services market for 2006, and announces a new class of business applications called service-oriented business applications. The merging of Web, IT and object/component technologies to form SOA and Web services is announced as the next stage of evolution for e-business (knowing that grid computing and autonomic computing will add their contributions too, but this is another story).

There is no doubt that the scientific field will get many benefits from this trend. As for CAPE, one can foresee several applications of SOA and Web services. However, it is sure that innovations will probably go beyond what is predictable at this stage of development. Here are a few examples:

- Sama et al. (2003) presents a Web-based process engineering architecture where simulator components can be executed over the Web.
- Many front-end engineering companies share design data over communications network. Access to this design data could be made easier through an SOA.
- Physical properties databases can be made available through Web services; a good example of such a service is Dechema's "DETHERM ... on the Web" on-line service (Westhaus 2004). This service is currently available through conventional technology (PhP requests on database) and could be made into a Web service, therefore directly interoperable with other programs.
- In the long run, process engineering software could interoperate with equipment manufacturers services not only to develop better simulation models by using the manufacturer's specific unit operation model, but as well to link into manufacturers' supply chain when moving into detailed design, procurement and commissioning.

As can be seen from these examples, the advent of service-oriented architectures brings many opportunities to the CAPE professional. Now let us move even further and come to semantic interoperability.

---

12 Even if a SOA does not imply the use of web services technology and vice versa.

5.3.2
W3C's Semantic Web Standards

The current World-Wide Web is very rich in terms of content, but is essentially *syntactic* or even *lexical*. Looking for information on the Web, using search engines, is done by finding groups of terms in the pages and in the documents, without taking consideration of the meaning of those terms.

For example, using the most popular search engine, Google, to look for information about the ESCAPE-15 conference, the first page brings the results seen in Table 5.1.

Thanks to the referencing work done by the conference organizers, the first hit is the conference's Web site. However, in the first page, together with the correct hit, Google reports a ski bag, a motor racing wheel, and a tour in New Zealand. One might wish to go to the "advanced search" page and specify that only Web sites about conferences should be returned. This is not possible, since Google does not allow this restriction. As a matter of fact, none of the most popular search engines currently used could restrict the search to a category of pages, as the *semantics* of the pages are unknown to them.

Supported by the W3C, of which it is a priority action, many projects aim at developing the semantic level, where information is annotated by its meaning. A necessary stage is to define consensual representations of the terms and objects used in the applications-these consensual representations are called *ontologies*. These ontologies will be expressed in OWL (Ontology Web Language), which itself is based on XML and RDF (Resource Description Language), a specialization of XML. Programs in the whole world support this movement towards the semantization of informa-

Table 5.1 ESCAPE-15 search with Google on 18 July 2004

ESCAPE 15
The ESCAPE (European Symposium on Computer Aided Process Engineering) series brings the latest innovations and achievements ...
www.ub.es/escape15/escape15.htm

Thule Escape 15 Cubic Foot Rooftop Cargo Bag
Buy Thule Escape 15 Cubic Foot Rooftop Cargo Bag here, one of many top quality Ski Rooftop Storage products ...
www.sportsensation.com/skiing/r/Ski-Rooftop-Storage/Thule-Escape-15-Cubic-Foot-Rooftop-Cargo--Bag-1330418.htm

---

Motegi Racing
Escape, 15" Wheels 01-On
info.product-finder.net/motegi/ Escape--15--Wheels-01-On-154.html

---

Grand **Escape**
15 Days Auckland to Christchurch
This morning we journey across the Auckland Harbour Bridge traveling through small rural farming communities. Visit the Matakohe Pioneer Museum ...
www.newzealandtours.net.nz/auckland/guided/akguid66x.html

---

tion. In Europe, the EC strongly supports through the Information Society Technologies (IST) program. A few ontology development projects have taken chemical engineering as their application domain. A good definition of ontologies is provided in the *Web Ontology Language Use Cases and Requirements* document published by W3C (2004):

> *Ontology defines the terms used to describe and represent an area of knowledge. Ontologies are used by people, databases, and applications that need to share domain information. Ontologies include computer-usable definitions of basic concepts in the domain and the relationships among them. They encode knowledge in a domain and also knowledge that spans domains. In this way, they make that knowledge reusable.*
>
> *The word ontology has been used to describe artifacts with different degrees of structure. These range from simple taxonomies to metadata schemes, to logical theories. The Semantic Web needs ontologies with a significant degree of structure. These need to specify descriptions for the following kinds of concepts:*
>
> - *classes (general things) in the many domains of interest,*
> - *the relationships that can exist among things,*
> - *the properties (or attributes) those things may have.*

The definition of ontologies is a multidisciplinary work, which requires competence (1) in the application area: processes, chemistry, environment, etc., (2) in the modeling of knowledge into a form exploitable by machines. It is also an important stake for the actors of the field, who will use the standards defined to annotate and index their documents, their data, their codes, in order to facilitate the semantic retrieval.

Applications of the Semantic Web are many. The last section of this chapter presents an example in intelligent reconfiguration of process simulations using software agents. Before this, it is worth listing the main use cases selected by the W3C working group on the definition of OWL that have guided its development before its official release as a standard:

- *Web portals.* A Web portal powered by ontologies will bring more relevant content by applying inferences on its content (e.g., a distillation column is a separation process, therefore information about distillation would be useful to readers interested in separation).
- *Multimedia collections.* Semantic annotation of large multimedia collections will help in the retrieval among these collections, e.g., a section of a video presentation about operating special equipment.
- *Corporate Web site management.* This is the same as above, with specific functionality for company personnel, such as finding competences among employee directories etc.
- *Design documentation.* The problem of documenting designs has been identified in the chemical engineering field as in other fields where design is a key phase; it is interesting to note that this problem has been outlined by the W3C as one which could most benefit of semantic annotations, allowing to retrieve design chunks in a structured manner.

- *Agents and services.* Ontologies will be used by software agents to discover and analyze service offers and select the most relevant one; the next section presents such a system developed in the COGents EC-funded project.
- *Ubiquitous computing.* New information and technical systems will be configured at runtime by appropriate selections of services in *unchoreographed* ways, that it, in configurations which were not predicted at the time of setting up the services; annotation of ubiquitous services by ontologies will help in interoperating such combinations.

5.3.3
Use of Ontologies by Software Agents

The IST COGents developed an agent-based architecture for numerical simulation, with a concrete implementation in the process simulation domain relying on the CAPE-OPEN interoperability standard. The project, which lasted two years (April 2002–March 2004), proposed and implemented a framework, designed the Onto-CAPE domain ontology of modeling knowledge, and demonstrated its benefits through case studies. COGents was funded by the European Community under the Information Society Technologies program, contract IST-2001-34431.

As before, the CAPE-OPEN standard facilitates process simulation software interoperability and can be the foundation for Web services in this domain. The COGents project pushed the technology further: we used cognitive agents to support the dynamic and opportunistic interoperability of CAPE-OPEN compliant process modeling components over the Internet. The result is an environment which provides automatic access to best-of-breed CAPE tools when required wherever situated.

For this purpose the COGents project:

- defined a framework allowing simulation components to be distributed and referenced on the Internet and intranets,
- defined representations of requirements and services in form of an ontology of process modeling, "OntoCAPE",
- designed facilities for supporting the dynamic matchmaking of modeling components,
- demonstrated the concepts through software prototypes and test cases.

The project was supported by case studies serving as examples: nylon-6 process modeling; HDA process synthesis and simulation. The nylon-6 process case study poses challenges to the component set-up and configuration: the choice on how a simulation shall be performed depends on the availability of solvers and discretization methods. The HDA process has been used as a case study in process design, process optimization and heat exchanger network synthesis. The availability of published results provides a benchmark for the agent-based design and optimization tools.

The architecture of the COGents framework is illustrated in Fig. 5.4.

The extended functionality of COGents is provided by a multi-agent system (MAS), represented by the DIMA block in the above figure. MAS aims to model com-

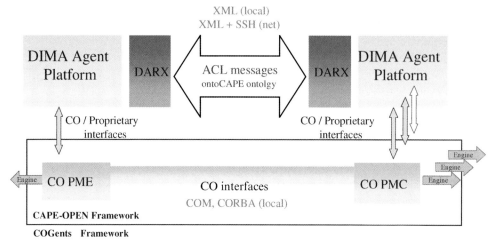

Figure 5.4 The COGents framework

plex systems as collections of interactive entities called agents. Each agent is autono-mous and proactive and can interact with others and act upon its environment, applying its individual knowledge, skills, and other resources to accomplish goals. In COGents the key role of the MAS is to conduct negotiation mechanisms for compos-ing the simulation during the design phase, as well as providing runtime facilities such as diagnostics and guidance to the users. The communication between individ-ual agents is done with messages exchanged using an Agent Communication Lan-guage (ACL), whose content is expressed using the OntoCAPE ontology. DIMA is complemented with DARX, which provides a global naming and location service on a network. COGents integrates a security layer based on SSH, which provides strong authentication and secure communications over the Internet.

The advantages of agent-oriented approach are as follows:

- *Openness.* New Agentscan be dynamically and easily added and/or removed.
- *Heterogeneity.* The various components can be developed with different program-ming languages, they can be executed on different platforms.
- *Flexibility.* Interactions between the various entities are not rigidly defined.
- *Distribution/Mobility.* The agents can be executed on a set of distributed machines and can move from one machine to another.

In COGents, agents are used to improve the dynamic of simulations and to facilitate the design and development of distributed large-scale simulations. These distributed interactive simulations are built from a set of independent simulation components linked together by a network. They provide rich adaptive simulations with agents that can interact with humans and each other.

As any application where domain knowledge has to be explicitly represented, COGents calls for an ontology to support the knowledge representation and inter-agent communication. More specifically, this ontology of the process modeling domain defines concepts indispensable for describing process modeling tasks, modeling strategies as well as software resources, and is the foundation of a matchmaking between requirements of users (i.e., process engineers) and suitable software components. OntoCAPE supports reasoning for mapping user's requests into modeling strategies and for locating software resources to implement the identified modeling strategies. OntoCAPE was developed in DAML + OIL, a predecessor of the OWL language.

More details on the COGents project, including full access to OntoCAPE, can be obtained from COGents (2004).

5.4
Conclusion (Economic, Organizational, Technical, QA)

Interoperability standards such as CAPE-OPEN, OPC, Web Services and the Semantic Web's OWL supporting reference ontologies, open new opportunities for the process industries. Once these ideas gain wide acceptance by the process engineering community, we will find ourselves facing some very major changes in the ways process engineering software are designed, developed, marketed, distributed and used, for the mutual benefit of users and vendors.

The market now has access to robust, reliable, commercial simulators that have standard software component interfaces. Process industries will be able to enjoy the lower cost and lower maintenance of commercial software, but this will be combined with an abundant flexibility. This combination will allow those companies to predict and manage process performance as never before. The number of potentially affected products is in the hundreds, due to the numerous application areas, components and suppliers. We will see many innovative combinations of process modeling components and services from large and small suppliers, used in opportunistic and changing ways depending on the modeling task at hand.

This new collaboration framework is called "co-opetition" as defined by Brandenburger and Nalebuff (1996): "Business is cooperation when it comes to creating a pie and competition when it comes to dividing it up." Plug-and-play capacity stimulates the market and creates new opportunities that could never have happened before. New value nets will be created with one supplier being another supplier's competitor, and at the same time the supplier's complement, as assembling components (or Web services in SOA) from several sources will provide more than just summing up the parts by operating them separately.

Be prepared for further innovations and business benefits in process and product engineering thanks to the increasing role of interoperability standards and to emerging information technologies.

Abbreviations

AIChE	American Institute of Chemical Engineering
API	Application Programming Interface
BPEL	Business Process Execution Language
BPEL4WS	Business Process Execution Language for Web Services
BPML	Business Process Markup Language
BPMI	Business Process Management Initiative
CAPE	Computer-aided process engineering
CCM	CORBA Component Model
CO	CAPE-OPEN
CO-LaN	CAPE-OPEN Laboratory Network
CORBA	Common Object Request Broker Architecture
(D)COM	(Distributed) Component Object Model
DTD	Document type definition
EAI	Enterprise Application Integration
ebXML	Electronic business XML
HTML	Hyper Text Markup Language
HTTP	Hyper Text Transfer Protocol
IDL	Interface Description Language
IIOP	Internet InterOrb Protocol
IS	Information system
IT	Information technologies
J2EE	Java 2 Platform Enterprise edition
MAS	Multi-agent system
OASIS	Organization for the Advancement of Structured Information Standards
OLE	Object linking and embedding
OPC	OLE for process control
OWL	Ontology Web Language
pdXI	Process Data Exchange Institute
pdXML	PlantData XML
OMG	Object Management group
RDF	Resource description framework
RPC	Remote procedure call
SAML	Security Assertions Markup Language
SQL	Structured Query Language
SOA	Service-oriented architecture
SOAP	Simple Object Access Protocol
UDDI	Universal Description, Discovery, Integration
UML	Unified Modeling Language
UO	Unit operation
WSDL	Web Services Description Language
WS-I	Web Services Interoperability Association
W3C	World-Wide Web Consortium
XML	Extensible Markup Language

Acknowledgements

The authors wish to express their thanks to colleagues of the CAPE-OPEN, Global CAPE-OPEN and COGents projects.

References

1 *Andrews W.* (2004) Predicts 2004, Gartner's predictions, www3.gartner.com/research/spotlight/asset-55117-895.jsp

2 BearingPoint, SAP and Sun Microsystems (2003), Livre blanc, Les services Web, Pourquoi? www.bearingpoint.fr/content/library/138-731.htm

3 *Bloomberg J.* Web services: A New Paradigm for Distributed Computing, The Rational Edge, September 2001, www-106.ibm.com/developerworks/rational/library/content/RationalEdge/archives/sep01.html

4 *Belaud J. P. Pons M.* Open Software Architecture for Process Simulation, Computer-Aided Chemical Engineering, 10, May 2002, Elsevier, Amsterdam, pp 847–852

5 *Belaud J. P. Braunschweig B. L. Halloran M. Irons K. Pi~nol D. Von Wedel L.* Processus de standardisation pour l'interopérabilité des composants logiciels de l'industrie des procédés, Système d'information modélisation, optimisation commande en génie des procédés, October Toulouse, France 2002

6 *Belaud J. P. Roux P. Pons M.* Opening Unit Operations for Process Engineering Software Solutions, AIDIC Conference Series, Vol. 6, AIDIC & Reed Business Information S.p.A. (2003) pp 35–44

7 *Brandenburger A. Nalebuff B.* Co-opetition, Currency Doubleday, New York 1996

8 *Braunschweig B. L. Britt H. Pantelides C. C. Sama S.* Process Modeling: the Promise of Open Software Architectures, Chemical Engineering Progress, September (2000) pp. 65–76

9 *Braunschweig B. L. Gani R. (eds.)* Software Architectures and Tools for Computer-Aided Process Engineering, Elsevier, Amsterdam 2002

10 COGents (2004), COGents project Web site, www.cogents.org

11 Cutter (2003) Consortium Web Services Terminology, Web Services Strategies, Vol. 2, No. 12, December 2003

12 *Fay S.* (2003), Standards and Re-Use, The Rational Edge, May 2003, www-106.ibm.com/developerworks/rational/library/2277.html

13 *Fieg G. Gutermuth W. Kothe W. Mayer H. H. Nagel S. Wendeler H. Wozny G.* A Standard Interface for Use of Thermodynamics in Process Simulation, Computers and Chemical Engineering, Vol. 19, Suppl., (2002) pp. S317–S320

14 Google Web APIs Beta (2004), www.google.fr/apis/index.html

15 *Heintzman D.* (2003), An Introduction to Open Computing, Open Standards, and Open Source, The Rational Edge, July 2003, www-106.ibm.com/developerworks/rational/library/content/RationalEdge/archives/july03.html

16 IBM Glossary (2004), Glossary of Computing Terms, www-306.ibm.com/ibm/terminology/goc/gocmain.htm

17 *Koch C.* (2003), The Battle for Web Services, CIO magazine, October 2003, www.cio.com/archive/100103/standards.html

18 *Iwanitz F. Lange J.* OPC-Fundamentals, Implementation and Application, Hüthig Fachverlag 2002

19 *Linthicum D. S.* Next Generation Application Integration: From Simple Information to Web services, September 2003, Addison Wesley, Boston 2003

20 *Mahalec V.* Open System Architectures for Process Simulation and Optimization, AspenTech Speech, ESCAPE'8 Conf., Belgium, 25 May 1998

21 *Manes A. T.* Web Services: a Manager's Guide, September 2003, Addison Wesley, Boston 2003

22 *Oellermann W.* Create Web Services with Business Value, .NET Magazine, November 2002, Vol. 2, Number 10, www.ftponline.com/wss/2002-11/magazine/features/wollermann/default.aspx

23 OPC foundation (1998), OPC Technical Overview, www.opcfoundation.org/01-about/OPCOverview.pdf

24 *Pons M. Belaud J. P. Banks P. Irons K. Merk W.* Missions of the CAPE-OPEN Laboratories Network, Proceedings of Foundations of Computer-Aided Process Operations, 2003, Coral Springs, Florida 2003

25 *Sama S. Piñol D. Serra M.* Web-based Process Engineering, Petroleum Technology Quarterly, 2003

26 *Sessions R.* What is a Service-Oriented Architecture (SOA)? ObjectWatch Newsletter, Number 45, October 2003, www.objectwatch.com/issue-45.htm

27 *Sherif M. H.* When is Standardization Slow ?, International Journal of IT Standards and Standardization Research, Vol. 1, Number 1, March 2003

28 Sim42 Foundation Simulator 42 Open Source Chemical Engineering Process Simulator, 2004 www.virtualmaterials.com/sim42

29 *Sprott D. Wilkes L.* Understanding Service-Oriented Architecture, Microsoft Architects Journal, EMEA edition, January 2004, www.thearchitectjournal.com/Journal/issue1/article2.html

30 *Teague T. L.* Electronic Data Exchange Using PlantData XML, AIChE Spring National Meeting, 10–14 March 2002, New Orleans Riverside, New Orleans 2002

31 *Teague T. L.* PlantData XML, Section 4.3 of Software Architectures and Tools for Computer-Aided Process Engineering, Elsevier, Amsterdam 2002b

32 W3C (2004) W3C, World-Wide Web Consortium, Web Ontology Language Use Cases and Requirements, www.w3.org/TR/2004/REC-webont-req-20040210/

33 *Warner A. G.* Block Alliances in Formal Standard Setting Environments, International Journal of IT Standards and Standardization Research, Vol. 1, Number 1, March 2003

34 *Weiss M. Cargill C.* Consortia in the Standards Development Process, Journal of the American Society for Information Science, Vol. 43, Number 8 (1992) pp. 559–565

35 *Westhaus U.* DETHERM ... on the Web, an On-line Service from DECHEMA, http://i-systems.dechema.de/detherm/ 2004

36 *White M.* Working Together: Collaborative Competition Creates New Markets, Cap Gemini Ernst & Young Center for Business Innovation E-journal, Issue 5 (2000) pp. 33–35, www.cbi.cgey.com/journal/issue5/index.html

Section 5
Applications

The previous sections of this book have shown how process systems engineering has developed methods and tools to address the increasing complexity of the process industries; it seeks to foster the development of new products and processes, to achieve optimal operation of complex equipment, and to help in the complex management of the global enterprises. Section 5 illustrates some applications of CAPE techniques, and aims to demonstrate what their benefits are, their current limits, and their short and long-term perspectives.

The first chapter illustrates the issue of education and training: how to teach the students to efficiently use very powerful tools, in order to better understand the concepts, to appreciate how the theory can be put into practice, while avoiding the dangers of misusing the software by merely pushing buttons to generate results. The applications covered deal mainly with process and product design, and illustrate also the concept of tool integration, since results must be carried out from one calculation step to the next.

The second chapter concentrates on model-based process operation. It illustrates various industrial applications of data validation, and shows how the use of more detailed models can improve the accuracy of estimating plant parameters. Use of thermodynamic constraints besides component and overall mass balances is illustrated. Examples are taken from a range of industries: oil refineries, chemicals, fertilizers, nuclear power plants. The main benefits are more reliable plant monitoring, capability to operate closer to limits with a better efficiency, early detection of faults, and reduction of analytical and instrumentation cost.

The last chapter illustrates CAPE techniques applied to solving production-planning problems for a multiproduct plant. The goal is to optimize revenue by reacting swiftly to changes in product demand, market prices and feed stocks availability. The uncertainty aspect is modeled by means of a stochastic approach. Multiple objectives are considered: either maximizing the expected value of the final profit over the planning period, or maximisation of the first quartile of the profit (robust solution). All steps in the application of the method are illustrated by means of a case study taken from a food additives plant.

Thus, Section 5 illustrates the diversity of CAPE tools and methods, and shows examples of current practice in areas ranging from process design to plant operation and production planning under uncertainty.

1

Integrated Computer-aided Methods and Tools as Educational Modules

Rafiqul Gani and Jens Abildskov

1.1
Introduction

The CAPE community has been developing computer-aided methods and tools for several decades and these days it is common practice in teaching as well as in industrial problem solving to use one or more pieces of currently available software. Students are trained to solve process-product engineering problems with state-of-the-art software, which they also later use during their professional career. Process simulators and their use in process design education has become a standard tool for process design courses everywhere. While these tools are able to provide excellent training in analysis of problems that are well-defined and have sufficient information to completely solve the problem, it is questionable if they are also suitable for solving open-ended problems (Doherty et al. 2000), such as those related to process-product design. Also, use of these tools in process-product design encourages the use of the inefficient trial and error solution approach as opposed to a systematic generate and test approach where additional tools for synthesis and design may be used together with process simulators.

As computer-aided design becomes more prevalent in the process industry, according to Finlayson and Rosendall (2000), it is essential that graduating engineers know the capabilities of the computer-aided systems that are available as well as the scepticism to interpret the results wisely. At the same time, advances in computer-aided design and simulation tools and reduced computing costs have allowed new uses of computing in chemical engineering education. Indeed, chemical process industries remain one of the strongest segments of the world-wide economy due to the cost effectiveness of well-designed chemical processes as well as to innovative chemistry (Doherty et al. 2000).

Process simulators (ASPEN+, PRO-II, gPROMS, ChemCad, ProSim, etc.) together with modeling and simulation software (Mathematica, Maple, MATLAB, etc.) have become standard computer-aided tools in chemical process design, process control

Computer Aided Process and Product Engineering. Edited by Luis Puigjaner and Georges Heyen
Copyright © 2006 WILEY-VCH Verlag GmbH & Co. KGaA, Weinheim
ISBN: 3-527-30804-0

and, chemical process-operation modeling. As industries face major new challenges because of increased global competition, greater regulatory measures, uncertainties in prices for energy, raw materials and products, etc., it becomes more and more important to consider integrated solution approaches. Similar to process integration, tools and/or problem integration imply the solution of more than one problem simultaneously or use of more than one tool in the solution of the problem. Also, the introduction of new courses such as product design has led to new challenges in fields such as applied thermodynamics (Abildskov and Kontogeorgis 2004), which also requires new software.

This chapter highlights the use of a system viewpoint within an integrated approach to the solution of chemical engineering problems. As noted by Edgar and Rawlings (2004), the chemical engineer leverages knowledge of molecular processes across multiple length scales to synthesize and manipulate complex systems that encompass both processes and products. Several computer-aided educational modules that encourage the development of this viewpoint, are presented in this chapter. First, a brief overview of the integrated approach to CAPE is given, followed by a short presentation of an integrated computer-aided system that has been used as a basis for the development of a number of computer-aided education modules. Three examples of these educational modules are presented together with references of where other modules can be found. In conclusion, uses of these modules in courses are discussed.

1.2
Integrated Approach to CAPE

An integrated approach to CAPE, also known as concurrent engineering, simply means the solution of two or more problems in a single step, for example, make decisions in the early stages of design that also select the control structure and guarantee acceptable environmental impact. In this way, it is similar to process integration where two or more operations are performed through a single operation, for example, a heat exchanger combining a cooling operation with a heating operation. Application of the integrated approach, however, also needs an integration of tools. As illustrated through Fig. 1.1, most CAPE problems (synthesis, design, and analysis) are multitask by nature and require a number of different tools. To achieve integration of tools, it is necessary to establish the workflow and data flow with respect to the solution steps and the tools that would be needed in each of the steps. Tools integration, therefore, avoids duplication of work while providing efficient data transfer from one tool to another. Through a computer-aided framework that includes a collection of tools (and their associated subtools such as databases, models, solvers, etc.) and allows access of the tools according to specific workflow and data flow, typical chemical engineering problems can be solved in an integrated manner. More details on tools integration can be found in Fraga et al. (2002).

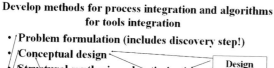

Develop methods for process integration and algorithms for tools integration

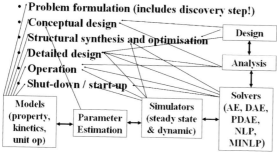

- Problem formulation (includes discovery step!)
- Conceptual design
- Structural synthesis and optimisation
- Detailed design
- Operation
- Shut-down / start-up

Figure 1.1 Multidisciplinary tools for nature process-product design problems

1.2.1
Integrated Computer-aided System

An integrated computer-aided system (ICAS) (Gani 2001) combines computational tools for modeling, simulation (including property prediction), synthesis/design, control and analysis in a single integrated system. These computational tools are presented as toolboxes. During the solution of a problem, the student moves from one toolbox to another in order to solve problems, which require tools from more than one toolbox. Typically, problems in process synthesis, process design, and process control require the use of more than one tool. For example, in process design/synthesis, one option is to define the system input stream, to analyze the mixture (use of analysis tools), to generate flow sheet alternatives (synthesis/design tools), to evaluate the alternatives (simulation and analysis tools), and finally, to optimize the flow sheet (design tools). Each toolbox has a number of tools associated with it and are connected to the necessary tools from other toolboxes. Figure 1.2 illustrates the architecture of ICAS.

ICAS has been developed specifically to solve problems in an integrated manner, which can be used to develop educational modules for different types of product-process engineering problems. It is currently used to solve industrial problems as well as research and teaching. In this chapter, only the teaching related features will be highlighted. As shown in Fig. 1.2, ICAS consists of a simulator (with steady state and dynamic simulation engines) having the same features as other process simulators. ICAS, however, also has tools that "add to the system" and "toolboxes" that help to solve some of the tasks typically found in different CAPE related problems (for example, designselection of solvents, synthesis of process flow sheets, environmental impact analysis, model parameter estimation, etc.). The "add to the system" helps to introduce new compounds into the database, new unit operation models into the simulation model library, and new property models into the property model library in an integrated manner and requiring no additional programming. Once all these additions are introduced to the system, all tools within ICAS will be able to use them. In this way, the "add to the system" and "toolboxes" help to define the problem

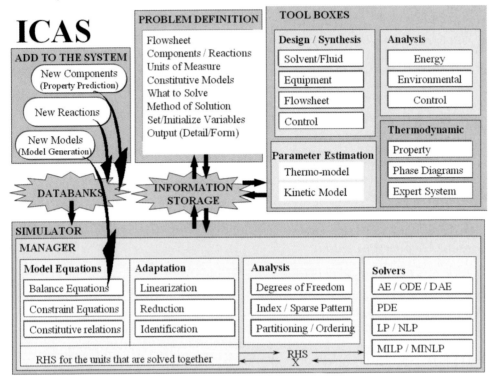

Figure 1.2 Multidisciplinary tools for nature process-product design problems

where most of the time is spent in software-based solution of problems. Often, the problems are not defined correctly or consistently, resulting in failure of the numerical solver. The different features of ICAS help to guide the students into defining/ formulating the problem correctly so that the numerical solver does fail, if a solution for the formulated problem exists.

1.3
Educational Modules

Three computer-aided educational modules involving property prediction (suitable for a course on thermodynamics or product design), extractive distillation-based separation (suitable for courses on separation processes, distillation, or process design), and model derivation and solution (suitable for courses on modeling, simulation and/or numerical methods) are presented. The objective of these educational modules is to highlight the solution strategies for typical chemical engineering problems (traditional as well as new) where software may be used (with clearly defined objectives) in some or all the solution steps (tasks). At the same time, it is emphasized that the software is just an efficient calculator (that is, it provides answers when asked) but it does not work as an engineer. Also, in addition to the above calculator service, the software plus the workflow and data flow provide insights to improve the solution efficiency (of the overall problem) and therefore, the productivity of the user (student).

1.3.1
Computer-aided Property Estimation

This computer–aided teaching module introduces the students to the workflow, data flow and the tools needed to perform phase equilibrium calculations (saturation point calculation and generation of various types of phase diagrams: vapor–liquid, liquid–liquid or solid–liquid). The students learn the importance of the property model selection, the need for property databases, the need for additional property models, the model equations, and finally, the important calculation steps for solving the problem. In this way, the students are able to appreciate not only the property related calculations but also understand their influence in other problems, such as process simulation and design. Two problems are presented here:

- Analyze the properties of a chemical called chemical fentanyl.
- Evaluate the binary mixture of ethanol–water.

The first problem could easily come from the product analysis step of a chemical product design problem, while the second problem could come from a bioprocess (downstream separation of a fermentation product or solvent based separation by distillation or even a solvent based crystallization process). In order to progress further in the process-product design problem, the pure component properties as well as the mixture properties need to be evaluated.

1.3.1.1
Analysis of Fentanyl

Here, we wish to analyze the properties of fentanyl (CAS number 000437-38-7) in terms of its state at the normal conditions of temperature and pressure, if it is toxic and its solubility properties in water and other solvents. The pure component properties that would be needed are: normal boiling point (T_b), normal melting point (T_m), the heat of fusion (ΔH_f), the heat of vaporization (ΔH_{vap}), the vapor pressure (P^{sat}), the Hildebrand solubility parameter (δ_S), the octanol–water partition coefficient (log K_{ow}) and a measure of toxicity (LC_{50}).

The following steps (workflow) could be performed:

1. Check databases to find properties of fentanyl (properties such as T_b, T_m, ΔH_f, (H_{vap}, P^{sat}, δ_S, log K_{ow} and LC_{50}).
2. If the properties cannot be found in the databases, use a property estimation package.
   a. Generate the needed properties through a property model by giving the molecular structural information.

3. Analyze the properties (estimated or retrieved from database).
   a. What is the state (solid, liquid or gas) at the normal condition of temperature and pressure?
   b. Is it a hazardous compound?
   c. Are there known solvents for fentanyl?
   d. How can solubility of fentanyl in solvents be quickly checked?

The methods and tools that are needed to perform the workflow shown above are the following: a fairly large database of pure component properties, a software package for prediction of pure component properties (with its resident model parameter tables), a software package for solvent search, a software package for solubility calculations (requires property models for mixture properties as well as algorithms for saturation point calculations). ICAS provides all the above in a single integrated system. Uses of ICAS (Gani 2001) for each of the above steps are highlighted below (information on all the ICAS tools can be found at www.capec.kt.dtu.dk/Software/ICAS-and-its-Tools/or in Gani (2001)):

Step 1: Search of the CAPEC database (Nielsen et al. 2001) in ICAS finds fentanyl but having only the molecular weight (336.48) and the normal melting point (360.65 K). This means all other properties need to be estimated. Databases in most process simulators will not have this compound or its properties.

Step 2: To generate the properties, the ProPred toolbox within ICAS is used. ProPred needs the molecular structure of the molecule (as a 2D drawing, as a 2D/3D mol.file or as a SMILES string). The CAPEC database has the SMILES string, from which ProPred is able to draw the molecule, identify the needed groups and estimate the properties (in the case of fentanyl, since all the group parameters were not available, it also needed to create the groups, which is an option available in ProPred). Table 1.1 gives the SMILES string for fentanyl, the 2D drawing of the molecule and the properties estimated by ProPred.

Step 3: Analyze fentanyl in terms of the generated properties.
   Step 3a: At the normal condition of 300 K and 1 atm, fentanyl is solid.
   Step 3b: It is hazardous as indicated by the $-\log(LC_{50})$ value. A high value (higher than 3) indicates a highly toxic compound. LC_{50} is the aqueous concentration causing 50% mortality in fathead minnow after 96 hours.
   Step 3c: The CAPEC database does not list any known solvents for fentanyl. However, as the Hildebrand solubility parameter is known (δ_S at 298 K), it can be used to obtain some idea of which compounds could be good solvents. A search of the database for compounds having similar δ_S (at 298 K) shows that hydrocarbons are likely to be good solvents while fentanyl will have very low solubility in water.
   Step 3d: To estimate the solubility, the needed properties can be seen from the following equation (condition for solid–liquid equilibrium where only one compound exists in the solid phase):

$$1 = x_S \gamma_S \exp\left[\Delta H_f / (RT_m)\{(T - T_m)\}\right] \tag{1}$$

Table 1.1 Estimated properties of fentanyl

Estimated properties of fentanyl	
SMILES string	CCC(=O)N(c1ccccc1)C2CCN(CCc3ccccc3)CC2
2D molecular structure	
T_b	703.47 K (estimated with created groups[a])
T_m	360.65 K
ΔH_f	44.66 kJ mol^{-1} (estimated with created groups)
ΔH_{vap} at 298 K	42.16 kJ mol^{-1}
δ_S at 298 K	20.78 MPa$^{1/2}$ (estimated with created groups)
Log K_{ow}	3.85
$-$Log (LC$_{50}$)	7.32
P^{sat} at 300 K	Fentanyl is solid at this temperature

a "Estimated with created groups" means that the need group parameters were not available and therefore, ProPred automatically generated the missing group parameters to estimate the needed properties.

where x_s is the saturation composition of solid s in solution, γ_s is the liquid activity coefficient of the solid compound in solution, R is the universal gas constant and T is the temperature at which solubility is to be calculated. From Eq. (1), it becomes clear that to estimate the solubility of fentanyl in a solvent, we need to estimate its liquid activity coefficient in the solvent (for which a property model is necessary) as well as the heats of fusion and melting point of fentanyl. A quick estimate may be obtained by setting $\gamma_s = 1$ (assuming ideal liquid). Note that if a liquid activity coefficient model such as UNIFAC (Hansen et al. 1991; Kang et al. 2002) is used, it will require the corresponding group interaction parameters, which for fentanyl are not available. Also, since γ_s is dependent on composition as well as temperature, an iterative solution technique would be necessary. The SoluCalc toolbox in ICAS, which has been specially developed for estimating solid solubility in solvents, can be used for this purpose. Figure 1.3 shows the calculated fentanyl saturation curve in hexane (solvent). Note that the students have the option to directly calculate the temperature versus composition diagram through ICAS utility toolbox, or, develop their own binary SLE phase diagram software using the property model (as a model object) from ICAS through modeling/simulation software (such as EXCEL, MATLAB, etc.).

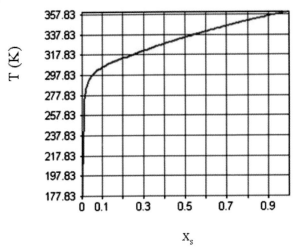

Figure 1.3 Estimated saturation solubility curve for fentanyl in hexane

The above workflow could be repeated for any new chemical being considered as a product, provided the needed property variables can be identified and their values measured or estimated through appropriate property models. Solving this type of problems with process simulators is not efficient and in most cases probably not also possible. Having all the necessary tools available in an integrated manner saves time and provides valuable insights to the problem and its solution.

1.3.1.2
Evaluation of Ethanol–Water Mixture

Here, we wish to evaluate the ethanol–water mixture with respect to its separation from a biofermentation reactor. Depending on the mixture characteristics, different separation schemes may be generated. Here, however, we will first look at the vapor–liquid equilibrium and confirm that it is indeed a minimum boiling azeotrope. Then we will introduce a solvent, for example benzene or ethylene glycol, to the system and evaluate the ternary mixture (in terms of ternary azeotrope and liquid–liquid miscibility). The methods and tools needed to perform these tasks (calculations) are available in most commercial simulators but they are not necessarily organized for an integrated approach. In this example, however, we will break down the problem into multiple tasks in order to understand the problem, to highlight the importance of property model selection, the need for accurate pure component properties (in this case, vapor pressure) as well as the associated workflow (calculation steps or tasks) and data flow.

Since ethanol–water is a nonideal mixture for which vapor–liquid (and possibly vapor–liquid–liquid when a ternary system is considered), equilibrium needs to be calculated, the properties, calculation methods and tools that are needed, are linked to the equilibrium model used. That is, if we select the equilibrium model as a two-model gamma–phi type, we may select an activity coefficient model for the liquid phase and the ideal gas model (equation of state) for the vapor phase. Neglecting the Poynting correction factor, the vapor–liquid equilibrium is represented by:

$$y_i = x_i \gamma_i P_i^{\text{sat}} / P \tag{2}$$

In the above equation, y_i is the vapor phase composition of component i, x_i is the corresponding equilibrium liquid phase composition, γ_i is the liquid phase activity coefficient of component i, P_i^{sat} is the vapor pressure of component i at the equilibrium temperature and P is the corresponding system pressure. Since γ_i is a function of composition and temperature and P_i^{sat} is a function of temperature, an iterative solution scheme needs to be devised to obtain the equilibrium temperature and the corresponding vapor composition for given liquid composition and pressure. Repeating the calculations for different values of liquid composition within the limit zero to one and keeping the pressure fixed at the original value, generates the entire so-called PT-xy diagram. Now consider the following three options:

- Given models for γ_i (for example, UNIFAC) and P_i^{sat} (for example, the Antoine correlation), develop a computer program to generate the phase diagram. In principle, any modeling software (EXCEL, MATLAB, etc.) can be used to develop the software with the ICAS supplied property model object.
- Given a program to calculate the saturation temperature and vapor composition for specified liquid composition and pressure (and for a selected set of property models), repeat the calculations to generate the phase diagram. In principle, any modeling software (EXCEL, MATLAB, etc.) can be used to develop the software with the ICAS supplied property-utility object.
- Given software with built-in models and calculation options, select the appropriate model and calculation option to generate the needed phase diagram. In principle, any process simulator and/or ICAS utility toolbox may be used. If, however, the compounds are not ethanol and water, then the available options in the software need to be checked.

All three options will give the required solution, while the first option will be time-consuming, the student will gain more insight on the needed workflow and data flow than the last option, which will be efficient in terms of problem solution but will provide little insight. An interesting approach could be to get the students to use all options, that is, use the first option at the beginning and use the last option when they are experienced in phase equilibrium calculations.

This example also provides insights on property model selection (assuming an ideal system, which is usually the default selection for many software, the azeotrope will not be found) and accuracy of the needed properties (that is, the predicted azeotrope location may be highly sensitive to the accuracy of the vapor pressure model). Also, the parameters of the liquid activity coefficient model are important as they may give different values of the location of the azeotrope.

Having analyzed the binary system, the next step is to analyze the mixture when a third component (for example, a solvent) is introduced. What happens to the azeotrope? Are there still a vapor and a liquid phase in equilibrium? If not, is there an additional liquid phase? If yes, how to calculate the phase compositions and is there also a ternary azeotrope?

As in the case of the binary mixture calculations, most commercial simulators also provide options to perform the calculations so that the above questions can be answered. The important points, however, are the following: Have the correct property model selections been made? What are the workflow and data flow? Are the results acceptable? Again, by breaking down the problem into multiple tasks, the students will gain more insights to the solution of the problem. In terms of calculations, in addition to the vapor–liquid equilibrium, the liquid–liquid equilibrium also needs to be computed:

$$x_{1i}\gamma_{1i} = x_{2i}\gamma_{2i} \tag{3}$$

The subscript 1 and 2 in the above equation indicates liquid phase 1 and liquid phase 2, respectively. The same liquid phase activity coefficient model will now be used in Eq. (2) and (3). However, the liquid composition from Eq. (2) needs to be checked for phase stability and if found unstable, Eqs. (2) and (3) will need to be solved simultaneously. The following calculation steps (workflow) could be used:

- Use Eq. (3) to identify any binary pair (there are 3 binary pairs in the ternary mixture), which splits into two liquid phases:
  - Check also if this is the vapor–liquid azeotrope point (for a vapor–liquid–liquid system, one pair will satisfy this condition).
  - For the binary system showing both an azeotrope as well as liquid–liquid phase split, add incremental amounts of the third component and perform the vapor––liquid–liquid calculations until there is only one liquid phase (note that for each calculation, the pressure is fixed at a constant value but the temperature is also calculated). The ethanol–water system with benzene as the solvent will show a vapor–liquid–liquid ternary system with the benzene–water pair showing the binary liquid–liquid phase split.
- If none of the binary pairs split into two liquid phases, there will only be a vapor in equilibrium with liquid and the calculations for the binary mixture can be repeated in the same way as before. The ethanol–water system with ethylene glycol as the solvent will show only a vapor–liquid system.

The three options listed above for the binary mixture calculations can also be repeated now for the ternary mixture calculations. Again, initially, it is better for the students to develop their own calculation program but later, they can use standard software (process simulator and/or ICAS utility toolbox).

Further extension to this problem could be considered by adding an inorganic salt to the ethanol–water mixture (to study the salting-in or salting-out effect). Here, again the workflow will basically remain the same but the data flow will be significantly different because of the different property models and their corresponding property parameters.

The above modules will prepare the students to solve all types of phase equilibrium problems in a systematic way, even when not all data and/or property model parameters are available. One advantage of allowing the students to develop their own software is that in the case of new compounds or systems, they will be better prepared to perform the necessary tasks. Note that since developing their software will not need much programming effort, they will actually concentrate on learning the calculations involved in each task.

1.3.2
Separation of Azeotropic Mixtures

The second computer-aided educational module introduces the students to aspects of design and simulation of solvent-based extractive distillation. The students use the phase diagrams that they have learned to generate. The main feature of this module is that it encourages the students to make the design decisions based on simple thermodynamic calculations rather than use the simulator on a trial and error basis to find the solution. That is, all the important design decisions are made through the generated phase diagrams and thermodynamic insights (for example, sequencing of distillation columns, selection of solvents, design of individual columns, etc.), which also generates an initial estimate for a detailed simulation of the process. In the final step, the initial estimate is passed to the simulator and the solution is obtained without too many iterations (by the rigorous model solver). Therefore, the student spends less time with the simulator and more time generating information (knowledge) that can be used for problem solution.

1.3.2.1
Problem Description
The problem which we will consider is the separation of a binary mixture of acetone and chloroform into high purity products. As this binary mixture forms an azeotrope (to be verified), solvent-based extractive distillation is an option. Benzene is a well known solvent that has been reported as a suitable solvent. Benzene, however, is not acceptable for environmental, health and safety (EHS) reasons and an alternative solvent needs to be found and verified.

1.3.2.2
Problem Solution
The following steps provide a solution to this binary mixture problem.

Step 1: Perform a mixture analysis and verify that an azeotrope exists and check for its dependence on pressure. Make decisions/choice of property model and calculation steps. An ideal system cannot be assumed since the binary mixture forms an azeotrope. A VLE-based phase diagram (using Eq. (2)) needs to be generated. Figures 1.4a and 1.4b show the binary azeotropes as a function of pressure (ICAS utility toolbox is used to generate these diagrams).

Step 2: Find solvents that perform as well as benzene but without the negative EHS properties of benzene. Using the ProCAMD tool in ICAS, a large number of candidate solvents are generated. The problem definition is as follows: find solvents that

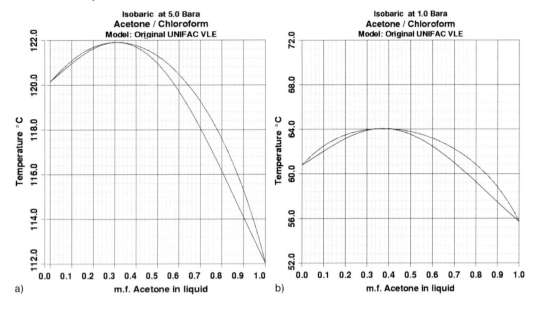

a)

b)

Figure 1.4 Acetone-chloroform VLE calculated at 5 bar (a). Acetone-chloroform VLE calculated at 1 bar (b)

are acyclic organic compounds having 320 K $< T_b <$ 420 K, $T_m <$ 250 K, that is more selective to chloroform than acetone (selectivity > 1.7), is totally miscible with acetone-chloroform (therefore a vapor–liquid system) and does not form azeotrope with either acetone or chloroform.

The solution statistics from ProCAMD is shown in Fig. 1.5. It can be noted that 5614 candidate molecules were generated, out of which 59 satisfied all constraints. From these 59 molecular structures, 133 isomers were generated and a more refined property estimation was made to identify 111 compounds that satisfied all constraints. Note that to identify the solvents, pure component properties (T_b, T_m) as well

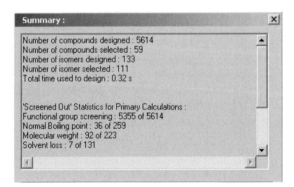

Figure 1.5 Solution statistics from ProCamd for the solvent selection/design task

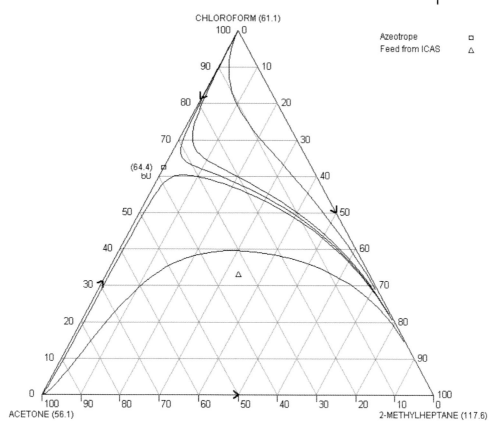

Figure 1.6 Calculated distillation boundaries for the acetone-chloroform solvent

as phase equilibrium calculations (selectivity, azeotrope calculation and liquid miscibility) needed to be performed. Therefore, an integrated system capable of doing these steps automatically for the user is very useful for this type of problems. Two of the alternatives found are methyl-n-pentyl-ether and 2-methylheptane.

Step 3: Analyze the alternative solvent candidates in terms of distillation boundaries. For this step, the PDS tool in ICAS is used. PDS performs, among others calculations of distillation boundaries, residue curves and distillation column design for a specified ternary system (reacting or nonreacting). This problem is nonreacting and the distillation boundaries calculated through PDS are shown in Fig. 1.6.

Step 4: Generate the process flow sheet and design the corresponding distillation columns. This can be done interactively. Design the first distillation column with the feed mixture and the fresh solvent added. The combination of the feed mixture and solvent places the total feed in the region where acetone can be obtained as the top product, and a mixture of mainly solvent and chloroform will be obtained as the bot-

All temperatures are displayed in Celcius

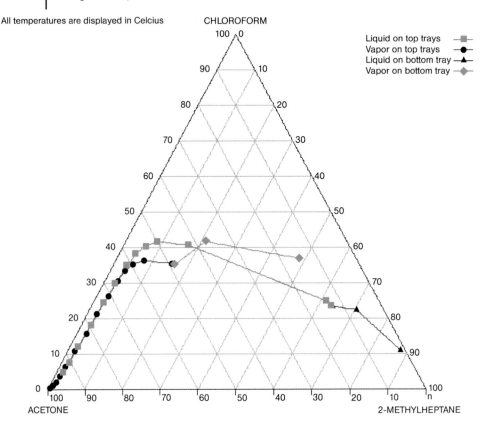

Figure 1.7 Design of the distillation column that is consistent with the distillation boundaries

tom product. This feed will then be sent to a second column, from where the chloroform will be obtained as a top product and the solvent will be recovered and recycled. PDS and ICAS-SIM (steady state simulation engine) can be used interactively to design the first column (number of stages, feed location, product purity, reflux ratio, etc.) and verified through steady state simulation. Then the second column is added and the procedure repeated, which also generates the total flow sheet. Figure 1.7 shows an output from PDS, highlighting the design calculations for the distillation column.

Step 5: Perform simulation and optimization to determine the optimal design of the separation process. In this case, all solvents can also be included in the same calculations and the optimization problem will find the optimal flow rate for each of the solvents. The flow sheet used in problem formulation and the sample simulation of the flow sheet are shown in Figs. 1.8a and 1.8b. This last step can be performed with the steady state simulation engine of ICAS or with any process simulator. ICAS has a

a)

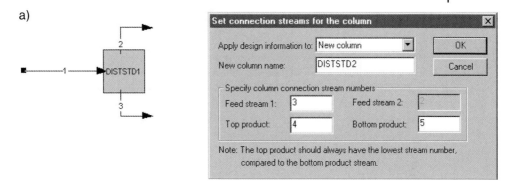

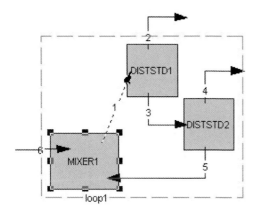

Figure 1.8a Generation of flow sheet (synthesis and design) with PDS-ICAS (a).

direct interface for the PRO-II steady state simulator. This means that any flow sheet synthesized and designed can be directly simulated in PRO-II without adding any further information in PRO-II.

Several variations of the problem are possible. For example, provide a binary azeotropic mixture that has significant pressure dependence so that the solvent-based separation can be compared with pressure-swing distillation. If a solvent that introduces a phase split is selected or specified, then at least one distillation column will have vapor–liquid–liquid phase equilibrium and the design calculations will have to be consistent with the resulting distillation boundaries. Also, a reacting system may be introduced. Basically, the tasks (steps in workflow) shown above would be similar in all cases but the data flow would be different because of specific choices of the models used to perform the tasks. If the student is given the calculation steps (workflow) and a corresponding set of integrated tools, the students will not only be able to solve these problems without too much difficulties but also gain valuable insights with each solution step.

b)

STREAM NUMBER	1	2	3	4
TEMPERATURE (K)	350.00000	332.04270	373.15131	342.63881
PRESSURE (atm)	1.00000	1.00000	1.00000	1.00000
ENTHALPY (K/Kmole)	-27615.86917	-24505.51626	-26910.01571	-14287.60495
ENTROPY (1/Kmole)	41.50294	35.62145	44.69363	37.41181
U-ENERGY (K/Kmole)	-27616.02699	-24532.76285	-26910.18654	-14315.72103
DENS. (Kmole/m^3)	6.33628	0.03670	5.85403	0.03557
VAPOUR FRACTION	0.00000	1.00000	0.00000	1.00000
LIQUID FRACTION	1.00000	0.00000	1.00000	0.00000
-------- (Kmole/hr)				
ACETONE	10.00039	9.05476	0.94563	0.94524
CHLOROFORM	10.98840	0.85620	10.13220	9.14408
2-METHYLHEPTANE	89.89447	0.06853	89.82593	0.93118
TOTAL	110.88325	9.97949	100.90376	11.02051

STREAM NUMBER	5	6
TEMPERATURE (K)	389.00747	350.00000
PRESSURE (atm)	1.00000	1.00000
ENTHALPY (K/Kmole)	-27545.64659	-19114.10880
ENTROPY (1/Kmole)	47.80138	36.69383
U-ENERGY (K/Kmole)	-27545.83081	-19142.82893
DENS. (Kmole/m^3)	5.42829	0.03482
VAPOUR FRACTION	0.00000	1.00000
LIQUID FRACTION	1.00000	0.00000
-------- (Kmole/hr)		
ACETONE	0.00039	10.00000
CHLOROFORM	0.98811	10.00000
2-METHYLHEPTANE	88.89475	1.00000
TOTAL	89.88325	21.00000

Figure 1.8b Verification by simulation of the generated process flow sheet (b)

1.3.3
Integrated Computer-aided Modeling

The third educational module deals with modeling issues. Here, the importance of model analysis before attempting to solve the model equations is emphasized together with model reuse in an external modeling/simulation environment. The degrees of freedom, the ordering of equations, the method of solution are all interrelated and through a computer-aided modeling toolbox, the students are encouraged to use these features whenever they have to solve problems represented by a set of equations. As shown in Fig. 1.9, the objective is to transform the model equations into a program code that can be used by other simulation engines and/or solvers. At the same time, the programming effort should be a minimum. But, before going to the solution phase, the model equations must be thoroughly analyzed. Although the modeling/simulation problems in most cases can be solved through a number of existing programs (MATLAB, Maple, Mathematica, etc.), in the current example, the use of ICAS-MoT is highlighted for the reasons given above.

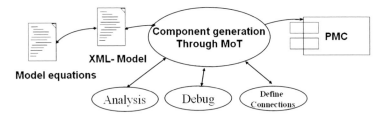

Model equations

Figure 1.9 Import of model equations to MoT and after transformation and analysis, export to a process modeling component (or external simulation engine)

The use of MoT, a tool in ICAS, as an educational tool will be highlighted through a simple reactor modeling exercise.

1.3.3.1
Modeling Problem Description

The series reactions:

$$2A \xrightarrow[(1)]{k_{1a}} B \xrightarrow[(2)]{k_{2b}} 3C \qquad (4)$$

are catalyzed by H_2SO_4. All reactions are first order in the reactant concentration. The reactions are carried out in a semi-batch reactor that has a heat exchanger inside with $UA = 35\,000$ cal h^{-1} K and ambient temperature of 298 K (see Fig. 1.10). Pure A enters at a concentration of 4 mol dm^{-3}, a volumetric flow rate of 240 dm^3 h^{-1}, and a temperature of 305 K. Initially there is a total of 100 dm^3 in the reactor, which contains 1.0 mol dm^{-3} of A and 1.0 mol dm^{-3} of the catalyst H_2SO_4. The reaction rate is independent of the catalyst concentration. The initial temperature of the reactor is 290 K.

The objective of this exercise is to highlight the basic features of ICAS-MoT and at the same time, the modeling steps needed to obtain the dynamic evolution of all concentrations in the semi-batch reactor for the given operating conditions.

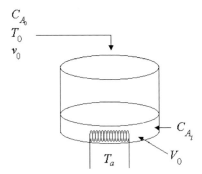

Figure 1.10 Semi-batch reactor scheme

1.3.3.2
Description of the Mathematical Model

Mole Balances

$$\frac{dC_A}{dt} = r_A + \frac{(C_{A0} - C_A)}{V} v_0 \tag{5}$$

$$\frac{dC_B}{dt} = r_B - \frac{C_B}{V} v_0 \tag{6}$$

$$\frac{dC_C}{dt} = r_C - \frac{C_C}{V} v_0 \tag{7}$$

where the rate laws are:

$$-r_{1A} = k_{1A} C_A \tag{8}$$

$$-r_{2B} = k_{2B} C_B \tag{9}$$

and the kinetic constants are Arrhenius-type

$$k_{1a} = k_{1A} \exp\left\{\left(\frac{E_{1A}}{R}\right)\left[\frac{1}{T_{1A}} - \frac{1}{T}\right]\right\} \tag{10}$$

$$k_{2b} = k_{2B} \exp\left\{\left(\frac{E_{2B}}{R}\right)\left[\frac{1}{T_{2B}} - \frac{1}{T}\right]\right\} \tag{11}$$

The relative rates are obtained using the stoichiometry (liquid phase) for the reaction series (Eq. (4)):

$$-r_{2B} = \frac{r_{2C}}{3} \tag{12}$$

$$r_{2C} = -3r_{2B} \tag{13}$$

So that the net rates are as follows:

$$r_A = r_{1A} = -k_{1A} C_A \tag{14}$$

$$r_B = r_{1B} + r_{2B} = \frac{-r_{1A}}{2} + r_{2B} = \frac{k_{1A} C_A}{2} - k_{2B} C_B \tag{15}$$

$$r_C = 3k_{2B} C_B \tag{16}$$

$$N_i = C_i V \tag{17}$$

$$V = V_0 + v_0 t \tag{18}$$

$$N_{H_2SO_4} = \left(C_{H_2SO_4,0}\right) V_0 = \frac{1\,\text{mol}}{\text{dm}^3} \times 100\,\text{dm}^3 = 100\,\text{mol} \tag{19}$$

$$F_{A_0} = \frac{4\,\text{mol}}{\text{dm}^3} \times 240\,\frac{\text{dm}^3}{h} = 960\,\frac{\text{mol}}{h} \tag{20}$$

Energy Balance

$$\frac{dT}{dt} = \frac{UA(T_a - T) - \sum_{i=1}^{NC} F_{i0} Cp_i(T - T_0) + \sum_{i=1}^{NR} \Delta H_{Rx_{i,j}} r_{i,j} V}{\sum_{i=1}^{NC} N_i Cp_i} \tag{21}$$

is equivalent to

$$\frac{dT}{dt} = \frac{UA(T_a - T) - F_{A0} Cp_A(T - T_0) + \left[(\Delta H_{Rx_{1A}})(r_{1A}) + (\Delta H_{Rx_{2R}})(r_{2B}) \right] V}{\left[C_A Cp_A + C_B Cp_B + C_C Cp_C \right] V + N_{H_2SO_4} Cp_{H_2SO_4}} \tag{22}$$

Summarizing, the process model is given by: four ordinary differential equations (Eqs. (4–6, 22)) and fourteen algebraic equations (Eqs. (8–20); note that Eq. (14) is actually two equations). This differential-algebraic system can be solved simultaneously using ICAS-MoT. The data for this problem can be found in Table 1.2.

Table 1.2 Data for the differential-algebraic system

Variable	Value	Units	Description	MoT-variable
C_{A_0}	4.0	mol dm^{-3}	Initial concentration of compound A	CA0
$C_{H_2SO_4,0}$	1.0	mol dm^{-3}	Initial catalyst concentration	CH2SO40
v_0	240.0	dm^3 h^{-1}	Initial flow rate	vo
V_0	100.0	dm^3	Initial reactor volume	V0
UA	35´000.0	cal h^{-1} K	Heat transfer coefficient	UA
T_a	298.0	K	Ambient temperature	Ta
T_0	305.0	K	Inlet Temperature	T0
T_{1A}	320	K	Reaction temperature (reaction 1)	T1A0
T_{2B}	300	K	Reaction temperature (reaction 2)	T2B0
k_{1A}	1.25	h^{-1}	Kinetic reaction constant(reaction 1)	k1A0
k_{2B}	0.08	h^{-1}	Kinetic reaction constant (reaction 2)	k2B0
E_{1A}	9500.0	cal mol^{-1}	Activation energy (reaction 1)	E1A
E_{2B}	7000.0	cal mol^{-1}	Activation energy (reaction 2)	E2B
Cp_A	30.0	cal mol^{-1} K	Thermal heat capacity of compound A	CpA
Cp_B	60.0	cal mol^{-1} K	Thermal heat capacity of compound B	CpB
Cp_C	20.0	cal mol^{-1} K	Thermal heat capacity of compound C	CpC
$Cp_{H_2SO_4}$	35.0	cal mol^{-1} K	Thermal heat capacity of catalyst	CpH2SO4
ΔH_{Rx1A}	−6500.00	cal mol^{-1}	Reaction enthalpy (reaction 1)	DHRx1A
ΔH_{Rx2B}	+8000.00	cal mol^{-1}	Reaction enthalpy (reaction 2)	DHRx2B
R	1.987	cal mol^{-1} K	Universal gas constant	R

1.3.3.3
Modeling Steps in MoT

The model developer does not need to write any programming codes to enter the model equations. Models are entered (imported) as text files or XML files, which are then internally translated.

Step 1: Type the model equations in MoT or transfer a text file or an XML file. In Fig. 1.11, the model (Eqs. (4–22)) has been typed into MoT.

```
#***********************************************
#*Nonisothermal Multiple Reaction            *
#*                                            *
#*CAPEC, Department of Chemical Engineering*
#*Technical University of Denmark            *
#*MSC, April, 2004                           *
#***********************************************

#The series reactions:
#              k1A        K2B
#       2A ------> B -----> 3C
#           (1)        (2)

#**************
#Solution    *
#**************

#Kinetic
  k1A = k1A0*exp((E1A/R)*(1/T1A0 - 1/T))
  k2B = k2B0*exp((E2B/R)*(1/T2B0 - 1/T))

#Rate Laws
  r1A = -k1A*CA
  r2B = -k2B*CB
  rA = r1A
  rB = k1A*CA/2-k2B*CB
  rC = 3*k2B*CB

#Reactor volume
  V = V0 + vo*t

  FA0 = CA0*vo
  Cpmix = CA*CpA + CB*CpB + CC*CpC

#Mol balances
  dCA = rA + (CA0 - CA)*vo/V
  dCB = rB - CB*vo/V
  dCC = rC - CC*vo/V

#Energy Balance
  dT = (UA*(Ta-T) - FA0*CpA*(T-T0) + (DHRx1A*r1A +
        DHRx2B*r2B)*V)/(Cpmix*V + CH2SO40*V0*CpH2SO4)
```

Figure 1.11 MoT model

a)
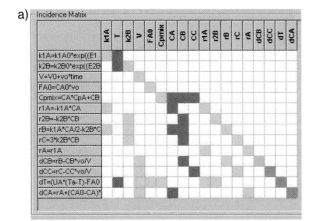

Figure 1.12a Incidence matrix (a). Equation partitioning and incidence matrix comparison (b)

b)
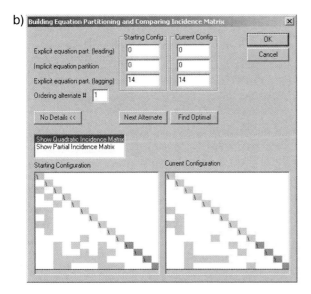

Step 2: Model translation. MoT translates the imported model, lists all the equations and variables found in the translated model. It has built-in knowledge to distinguish between algebraic equations (explicit and implicit), ordinary differential equations and partial differential equations. The variables are classified as parameters, known, unknown (implicit), unknown (explicit), dependent and dependent prime (used for differential equations only). MoT automatically identifies the unknown (implicit), unknown (explicit) and dependent equations. The user needs to classify the known variables as either parameters (which could then be selected for model parameter estimation) or known variables (which could be used as design variables for optimization). Also, the user needs to link the dependent variables to the dependent prime (that is, in dy/dt, y is the dependent variable and dy is the dependent prime).

Step 3: Incidence matrix analysis. As the variables are assigned, MoT is able to generate a corresponding incidence matrix (equations are placed in rows and variables in columns) and order the equations as near as possible to a lower tridiagonal form. Also, unless the degrees of freedom are matched (that is, a square matrix of equations and unknown variables and the known variables equal the degrees of freedom), MoT does not allow the solver to be called. A sample of the incidence matrix is shown in Figs. 1.12a and 1.12b.

Step 5: Define the independent variable. In this case, the independent variable is time t. MoT is now able to find the 14 algebraic equations and 4 ordinary differential equations and is therefore ready to start the solver. However, before the solver is called, the initial values for the dependent variables need to be specified together with the parameters and the known variables.

Step 6: Set variable values. Figure 1.13 shows the specified values for the variables.

Model Solution
Step 7: Select variables for output. The user may select the variables whose values can be stored and visualized as the numerical solver progresses towards the solution.

	Dependent	Dependent Prime	Unknown	Explicit	Known	Parameter
T	298					
CA	1					
CB	0					
CC	0					
dCA		0				
dCB		0				
dCC		0				
dT		0				
k1A				0		
k2B				0		
r1A				0		
r2B				0		
rA				0		
rB				0		
rC				0		
V				0		
FA0				0		
Cpmix				0		
k1A0					1.25	
E1A					9500	
R					1.987	
T1A0					320	
k2B0					0.08	
E2B					7000	
T2B0					300	
V0					100	
vo					240	
CA0					4	
CpA					30	
CpB					60	
CpC					20	
UA					35000	
Ta					298	
T0					305	
DHRx1A					-6500	
DHRx2B					8000	

Figure 1.13 Initial condition and values of known variables

a)
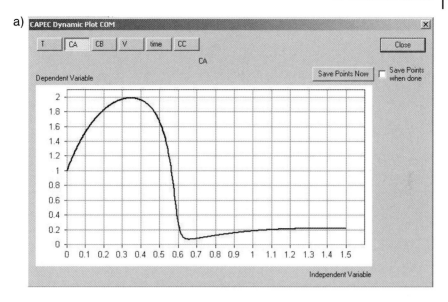

b)
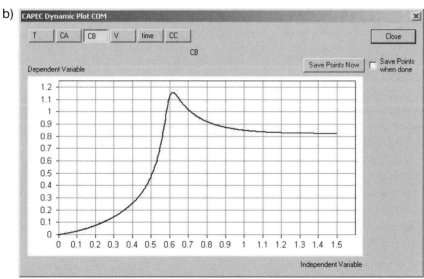

Figure 1.14 Component A: dynamic concentration motion. Component B: dynamic concentration motion

Step 8: Select the solver. In this case, dynamic option and forward integration with the BDF method is chosen and an end-time of 1.5 hours may be specified.

Step 9: Simulation results. As the numerical solver integrates, the selected variables values from Step 7 will be shown in dynamic plots and their values will be stored in files for later use. Figures 1.14a and 1.14b shows two samples of the dynamic plots.

Step 10: Saving the MoT file for reuse as well as export to other simulation engines. When the user is satisfied with the model and its solution, the MoT file can be saved for use within the ICAS simulation engine, for use from EXCEL or for use from any external simulation engine (with or without the CAPE-OPEN interface).

Further expansion of the exercise includes providing experimental data and regressing the kinetic model parameters and process optimization. Other exercises may also be developed where the generated model object is used to simulate the reactor, which is part of a process flow sheet. The same procedure can also be repeated to generate model objects for new property models, kinetic models and unit operation models. Using these model objects and a simulation environment such as EXCEL, the student can develop their own process simulator. For process design tasks involving new chemical products, this option is very practical and useful as the available process simulators do not have the chemicals and/or the models to handle them.

1.3.4
Other Educational Modules

A number of ICAS-based educational modules have been developed and can be downloaded from the following address: www.capec.kt.dtu.dk/Software/ICAS-Tutorials/ICAS-Tutorials-Workshops.

These tutorials cover problems related to:

• computer-aided property estimation
• computer-aided modeling
• computer-aided product design
• computer-aided separation process design
• computer-aided batch process modeling
• integrated computer-aided process engineering.

The objective of all these exercises is to highlight a systematic solution procedure and the use of an integrated set of methods and tools.

Other useful information can be found in the document from CACHE Corporation (2004) on *Computing through the curriculum: an integrated approach for chemical engineering* (www.che.utexas.edu/cache/newsletters/fall2003_computing.pdf). See also *Strategies for creative problem solving* by Fogler and LeBlanc (www.che.utexas.edu/cache/strategies.html); *The frontiers in chemical engineering education* (web.mit.edu/che-curriculum/); the EURECHA Web site at (www.capec.kt.dtu.dk/eurecha/); the EFCE working party on Education (www.efce.org/wpe.html).

1.4
Conclusion

The educational modules are documents containing the problem definition together with a detailed step by step solution strategy where the calculations for each step are highlighted with possible use of specific software (data flow and workflow). They can be used by the teacher to highlight an algorithm or methodology or technique (theory). They can also be used by the student to learn how to apply the theory (algorithms/methods) to solve problems.

Through the education modules, some of the important issues (including the danger of misuse) related to the use of computers in chemical engineering education have been highlighted. They have been prepared such that the user is in charge of the navigation and decisions while the computer does the calculations, data transfer, code generation, etc., which it is supposed to perform very efficiently. One of the principal experiences from the use of the presented educational modules has been that once the students understood the main ideas and got familiar with the workflow and data flow, they are able to tackle a wide range of similar problems without much help and in a very short time. In this way, they also learn to appreciate that the computer-aided tools are there to help them but they are the ones who need to make the right decisions and drive the use of the software in the appropriate direction. Finally, it was found that the students were able to appreciate the concepts better, were able to solve more challenging problems in a shorter time, resulting, thereby, an increase in productivity. They were able to use the same methods and tools also for problem solution in other courses. The feedbacks from the students have also helped to improve the software as well as the workflow and data flow of the educational modules. Finally, we would like to emphasize that software should not be used as a replacement of the process-product engineer; it should be used to do what it was designed for, with the user always in charge of directing it.

References

1 Abildskov J. Kontogeorgis G. M. Chemical product design: a new challenge of applied thermodynamics. Chemical Engineering Research and Design 82(A11) (2004) p. 1494–1504

2 Doherty M. F. Malone M. F. Huss R. S. Decision-making by design: experience with computer-aided active learning. AIChE Symposium Series 323(19) (2000) p. 163–175

3 Edgar T. F. Rawlings J. B. Frontiers of chemical engineering: The systems approach. DYCOPS Conference. Paper No. 206, Boston, MA, July 2004

4 Finlayson B. A. Rosendall B. M. Reactor transport models for design: how to teach

students and practitioners to use the computer wisely. AIChE Symposium Series 323 (96) (2000) p. 176–191

5 Fraga E. S. Gani R. Ponton J. W. Andrews R. Tools integration for computer-aided process engineering applications, in B. Braunschweig and R. Gani (eds.) Software Architectures and Tools for Computer-aided Process Engineering. CACE-11, Elsevier Science, Amsterdam, (2002) pp. 485–514

6 Gani R. ICAS Documentations CAPEC Internal Report, Technical University of Denmark, Lyngby, Denmark 2001

7 Hansen H. K. Rasmussen P. Fredenslund A. Schiller M. Gmehling J. Vapor–liquid––equilibria by UNIFAC group contribution.

Revision and extension. Industrial Engineering Chemistry Research 30 (1991) p. 2352–2355

8 *Kang J. W. Abildskov J. Gani R. Cobas J.* Estimation of mixture properties from first and second-order group contributions with UNI-FAC models. Industrial Engineering Chemical Research 41(13) (2002) p. 3260–3273

9 *Nielsen T. L. Abildskov J. Harper P. M. Papaeconomou I. Gani R.* The CAPEC Database, Journal of Chemical Engineering Data 46 (2001) p. 1041–1044

2
Data Validation: a Technology for Intelligent Manufacturing

Boris Kalitventzeff, Georges Heyen, and Miguel Mateus

2.1
Introduction

This document is intended to progressively demonstrate the technical assets of the data validation technology. Most of the technical features of the technology will be enlightened by specific process systems. However, validation technology can be and is implemented in various industrial sectors. Namely, it covers chemical, petrochemical and refining process plants, thermal and nuclear power plants, upstream oil and gas exploitation fields. Data validation is an extension of data reconciliation. Before demonstrating the technical assets of the validation, the reconciliation concept will be reviewed.

2.2
Basic Aspects of Validation: Data Reconciliation

Data reconciliation (DR) is the first mathematical method that addressed the concept of data validation for linear problems. It exploits information redundancy and (linear) conservation laws to extract accurate and reliable information from measurement data and from the process knowledge. It allows for the production of a single consistent set of data representing actual process operations, assuming the plant is operated in a steady state.

To understand the basic principles of data reconciliation, one must first recognize that plant measurements (including lab analyses) are not 100% error free. When using these measurements without correction to generate plant balances, one usually gets incoherence in these balances.

Some sources of errors in the balances directly depend on sensors themselves:

- intrinsic sensor accuracy,
- sensor calibration,
- sensor location.

Computer Aided Process and Product Engineering. Edited by Luis Puigjaner and Georges Heyen
Copyright © 2006 WILEY-VCH Verlag GmbH & Co. KGaA, Weinheim
ISBN: 3-527-30804-0

A second source of error when calculating plant balances is the small variations in the plant operating conditions and the fact that samples and measurements are not exactly taken at the same time. Using time averages for plant data partly reduces this problem. However, lab analyses are usually carried out at a low frequency, and thus can seldom be averaged.

Finally, one must also realize that in some parts of a plant too many measurements are available, whereas in other parts some measurements are missing and must be back-calculated from other measurements.

As shown in detail in Section 3, Chapter 3 of this book, data reconciliation can be expressed mathematically as:

$$\text{Min} \sum_i \left(\frac{y_i^* - y_i}{\sigma_i} \right)^2 \tag{1}$$

subject to
$$F(x, y^*) = 0$$
$$G(x, y^*) \geq 0$$

where	y_i^*	is the reconciled value of measurement i,
where	y_i	is the measured value of measurement i,
	x_j	is the unmeasured variable j,
	σ_i	is the standard deviation of measurement i defining its confidence interval.
	$F(x,y^*) = 0$	corresponds to the process equality constraints.
	$G(x,y^*) \geq 0$	corresponds to the process inequality constraints.

The term $\left(\dfrac{y_i^* - y_i}{\sigma_i} \right)^2$ is called the *penalty* of measurement i.

In early publications on DR, equality constraints were considered linear. Thus, one obtains a quadratic formulation, where the Jacobian matrix of F is constant. It is a Gaussian regression problem: given a set (y, σ_y), the algorithm provides x and y^* vectors together with their standard deviation σ_{y^*} (when computed).

When inequality constraints were not considered, some values y or x could be negative, what had no physical meaning in chemical or mechanical processes, where most variables must be positive (e.g., pressure, flow rate, mole fraction). It was considered as a source of information because one had to find which measurement was responsible for that negative value. Later on, simple inequalities ($y \geq 0$, $x \geq 0$) were considered.

When F or G is nonlinear, the DR problem can be solved by sequential linearization. The minimization problem is solved iteratively, using algorithms such as SQP (sequential quadratic programming). It is now possible in some commercial codes to calculate not only the reconciled values of measurements (y^*, σ_{y^*}) but also unmeasured state variables (x, σ_x) and some key performance indicators (KPIs) related to measured and unmeasured state variables (y^*, x), as well as their uncertainty σ_{KPI}:

measurements	y		y^*	$\}$	
a priori accuracy	σ_y	VALIDATION	σ_{y^*}		KPI
first principle modeling		$\longrightarrow$	x	$\}$	σ_{KPI}
statistical laws			σ_x		

Another problem is the identification and elimination of gross errors in the measurement data. This is a key asset of modern validation tools. We use the term data validation instead of DR, when all these features are exploited in the corresponding software tool, together with the use of thermodynamic conservation and equilibrium laws besides the mass balance equations.

The validation problem is solved using an interior point SQP solver in the VALI software that will be considered as an example in the subsequent applications [1].

2.2.1
Redundancy Analyses: Local/Overall

The level of redundancy is the number of measurements, which are available beyond the absolute minimum needed to calculate the system. Three different cases can be encountered:

- If a system's redundancy is negative then there is not enough information to determine the state of the system. Additional measurements need to be introduced.
- A redundancy equal to zero means that the system is globally just calculable.
- And finally, if a system has a positive redundancy, DR can use it as a source of information to correct the measurements and increase their accuracy. In fact, each measurement is corrected as slightly as possible but in such a way that the reconciled measurements match all the constraints of the process model.

However, overall redundancy is not enough. It must also be achieved at the local scale. Indeed, redundancy can be positive at the global scale, but negative locally; consequently, information is lacking to completely describe the whole process.

This point is illustrated with Fig. 2.1, based on a typical synthesis loop. Components A and B are introduced into the process feed, and converted into component C in the reactor unit SYNTHES ($2C = 3A + B$). Afterwards, the product ABC is separated in three distinct streams. One is recycled upstream in the process, another represents a purge, and finally an outlet stream contains only the compound C.

Let us consider a process model restricted to mass balances. Measured variables are shown on Fig. 2.1. This simple process model presents a global redundancy level of 2 (20 equations for 18 unmeasured variables). However, local redundancy of unit SEP-2 is equal to zero. If one of the measurements around this unit was missing then global redundancy of the model would still be 1 but local redundancy of unit SEP-2 would be -1. Therefore, the system would not be reconcilable until a supplementary measurement around the mentioned unit has been provided.

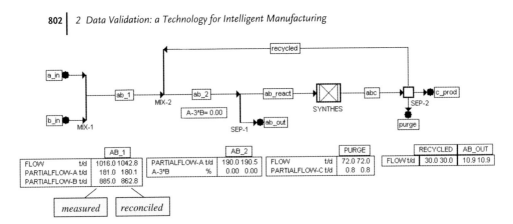

Figure 2.1 Process flow diagram (PFD) of a synthesis loop

2.2.2
If Complementary Measurement(s) are Needed: Which One(s)?

If the available measurement set is not enough to calculate all required process performance parameters, how do you propose an extra set from which complementary
measurements can be chosen? Thus, the system becomes either just calculable or
locally redundant, but necessarily globally redundant, as illustrated before.

Consider the previous example, but here we would remove the total flow rate measurement of stream "purge". Reconciliation software would then propose a set of variables from which possible complementary measurements ought to be chosen.
Namely, the software would purpose in this case a choice between partial flow rates
of compounds A and B in either stream "abc" or "purge", or compound C partial flow
rates in either stream "abc" or "c_prod".

If it is not possible to add any measurement to the system (because of economical
constraints for example), another way of avoiding negative redundancy is to aggregate some units in the model as a more global "black box" (that simply ensures
global balances to be satisfied). Less information will be obtained locally, but this
may allow estimating the required KPIs.

2.2.3
Increased Accuracy on Measured Data: Why?

As explained before, data reconciliation is based on measurement redundancy. This
concept is not limited to replicate measurements of the same variable by separate
sensors; it includes the concept of topological redundancy, where a single variable
can be estimated in several independent ways, from separate sets of measurements.
Therefore, *a posteriori* accuracy of validated data will be better than *a priori* accuracy
of measured data. *A priori* and *a posteriori* means before and after consistency treatment, or in other words before and after validation and reconciliation.

Table 2.1 DR back corrects measurements and increases their accuracy

			Meas.	Meas. Acc.	Reconc.	Reconc. Acc.
AB_1	Flowrate	ton/d	1016,0	3,00 %	1042,8	1,64 %
	Partial Flowrate (A)	ton/d	181,0	3,00 %	180,1	2,98 %
	Partial Flowrate (B)	ton/d	885,0	3,00 %	862,8	2,00 %
RECYCLED	Flowrate		30,0	3,00 %	30,0	3,00 %
AB_2	Partial Flowrate (A)		190,0	3,00 %	190,5	1,60 %
	A−3*B		0,0	0,00 %	0,0	0,00 %

In the previous example, unit MIX-2 presented a level 2 redundancy. Indeed, for 5 equations and 9 variables (and thus 4 degrees of freedom) we have 6 measurements (6 − 4 = 2). Table 2.1 shows the *a priori* and the *a posteriori* accuracy of those measurements around unit MIX-2.

Reconciled measurements are more accurate than raw data when measurement redundancy is available. But when no redundancy is available locally, no improvement can be expected. This is the case for the estimation of the recycled flow rate: the measured value is not corrected, and its accuracy is not improved. When some measurement is not corrected that does not imply it can be trusted; this would only be the case if the standard deviation would decrease.

2.2.4
DR Avoids Error Propagation

Progress in automatic data collection has presented plant operators with a flood of data. Tools are needed to extract and fully exploit the relevant information it contains. Furthermore, most performance parameters are often not directly measured, but calculated from measured values. Thus, random errors on measurements also propagate in the estimation of KPIs. Data reconciliation, on the contrary, allows state estimation and measurement correction problems to be addressed in a global way. As a result, validation technology avoids error propagation, and provides the most likely estimate of the actual operating point of the process. Thus, the plant can be safely operated closer to its limits. Illustration of error propagation is addressed in Table 2.2 for the example considered in Fig. 2.1. The goal is to estimate the flow of component C in the process output.

Because raw measurements are not error free, mass balance equation around mixer MIX-1 is not respected (fourth row of Table 2.2). Cases 1 to 3 show what happen when each of the three (process inlet) flow rates are manually corrected to close the mass balance, the flow rate of stream C being computed afterwards. In the last case DR is used to provide a consistent and accurate set of reconciled measurements. Indeed, Table 2.2 shows a balance value equal to zero. Note that measurements may be considered as correct since reconciled values are inside their confidence limits.

Table 2.2 Error propagation

		Measured	Accuracy	Case 1	Case 2	Case 3	Reconcilied	Accuracy
A in	ton/d	181.0	3.00 %	181	181	131	180.1	2.98 %
B in	ton/d	885.0	3.00 %	885	835	885	862.8	2.00 %
AB in	ton/d	1016.0	3.00 %	1066	1016	1016	1042.8	1.64 %
Balance in	*ton/d*	−50.0	/	0	0	0	0	/
ABC Purge	ton/d	72.0	3.00 %	72	72	72	72.0	3.00 %
C out	ton/d	/	/	994	994	944	959.9	1.80 %

Knowing that the standard deviation of flow measurements is 3 % of the measured value, one obtains for outlet compound C the flow rate:

- with DR: a standard deviation equal to 1.80 % with an estimate of 960 ton/d;
- with manual correction: a spread of estimates equal to 5.03 % (from 944 to 994 ton/d).

Thus, DR avoids error propagation and so provides more accurate computed parameters than those calculated by less rigorous or *ad hoc* correction modes. Plant engineers have to solve that type of problem regardless of if they have the appropriate tools or not.

2.2.5
Process Measurements to be Exploited

Key performance indicators (KPIs) can be determined accurately by validation of process measurement data. They are very useful for many purposes, e.g., revamping, energy integration, improved follow-up of the plant, possibility of working closer to specifications, detecting degradation of equipment performance, etc.

A hydrogen plant process is used to illustrate the determination of accurate and reliable KPI. Namely, this example concerns the steam to carbon ratio (S/C) in the steam reformer feed, that is, one of the key control parameters in such plants. It allows controlling the conversion of methane to carbon oxide and hydrogen while avoiding carbon deposition on the catalyst. Two different cases were studied to compute this ratio:

- First, DR was not considered. Ratio S/C was calculated from raw measurements of flow rates and compositions of process inlets (steam and natural gas) and reforming gas recycled.
- Afterwards, the same KPI was determined by means of DR.

Each of these two cases were reassessed, considering a measurement error on the steam flow rate (e.g., due to a leak). Namely, the steam flow rate is measured at either 72 ton/h or 78 ton/h.

Results shown in Table 2.3 demonstrate that the uncertainty on the S/C ratio is reduced when data reconciliation is performed. Also, reconciled S/C ratio is less sensitive to the flow rate measurement error, which is detected and corrected by data reconciliation. Thus, reconciliation detects errors in available measurements and yields accurately consistent and complete estimates of measured as well as unmeasured process parameters. Furthermore, in industrial practice one must take a safety margin for the S/C ratio to avoid carbon deposition in the catalyst. With DR, safety margins can be thinner, steam consumption is reduced and therefore plant operation costs less.

Table 2.3 KPI computation

	Without meas. errors		With meas. errors	
	S/C ratio	rel. error	S/C ratio	rel. error
without DR	3.545	4.24 %	3.840	4.24 %
with DR	3.514	3.52 %	3.673	3.53 %

Here a real industrial case encountered in a hydrogen plant is described, for which validation technology was applied. In a hydrogen plant (operated by ERE company), the feed gas composition was not monitored accurately; measurement errors were leading to an approximate knowledge of the steam/carbon ratio [2], uncertainty being on the order of 30 %. However, the hydrogen production efficiency and cost are strongly related to this ratio. Indeed a low S/C ratio decreases energy consumption. Therefore, a potential return of 500,000 euro per year had been identified. On the other side, a low S/C ratio could lead to carbon deposition (see Fig. 2.2) entailing a risk of catalyst damage (shut down for replacement costs five million euro).

With on-line validation software the steam/carbon ratio is determined nowadays with a precision of 1 %. This allows operating at the optimal point where energy costs are mastered and carbon deposition is avoided. This example shows how validation software allows for operation closer to the limits, taking care of safety constraints.

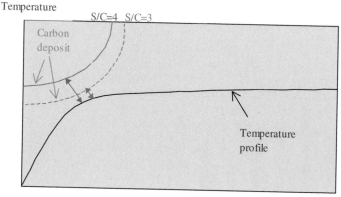

Figure 2.2 Profile of reformer reactor (courtesy of BP-ERE [3])

2.3
Specific Assets of Information Validation

Data validation is an extension of DR. In that case the set of corrected measurements and other calculated data respect linear and nonlinear constraints (mass, components and energy balances, reaction constraints as well as physical and chemical thermodynamic equilibrium constraints). Furthermore the technology includes data filtering, gross error detection/elimination, and it also provides the *a posteriori* accuracy of all the calculated data. Therefore, accurate and reliable KPIs are determined, as well as their accuracy. Moreover, validation software detects faulty sensors and pinpoints degradation of equipment performance (heat rate, compressor efficiency, etc.).

2.3.1
Accuracy of Nonmeasured but Calculated Data

Unmeasured variables of the system are calculated and their accuracy is quantified on basis of the measurements that are related to them. Therefore, in addition to providing substitution values for failed instruments, data validation software also calculates values that are not directly measured. Validation acts as a set of "soft sensors" that are robust and accurate because they are based on the reconciled values of all the measurements. Typically, validation technology provides three times more calculated data (and their accuracy), than the number of effectively measured data.

Benefits are undeniable, costly lab analyses can be avoided. For instance, on the chemical site of Wacker Chemie (Germany) an on-line implementation of validation software reduced the number of routine analyses up to 40% (see Fig. 2.3) [3].

Wacker considered validation as a revolutionary way for quality follow-up of their plants: f_{obj}, the sum of weighted squares of measurement corrections were checked for three years (see Fig. 2.4) [3]. They showed a reduction of the objective function (f_{obj}) from 30,000 to 1000, demonstrating a better quality of sensors tuning. Any increase of that validation criterion alerts operators on possible plant upset.

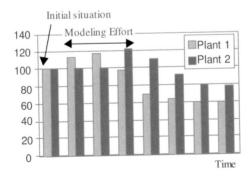

Figure 2.3   Reduction of lab analysis cost (courtesy of Wacker Chemie [3])

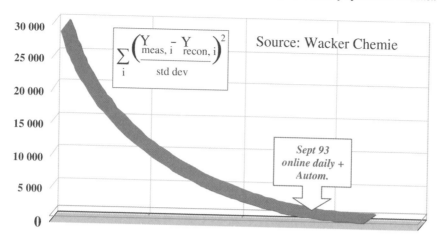

Figure 2.4 Sum of weighted squares of measurement corrections (courtesy of Wacker Chemie [3])

Furthermore, Wacker also follows the ratio $\frac{\chi^2}{f_{obj}}$, based on the chi-square statistical test (χ^2). The chi-square test value depends on the number of redundancies of the system and on the statistical threshold of the test, typically 95 %. Active bounds are considered as adding new levels of redundancy.

Two different cases are possible, whether the ratio is higher or lower than 1:

- If $\frac{\chi^2}{f_{obj}} > 1$: no presence of gross errors in the set of measurements can be expected.

- If $\frac{\chi^2}{f_{obj}} \leq 1$: presence of at least one gross error in the set of measurements is expected.

A data reconciliation result can only be exploited if the chi-square test is satisfied.

Gross error detection and elimination is a feature of validation software that will be detailed next.

2.3.2
Key Performance Indicators and Their Accuracy

Key performance indicators (KPIs) are identified in the same way as nonmeasured state variables. Because measurement errors have been withdrawn from the set of reconciled data, the best possible estimate of the plant performance is delivered. Thus, KPIs can be accurately determined.

Typical KPIs include:

- global plant efficiency
- yields

- steam/carbon ratio, oxygen/carbon ratio, H_2/N_2 ratio, etc.
- specific energy consumption
- specific energy cost
- equipment duty and efficiency
- catalyst activity, etc.

Table 2.4 shows S/C ratio values and accuracy using data validation technology. In the third case, thermodynamic constraints were taken into account. KPI accuracy is more improved with data validation than with data reconciliation. This is due to the fact that data validation considers all available process information (temperatures, pressures, chemical reactions, equilibrium constraints, etc.), the redundancy level being thus higher. Moreover, S/C ratio is much less sensitive to measurement bias, as demonstrated with the introduction of a measurement error on the steam flow rate entering the reformer (see Table 2.4). The additional assets of data validation are described here after.

Table 2.4 S/C ratio

	Without meas. errors		With meas. errors	
	S/C ratio	rel. error	S/C ratio	rel. error
without DR	3.545	4.24 %	3.840	4.24 %
with DR	3.514	3.52 %	3.673	3.53 %
with data validation	3.423	0.63 %	3.432	0.63 %

2.3.3
Nonlinear Thermodynamic-based Data Validation

2.3.3.1
The Limitation of (Linear) Mass Balance-based Reconciliation
Most commercial data reconciliation packages are based on a linear solver and reconcile measurements on the basis of overall mass balances. Moreover, bounds on variables are seldom considered, meaning that negative flow rates or negative inventories can appear in the results. Additionally, mass balance-based systems only offer a low level of redundancy: at the most one gets one level of redundancy around each node where all incoming and outgoing rates are measured. As a consequence, the improvement in data quality is low and the results are very sensitive to gross errors in the measurements.

On the contrary, thermodynamic-based data validation software provides additional equations increasing consequently the redundancy of the system, making it more accurate and less sensitive to measurement errors. At the same time, key performance indicators can be directly derived with a high level of accuracy and reliability. Of course, using thermodynamic properties has its drawback: most of the equations become nonlinear making linear solvers useless. Therefore, one must then use a nonlinear algorithm as large scale SQP-IP (sequential quadratic programming-

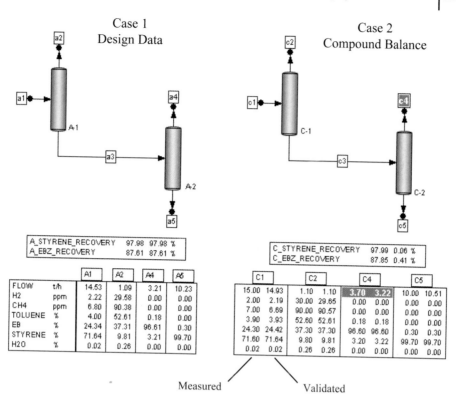

Case 1 — Design Data

A_STYRENE_RECOVERY	97.98	97.98	%
A_EBZ_RECOVERY		87.61	87.61 %

		A1	A2	A4	A5
FLOW	t/h	14.53	1.09	3.21	10.23
H2	ppm	2.22	29.58	0.00	0.00
CH4	ppm	6.80	90.38	0.00	0.00
TOLUENE	%	4.00	52.61	0.18	0.00
EB	%	24.34	37.31	96.61	0.30
STYRENE	%	71.64	9.81	3.21	99.70
H2O	%	0.02	0.26	0.00	0.00

Case 2 — Compound Balance

C_STYRENE_RECOVERY	97.99	0.06 %
C_EBZ_RECOVERY	87.85	0.41 %

C1		C2		C4		C5	
15.00	14.93	1.10	1.10	3.70	3.22	10.00	10.51
2.00	2.19	30.00	29.65	0.00	0.00	0.00	0.00
7.00	6.69	90.00	90.57	0.00	0.00	0.00	0.00
3.90	3.93	52.60	52.61	0.18	0.18	0.00	0.00
24.30	24.42	37.30	37.30	96.60	96.60	0.30	0.30
71.60	71.64	9.80	9.81	3.20	3.22	99.70	99.70
0.02	0.02	0.26	0.26	0.00	0.00	0.00	0.00

Measured   Validated

Case 3 — Mass Balance

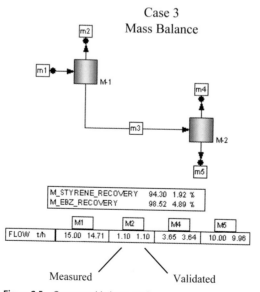

M_STYRENE_RECOVERY	94.30	1.92 %
M_EBZ_RECOVERY	98.52	4.89 %

	M1		M2		M4		M5	
FLOW t/h	15.00	14.71	1.10	1.10	3.65	3.64	10.00	9.96

Measured   Validated

Figure 2.5 Compound balances influence on accuracy

interior point), which has been implemented to solve complex nonlinear data reconciliation problems.

2.3.3.2
Example: Reconciliation of Two Distillation Columns

Two consecutive distillation columns are used to separate styrene (the final product) from unreacted ethylbenzene (EB), which is recycled to the reaction section (see Fig. 2.5).

Case 1 presents the design mass (and compound) balance of the plant. Case 2 presents typical measured values with a significant bias on the flow rate of recycled EB (stream c4) as reconciled in a compound-based data reconciliation system. The bias is clearly identified (3.70) and corrected (3.32) so that the styrene and EB recovery are accurately determined (87.86 ± 0.41 %). Case 3 then presents the same flow rates reconciled using a simple mass balance system, which is unable to detect the measurement error and therefore calculates a wrong recovery of EB and styrene. One can see that the accuracy of the computed recoveries is considerably better when performing a compound balance than with a simple mass balance (in this case, more than ten times better).

2.3.4
Exploiting LV and LLV Equilibria as Source of Information

Variables describing the state of a process must be reconciled to verify consistency constraints representing basic laws of physics: dew point and boiling point constraints in condensers, evaporators, or distillation columns are a source of information exploited by thermodynamic-based validation software.

The process of industrial ammonia production may be subdivided into three distinct parts: synthesis gas production, compression section and ammonia synthesis loop. Process natural gas (PNG) and steam enter the primary reformer reactor, after sulfur removal of PNG. High temperature and low temperature shift sections follow the secondary reforming, where compressed air is also introduced. After the methanator section, synthesis gas is partially recycled upstream in the process and partially introduced in the hyper compressor section. Finally, gas enters the ammonia synthesis loop. Figure 2.6 represents an ammonia synthesis loop process flow diagram (PFD), which can be considered as having an 8-digit structure with a heat exchanger in the middle.

The synthesis gas enters the hyper compressor as well as the recycle gas, then the outlet (process gas) is cooled and partially condensed (106F) to recover ammonia. Afterwards, gas is heated through a counter-current heat exchanger, goes to the reactor section, then again to the same heat exchanger (at lower pressure than the cold process gas) before closing the synthesis loop.

Condenser temperature (see Table 2.5) reflects a compromise between ammonia content and flow rate of the gas entering the reactor section. Considering condenser pressure as constant (158 bar) to simplify the following illustration, and condenser

Table 2.5 Condenser 106F measurements

		Raw measurements	Validated measurements
Condenser 106F	T	−14°C	−16.50°C
	Vapor flow rate (Nm³ h⁻¹)	456,890	455,040
Reactor	P_{in}	165 barg	165 barg
	T_{in}	185°C	181 C
	%mol NH₃	2.4	2.829
	%mol inerts	11.2	11.07

inlet composition and vapor flow rate specified, three different "what if" cases were studied (see Table 2.6). First, temperature was assumed equal to measured temperature −14°C. In the second column, temperature was considered the same as validated value −16.5°C. Finally, temperature is computed for ammonia content in vapor phase identical to raw measurement 2.40%.

Thus, a large amount of information can be extracted from the results:

- At 158 bar, hydrogen solubility rises slightly with temperature.
- If temperature is considered equal to the raw measurement (−14°C), ammonia vapor composition estimated is considerably different from measurement (3.1% instead of 2.4%). This proves inconsistence in the measurement set. On the contrary, vapor flow rate computed seems closer to that of the measurement value.
- In the second "what if" case, we reproduce validated data.
- To reach specified reactor inlet ammonia content (2.4%), temperature should be −20.8°C, instead of the −14°C measured. Therefore, vapor flow rate decreases.

This illustration shows the limitations of any partial "manual" validation.

Why is validated data so important in this particular case? The "what if" computations show the size of uncertainty of different data. The more NH₃ you condense in the condenser the better, but this has a direct cost, the energy spent in the cooling loop. How do you optimize any compromise if only nonvalidated data are available? Does it make sense?

Table 2.6 LV equilibrium calculation results

T (°C)	−14	−16.5	−20.8
Vapor fraction	0.9586	0.9558	0.9517
%mol NH₃ in vapor phase	3.10	2.83	2.40
Vapor flow rate (Nm³ h⁻¹)	456,330	455,039	453,049
%mol H₂ in liquid phase	0.38	0.36	0.33
Liquid flow rate	14.95	15.93	17.44

2.3.5
Exploiting Reactions and Chemical Equilibria as Source of Information

This point can be illustrated with the same ammonia process described previously (see Fig. 2.6), in particular its reactor section. Ammonia is produced in a two adiabatic catalytic stages reactor. Reactants are nitrogen and hydrogen, entering the reactor in a stoichiometric mixture. Ammonia formation reaction is exothermic and reversible; therefore, gas leaving the first adiabatic stage is cooled before entering the second stage. Furthermore, the model considers a performance equation, consisting in the introduction for both adiabatic stages of a ΔT_{eq} parameter, which takes into account deviation from chemical equilibrium. Because reaction is exothermic, ΔT_{eq} will be positive.

Thus, important information that can be extracted from data validation, considering reactions and chemical equilibrium, are performance parameters ΔT_{eq} (see Table 2.7). Results pinpoint a closer approach to equilibrium in the first catalyst bed. In addition, it is possible to visualize validated ammonia concentration profile together with equilibrium curve and plant measurements (see Fig. 2.7). The two vertical lines represent measured inlet and outlet temperatures of the heat exchanger between the two catalyst beds.

One cannot accept a measurement point above the equilibrium curve. This erroneous measurement set could not have been noticed any other way than exploiting reactions and chemical equilibria as an information source.

Table 2.7 Performance parameters

	ΔT_{eq} (°C)
First catalytic bed	6
Second catalytic bed	14

2.3.6
Exploiting Process Information

As explained before, data validation is based on measurement redundancy. The plant structure yields additional information, which is exploited to correct measurements. Consequently, considering a process at a global scale brings more accuracy to validated data than only taking into account a local section of the process. It is the same for the accuracy evolution of key performance indicators.

Considering the same ammonia process as before, the H_2/N_2 ratio in the synthesis loop was estimated in several ways. First, only a local section of the process was considered (the synthesis loop). Then, additional information of the plant was successively added until the whole process was taken into account. Results pinpoint a substantial reduction of the KPI inaccuracy when more and more process information is considered (see Fig. 2.8).

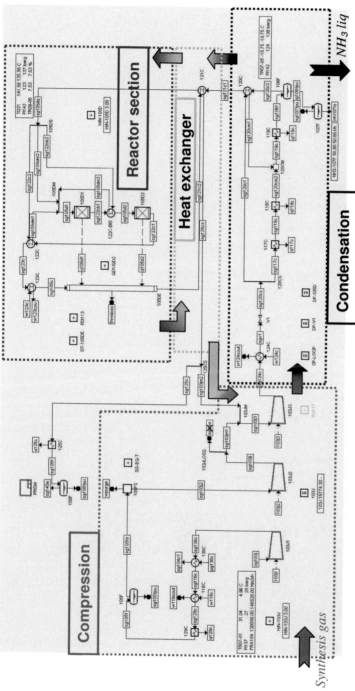

Figure 2.6 Ammonia synthesis loop PFD

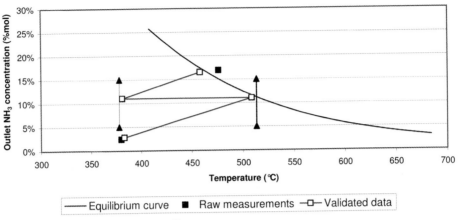

Figure 2.7   Synthesis reactor equilibrium curve

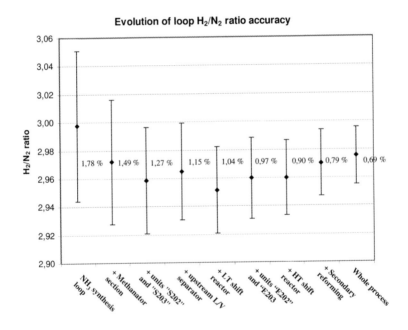

Figure 2.8   Evolution of a KPI imprecision according to process information taken into account

It was previously demonstrated that validation technology avoids error propagation. In fact, data validation software propagates accuracy. This technology combines process information and raw measurement data. The more information of the process taken into account, the more nonmeasured data (and so KPIs) will be accurate and reliable.

2.3.7
Detection of Leaks

Validation technology points out process performance degradation sources and helps to operate the plant closer to its ultimate performance. In particular, validation allows the detection of leaks. This can be illustrated by a practical case study related to a previous ammonia plant, where a leak in a NH_3 synthesis loop was discovered. It would have been hardly detected by tools other than validation technology.

A Carbochim plant operated in Belgium at 90 % of nominal capacity; a retrofit was studied to restore the expected capacity. Validation conveyed a leak in the heat exchanger in the middle of the 8-structure synthesis loop (see Fig. 2.6). Thus, part of the process gas was cycling around from the compressor and condenser section, to the heat exchanger and again to the compressor. That leak had not been suspected. It probably developed and increased smoothly, but the question is, how could it have been discovered in the absence of the appropriate tool? Plant was shut down for isolating leaking tubes in the exchanger and was reopened to easily achieve the expected production rate without any costly additional investment.

2.4
Advanced Features of Validation Technology

2.4.1
Trivial Redundancy

Trivial redundancy cases are met when the validated value of a measured variable does not depend at all upon its measured value but is inferred directly from the model.

This can occur in particular in L/V equilibrium drums, where complementary thermodynamic constraints must be respected. Indeed if, e.g., temperature, pressure, flow rate and composition of a condenser inlet stream were known together with the unit pressure drop, any complementary measurement (e.g., outlet temperature) would be considered as a trivial redundancy. Proper validation software detects trivial measurements, which then are no longer considered as measured. As a consequence, their measurement accuracy will not affect the accuracy of the respective validated variable.

2.4.2
Gross Error Detection/Elimination

Gross errors are detected by means of a chi-square (x^2) statistical test, which has been previously explained at Section 2.3.1.

2.4.2.1
Detecting Gross Errors
The x^2 statistical test enables the detection of gross errors in the sets of measurements. The x^2 value depends on the total number of redundancies of the system, active bounds being considered as adding new levels of redundancy, and on the sta-

tistical threshold of the test, typically 95 %. If the weighted sum of penalties is higher than the χ^2 threshold value, then there is a significant suspicion that gross errors exist. In such a case, all results obtained with that model are to be used with caution: validated values, identified performance factors and their reconciled accuracy.

2.4.2.2
Eliminating Gross Errors: The Highest Impact Method

Identifying the actual source of the gross errors is not always trivial and requires a careful analysis of the results. The conventional technique (highest penalty method) is to ignore the measurements for which the highest corrections are made. This method is known to be inadequate in detecting some gross errors, for example, when the corresponding measurement is specified with a high level of accuracy as compared to the other measurements.

On the contrary, the highest impact method detects the impact on the total sum of penalties by removing each of the measurements. This approach is in principle highly time-consuming and is therefore not used by most data validation packages. However, by means of specific algorithm, one can apply this technique in a calculation time of the same order of magnitude as a single validation run.

2.4.3
How to Validate with Petroleum Fractions

The modeling of a refinery process or a part of it is always confronted by the complexity of the petroleum and its products. Indeed, crudes and petroleum cuts are mixtures of a large number of chemical compounds, thus making it very difficult to model their properties without accurately knowing their composition. Therefore, it is common practice to model such streams by the well known pseudo-component concept.

2.4.3.1
Concept

A pseudo-component is a hypothetical molecule characterized by its density and its boiling temperature. Those parameters are then used to estimate the other thermodynamic properties (like critical properties or specific heat capacity) using empirical correlations as proposed for example by American Petroleum Institute (API). According to the crude type and origin, different pseudo-components must be used to get an accurate representation. The usual way of characterizing petroleum fractions is to generate a defined mixture of pseudo-components, with given boiling point, having the same properties as the Petroleum fraction. Namely, their composition and their density are identified in order to match all stream distillation curves and densities. Most common standards for distillation curves are true boiling point (TBP) and ASTM; each of them can be expressed on a weight basis or on a volume basis (see Fig. 2.9).

Several petroleum cuts involved in a distillation process can be modelled as a data validation system involving:

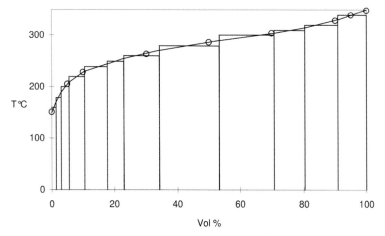

Figure 2.9 Decomposition in pseudo-components based on a TBP curve of a gas oil

- as variables, the pseudo-compounds densities and their measured volume fractions in each stream;
- as equations, the mass balances of the distillation column for each pseudo-component;
- as measurements, the densities and TBP or ASTM curves of all connected streams.

On this basis, data validation will generate calculated distillation curves from measured TBP or ASTM data, as it identifies the density of each pseudo-compound; this involves minimization of weighted deviation between measured and calculated distillation points, under density constraints, mass and thermal balance constraints. The other thermodynamic properties of the pseudo-compounds are also estimated.

2.4.3.2
Crude Oil Atmospheric Distillation Example

Following example concerns the modeling of a crude oil distillation unit (CDU), preceded by the preheating train (see Fig. 2.10) [4]. The crude oil is separated into six Petroleum cuts: naphtha, jet, kerosene, gasoline, diesel and residue.

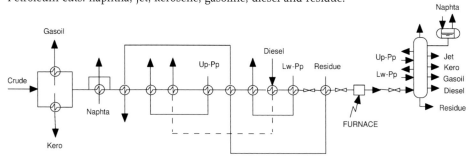

Figure 2.10 Preheat train and CDU

Measurements available to perform the modeling are:

- density and distillation curves (ASTM-D86) of the petroleum cuts,
- temperature, pressure and flow rates of the streams,
- design data of the exchangers.

These measurements are validated and the other thermodynamic properties of pseudo-compounds are subsequently computed. Furthermore, with several sets of measurements taken in one year it was also possible to confirm fouling problems for the exchangers at the end of the preheating train. Indeed, their heat transfer coefficient decreased by a factor of two after one year of operation. Thus, data validation uses a rigorous method integrating robustly complex distillation systems. This forms a sound basis for the analysis of refinery performance, and for instance, of a retrofit potential.

2.4.4
Advanced Process Control Benefits from Working with Data Validation

Nowadays plants face a market where margins are under pressure due to global competition, more stringent environmental regulations, a higher demand for flexible operation and more severe safety requirements. Control techniques are required to increase those margins. Advanced process control (APC) systems can help optimize control to deal with those challenges [5]. Data validation technique enhances the quality of information allowing APC systems to work more efficiently.

2.4.4.1
Data Validation-APC: How They Work Together

Figure 2.11 describes how data validation software works hand to hand with an APC system in order to improve its efficiency. A plant process is permanently submitted to disturbances, causing changes in operating conditions. APC systems use a reduced dynamic model to predict the plant behavior when submitted to disturbances, and thereafter take the adequate actions to counter them (multiple input multiple output, MIMO control). Without data validation, input and output stream measurements are introduced directly into a dynamic model with no ways of chec-

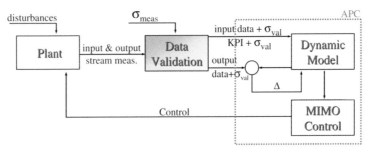

Figure 2.11 Data Validation working together with APC system

king raw information reliability and coherency. Some measurements could be erroneous and balances not be closed.

Data validation software uses input and output streams of raw measurements in order to provide one coherent and accurate data set. With data validation, APC systems are allowed to take actions on the process based on coherent and reliable measured and nonmeasured data. Validated data contains measurements, equipment parameters, KPIs, and many other nonmeasured but validated data. The *a posteriori* accuracy of measurements and KPIs is provided.

When a dynamic model is tuned according to validated data, benefits are generated as early as at the model design stage.

2.4.4.2
Benefits at Model Design Stage

Reduced dynamic model must be certified: dynamic model parameters are chosen and adjusted in order to produce results identical to measurements ($\Delta = 0$ on Fig. 2.11). Benefits using validation techniques are double:

- Measurements, to which dynamic model results are compared, are checked and corrected by data validation techniques. Measurements are much more reliable (they represent the actual process operation) and thus the model will be more reliable as well.
- Data validation technology reduces the number of principal directions needed to represent process variability, allowing the reduced dynamic model to represent the same level of variability using a model with a lower number of principal directions (see Fig. 2.12) [6].

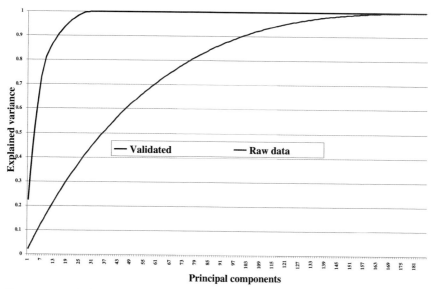

Figure 2.12 Level of variability according to number of principal directions

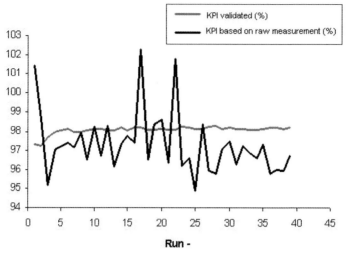

Figure 2.13 KPI follow-up and control

Figure 2.12 illustrates the number of principal directions (or components) necessary to represent the variability of a given system when the latter is based on validated data or on raw measurement data. Taking into account more principal components allows for the explanation of a higher fraction of the total process variability:

- When using raw measurements, a large number of components are needed to explain most of the process variability (upper limit is the number of original variables, 186).
- When using the validated data sets, the number of significant principal components tends to a much lower number than the number of variables (upper limit is the number of degrees of freedom of the data validation model).

This reduction in the problem size allows the dynamic model to be more reduced, when based on validated data (accuracy increased and noise reduced). Control of the process is made easier and the computing demand is decreased.

Furthermore, since data validation technique enforces the strict verification of all mass and energy balance constraints, use of this technology ensures that the principal components represent the proper process behavior.

2.4.4.3

Benefits at Operation Stage

Process control behavior can be very different whether APC is working together with data validation or not. Figure 2.13 presents the evolution of a process yield (KPI) versus time (run) whether data validation software is used or not:

- Without data validation, APC detects a KPI variation and tries to stabilize the process operation. Based on raw data with embedded errors, APC takes actions risking being unuseful, resources-expensive, and even process-disturbing.

- With data validation, APC considers actual process operation (validated, measured and nonmeasured, data are used as inputs to APC). APC can now use all of its resources on optimization of the process rather than on more stable operations.

2.5
Applications

2.5.1
On-line Process Performance Monitoring

The goal is to deliver on a periodic basis (typically each 10 to 60 minutes) a coherent heat and mass balance of a production unit. In addition to the compound balances, the laws of energy conservation are introduced in the form of heat balances. This more detailed modeling of the production unit allows validation software to work as an advanced process soft sensor and to determine reliable and accurate KPIs.

Typical benefits are:

- access to unmeasured data, which is quantified and related accuracy determined;
- early detection of problems: sensor's deviation and degradation of equipment performance are pinpointed;
- quality at process level: anticipate off-spec products by carefully monitoring the process;
- work closer to specifications: as the accuracy of measurement data improves, the process can be safely operated closer to the limits. This feature is reported as being financially the most productive.
- decreased number of routine analyses (up to 40 % in chemical applications);
- reduced frequency of sensor calibration (only faulty sensors need to be calibrated).

Improvement of Product Selectivity in a BASF Plant

This example shows how the operation of a production unit at a BASF operating division of performance chemicals can be improved using data reconciliation [7]. The product C is produced by conversion of component A with component B using 2 reactors. Several undesired by-products are generated, thus selectivity has to be maximized. Process model generated took into account only component mass and atomic balances. Several data sets at different process conditions were validated and from those the selectivity of product C was calculated. The diagram Fig. 2.14a (courtesy of BASF) shows this selectivity as a function of residence time in the first reactor, calculated from measured values; Fig. 2.14b shows the results from validated data.

The selectivity, calculated from crude data, is spread widely and in some cases selectivity values of more than 100 % were obtained, which is meaningless. The corresponding unfeasible area is marked on the charts. One could estimate in this case that a residence time of about 45 minutes is enough to maximize selectivity. However the selectivity based on reconciled data shows a clearer trend and does not

(a)

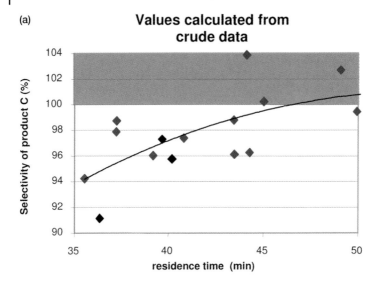

(b)

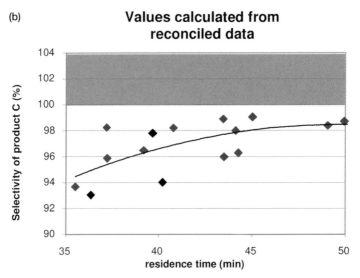

Figure 2.14 Nitrile selectivity as a function of residence time in the reactor: raw data versus reconciled data (courtesy of BASF [7])

exceed the 100 % boundary. One realizes that residence time should be larger than the one estimated without data validation, in order to achieve the product optimal selectivity (residence time of about 48 minutes). This example (considering only a restricted part of a process) shows that the evaluation of selectivity is meaningful only on the basis of validated operational data. These lead to a safe interpretation of measurements. By doing so, a selectivity close to 99 % can be obtained systematically, which is 2 % higher than the average figure obtained without data validation.

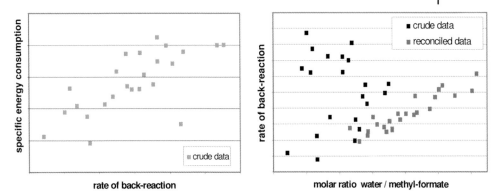

Figure 2.15 Evolution of the specific energy consumption as a function of the rate of back-reaction and parameters that influence this rate: crude data versus reconciled data (courtesy of BASF [7])

Reducing Energy Consumption in a Formic Acid Plant of BASF

A main problem at the formic acid production is the undesired back-reaction of formic acid during distillation, which increases the specific energy consumption [7]. This is shown in the left diagram of Fig. 2.15, on the basis of measured values within a time interval of 6 days. BASF looked for process parameters, which may influence the back-reaction, in order to decrease operation costs (specific energy consumption). One of them is the molar ratio of water to methyl formate, both educts of the formic acid synthesis.

The diagram on the right shows the influence of the mentioned molar ratio to the rate of the unwanted back-reaction:

- Without data validation (raw data , black symbols), no influence is visible but only a cloud of data.
- Using validated values (grey symbols) a clear trend is visible, which means that reducing the molar ratio decreases the rate of back-reaction.

Both parameters could be correlated only by data validation. Due to these results the specific energy consumption can be reduced by 5 %. Data validation allows the most effective command variables for the control of a process to be determined. This study led to the discovery of which control variable had a dominant effect on the said rate of back-reaction, and consequently on the specific energy consumption.

Performance Monitoring at KKL Nuclear Power Plant

On-line implementation of validation software in the nuclear power plant (NPP) of Leibstadt – Switzerland (KKL) generated substantial benefits (two million USD per year) over the past 10 years. The priority of NPP operators is to run their plant as close as possible to the licensed reactor power in order to maximize the generator power. To meet this objective, plant operators must have the most reliable evaluation of the reactor power. The definition of this power is based on a heat balance using several

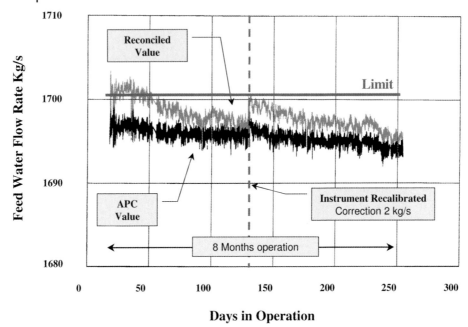

Figure 2.16 Operating closer to the limits: site feed water flow of the
NPP Leibstadt (courtesy of Kernkraftwerk Leibstadt [8])

measured process parameters, among which the total feed water flow rate is the most
critical value.

On-line implementation of validation software in the NPP of Leibstadt – Switzerland (KKL) has quantified the deviation between the actual and the measured feed
water flow rate (see Fig. 2.16). In Fig. 2.16, only one recalibration is illustrated. This
was used to convince the legal authorities about the reliability of the implemented
validation technique. Validation results were also compared to test runs.

In agreement with the authorities in charge of safety of NPP, KKL nowadays reca-
librates the measured flow rate based on the validated value, as soon as a deviation
becomes significant. This enables the power plant to work close to its maximum
capacity throughout the whole year (1145 MW). Prevention of losses due to heat balance errors increased the plant output by 5 MW. In addition, the use of this technology also made the annual heat cycle testing obsolete and significantly reduced the
cost for mechanical and instrumentation maintenance [8].

Performance Monitoring of Refinery Units at LOR (Lindsey Oil Refinery), U.K.
On-line validation software is used at LOR for the performance monitoring of refinery units for several years. One set of applications is about the follow-up of fouling
of the heat exchangers of several preheat trains. The main goal of the application is
to determine the appropriate amount of anti-fouling product in order to maintain an
adequate operation of the preheat trains and thus energy efficiency of the plant.

Another set of applications concerns the follow-up of furnaces and power plant boilers. The goal here is to determine with sufficient reliability and accuracy their energy efficiency. Any inappropriate operation can easily be detected and corrected when necessary.

Performance Monitoring of PE Plant at Gonfreville, France

The application enables any deviation within the instrumentation to be detected and provides guidance to the operators for the recalibration of the on-line analyzers. In addition, it ensures that the on-line soft sensors remain valid by counter-checking the quality of the instrumentation on which they rely.

2.5.2
On-line Production Accounting

2.5.2.1
Description and Benefits

This solution aims at providing a clear view of the production accounting, on a daily basis, of a whole industrial site: rigorous and automatic procedure for production accounting based on closed material balances. These material balances can be performed either:

- On a global mass balance basis: mass flow rates, in terms of tons entering and tons going out of each production unit, are reconciled to generate a coherent mass balance of the whole site. This approach is typically applied in refineries and covers the whole site including the tank farm.
- On a chemical compound basis: additional information is then required on the composition of the various streams and the reactions schemes. This approach is typically applied in chemical and petrochemical production plants.

Typical benefits are:

- Actual plant balances: closed balances are key elements as much for effective production accounting as for efficient performance monitoring.
- Decrease of unidentified losses and surpluses: abnormal conditions leading to losses and/or apparent surpluses are identified and can be corrected before they impact the economics of the plant.

Several real cases can be referred to, namely an adiponitrile plant and two refineries.

Production Accounting at ERE and Holborn Refineries

On-line validation software establishes the daily mass balance of the whole ERE refinery (BP refinery located at Lingen, Germany), covering about 150 tanks and about 50 production or blending units. Only a global mass balance (in tons) is made around each unit. The person in charge of the use (and maintenance) of the system spends about 30 minutes per day to generate all the validated reports and inputs for the production accounting. More recently the Holborn refinery in Germany has installed a similar system, which also automatically detects abrupt changes in measured data identifying possible changes in operation or instrumentation failure.

Production Accounting at Butachimie, France

In this application the modeling includes compound balances of each main piece of equipment of an adiponitrile production facility. Reconciled compound balances are provided on a daily basis. All main chemical compounds as well as the catalysts used in the system are rigorously tracked all over the process unit.

2.6
Conclusion

Data reconciliation and validation is nowadays a mature technology. However it is often confused with flow sheeting and process simulation. Still, much has to be done to inform engineers and managers who have not learned about this technology during their studies. We have tried to convey the importance of this technology, and the very high diversity of applications and benefits that it can provide for the process industry.

References

1 *Belsim* VALI 4 User's Guide, Belsim, Belgium, 2005

2 BP-ERE, AN-ERE-01.pdf, available at: www.belsim.com, 2004

3 WACKER AN-Wacker-01.pdf, available at: www.belsim.com, 2002

4 *Delava P. Maréchal E. Vrielynck B. Kalitventzeff B.* Modeling of a Crude Oil Distillation Unit in Terms of Data Reconciliation with ASTM or TBP Curves as Direct Input. Application: Crude Oil Preheating Train, Proceedings of ESCAPE-9 conference, Budapest, May 31-June 2 1999, *Computers and Chemical Engineering* Suppl., (1999) pp. 17–20

5 APC Systems www.ipcos.be, 2005

6 *Amand T. Heyen G. Kalitventzeff B.* Plant Monitoring and Fault Detection: Synergy between Data Reconciliation and Principal Component Analysis. Computers and Chemical Engineering 25 (2001) p. 501–507

7 BASF, AN-BASF-02.pdf, available at: www.belsim.com, 2002

8 Kernkraftwerk Leibstadt, AN-KKL-01.pdf, available at: www.belsim.com, 1995

3

Facing Uncertainty in Demand by Cost-effective Manufacturing Flexibility

Petra Heijnen and Johan Grievink

3.1
Introduction

This chapter deals with a case of flexible production planning for a multiproduct plant to optimize expected proceeds from product sales when facing uncertainty in the demands for existing and emerging new products over the planning period. The manufacturing capacities of the plant (that is, the nominal production rates) for the existing and new products are fixed by its design and these are not subject to adaptations by making changes to the plant. Hence, the flexibility refers exclusively to the planning problem and it is not coupled with a plant redesign.

As the inherent uncertainty in customers' demand forecasts is hard to defeat by a company, the industry's specific capabilities with respect to responding rapidly to new and changing orders must be improved. New technologies are required, including tools that can swiftly convert customer orders into actual production and delivery actions. On the production side, this may require new planning technologies or new types of equipment that are, for example, dedicated to product families, rather than to individual products. Many companies need to use medium term planning in their product development and manufacturing processes in order to sustain the reliability of supply and the responsiveness to changing customer requirements.

Flexibility is often referred to in operations and manufacturing research as the solution for dealing with swift changes in customer demands and requests for in-time delivery (Bengtsson 2001). The concept has received even more attention with the upcoming of e-business in the chemical industry. The actual meaning, interpretation and consequences of "operating flexibility" are, however, not instantly clear for a particular case or company (Berry and Cooper 1999). A number of uncertainties may induce organisations to seek more flexible manufacturing systems. Common sources of uncertainties are depicted in Fig. 3.1.

Computer Aided Process and Product Engineering. Edited by Luis Puigjaner and Georges Heyen
Copyright © 2006 WILEY-VCH Verlag GmbH & Co. KGaA, Weinheim
ISBN: 3-527-30804-0

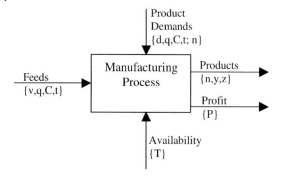

Figure 3.1 Types of uncertainties

On the input side, manufacturing systems have to deal with suppliers' reliability with respect to feed stock supply, involving quantities (v), quality (q), cost (C), and with uncertainties in time (t). Secondly, process inherent uncertainties exist, concerning equipment availability (T), and modeling uncertainties. On the product demand side, the same types of uncertainties can be found for each product, involving demand (d), quality (q), and cost (C), and time (t). For the products a distinction is made regarding two different sales conditions. At the beginning of the planning period some sales contracts can be secured, under which the amounts (y) that can be manufactured and sold. The excess manufacturing capacity of the plant can be used to make the amounts (z), which will capture market opportunities during the planning period. It is noticed that the number of products (n) can change over time. This change reflects a trend towards diversification in many production markets. To achieve this diversification and to cope with shorter product life span, it seems preferable for manufacturing systems to have *flexible* resources.

Extensive research has been done into the flexibility of (chemical) processes that are subject to uncertainties on the input side and with respect to the availability of the processing equipment, possibly influencing the feasible operating region of the plant (Bansal et al. 1998; Swaney and Grossmann 1985). Less research has been done, however, into flexibility that is characterized by the possibility to cope with changes in demand or product mix. The right way to respond to change is always system specific, and dependent on the system's flexibility. Many approaches for dealing with uncertainties exist (Corrêa 1994). As this study concerns product mix variations and demand variations, the monitoring and forecasting technique was selected. The uncertainty aspect is modeled by means of a stochastic approach.

In the development of a planning technique its applicability requires careful consideration. Firstly, the technique should be compatible with the work processes and the associated level of technical competence. Among others, this requires that the input and the output can be well understood and interpreted by those who will use it. Secondly, the cost of using the technique (time and money wise) should remain low. It would be very helpful to use input data that can be obtained without excessive efforts, while the results of the planning can be easily (re)produced with small computational effort.

3.2
The Production Planning Problem

A case study was developed based on experiences at a company that makes various types of food additives in a multiproduct batch plant. In this plant several groups of products are produced on a number of reactors. At the beginning of every new production period of one year, planning management agrees with the customers about the amount and price of products that will be produced to meet customer demands in the coming period. These agreements between company and customers are laid down in annual contracts. The demand for products that could be sold in these annual contracts is in general very large and the total capacity of the plant could have been sold out. However, planning management has strong indications that the demand for a new and very profitable product will increase during the coming production period and it could be very attractive to keep some of the capacity free for this newcomer on the market.

Not only that, but also for the current products it could be quite profitable to not sell all capacity beforehand, since the price for which the products can be sold during the production period is in general significantly higher than before by contract price. Unfortunately, the demand for the products during the production period cannot be assured. Planning management would like to establish in the production planning how much of the current products they should sell in annual contracts and how much capacity they should leave open for every individual current product and for the new one in such a way that the final profit achieved at the end of the production period is as high as possible. The plant production capacity acts as a restriction on the total amount that can be produced.

The next sections will introduce a simple probabilistic model for the product demands as well as a manufacturing capacity constraint (Section 3.3). The realised product sales are related to the corresponding profit over the planning period (Section 3.4). In order to optimize the manufacturing performance, two objective functions are chosen that take into account the distributive nature of the demands and product sales (Section 3.5). The first objective is the expected value of the final profit over the planning period. The second objective is a measure for the robustness of the planning; it involves maximization of the first quartile of the profit. The outcome of the modeling is a multiobjective, piecewise linear optimization problem (Section 3.6). Due to the discontinuities the problem is solved by means of a direct search method, the Nelder and Mead algorithm. The multiobjective problem is turned into two single objective problems. The solutions to these problems define the full range between maximum expected profit (with a high risk) and the robust profit (for a low risk scenario). This approach allows a production manager to take a preferred position between these two extremes. The result is a production planning and the associated profit.

Each step in the model development is illustrated by its application to the case study of the food additives plant, taking base case values for model parameters. To be able to make a good evaluation of the risks, the sensitivity of the profit and the optimal planning are studied for small changes from the nominal model parameters

(Section 3.7). Finally, the implementation aspects of the proposed planning method are discussed (Section 3.8).

3.3
Mathematical Description of the Planning Problem

To solve this production planning problem we need to formulate it in a more formal way. Assume that the current product portfolio consists of n products that can be produced on several exchangeable units. The decision about on which specific unit a product will be made is established in the production schedule and is considered to be outside the scope of the production planning. In the production planning, the planners take the overall production capacity into consideration without allocating products to specific units.

In the production planning the following decisions variables should be established:

- the amount of the current products sold in annual contracts in ton per year:

y_i, $i \in \{1, 2, ..., n\}$;

- the capacity left open for the current products and for the new one in ton per year:

x_i, $i \in \{1, 2, ..., n, n + 1\}$.

The information that is needed to make these decisions consists of the following parameters:

- the profit that can be made with the production of one ton of a certain product, depending on the retail price and on the production costs, divided into:
  - the profit made on the current products sold in annual contracts in dollars per ton:

  σ_i, $i \in \{1, 2, ..., n\}$;

  - the profit made on sold amounts of the current products and of the new one during the production period in dollars per ton:

  ϱ_i, $i \in \{1, 2, ..., n, n + 1\}$;

- the total production time available in hours per year: T;
- the production time needed to make the products in hours per ton:

τ_i, $i \in \{1, 2, ..., n, n + 1\}$;

- the demand for the current products that can be sold in the annual contracts:

δ_i, $i \in \{1, 2, ..., n\}$;

Since the total amount of product made during the production period cannot exceed the total available production time, the decision variables are restricted by:

$$T = \sum_{i=1}^{n} \tau_i y_i + \sum_{i=1}^{n+1} \tau_i x_i \tag{1}$$

The assumption is made that the planners have enough and correct information to make a good estimation of the values of these parameters. Therefore, these parameters are assumed to be the deterministic factors in the planning problem.

- the demand for the current products and for the new one during the production period in ton per year: d_i, $i \in \{1, 2, ..., n, n + 1\}$.

The uncertainty of the demand during the production period is quite large and therefore these factors are assumed to have a stochastic nature. The assumption is made that the planners have enough information to indicate the minimum and maximum demand that can be expected and the mode of the demand, that is, the demand for which the probability density function is maximized.

The demand for the products will therefore be modeled by a triangular distribution with the probability density function given in Eq. (2). This triangular form (see Fig. 3.2) corresponds with the shape used in fuzzy modeling.

$$f(d_i) = \begin{cases} \dfrac{2(d_i - \alpha_i)}{(\beta_i - \alpha_i)(\gamma_i - \alpha_i)} & \alpha_i \leq d_i \leq \beta_i \\[2ex] \dfrac{2(\gamma_i - d_i)}{(\gamma_i - \beta_i)(\gamma_i - \alpha_i)} & \beta_i < d_i \leq \gamma_i, \ i \in \{1, 2, \ldots, n, n+1\} \\[2ex] 0 & \text{otherwise} \end{cases} \tag{2}$$

in which α_i is the minimum, β_i the mode and γ_i the maximum of the demand d_i for a certain product i.

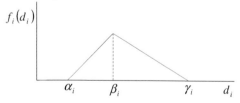

Figure 3.2 The triangular probability density function of the demand

The expected demand for product i will then be:

$$E(d_i) = \frac{\alpha_i + \beta_i + \gamma_i}{3}, \in \{1, 2, ..., n, n + 1\} .$$

The probability distribution of the demand d_i reads:

$$F(d_i) = \begin{cases} 0 & d_i \leq \alpha_i \\[2ex] \dfrac{(d_i - \alpha_i)^2}{(\beta_i - \alpha_i)(\gamma_i - \alpha_i)} & \alpha_i < d_i \leq \beta_i \\[2ex] 1 - \dfrac{(\gamma_i - d_i)^2}{(\gamma_i - \beta)(\gamma_i - \alpha_i)} & \beta_i < d_i \leq \gamma_i \\[2ex] 1 & d_i > \gamma_i \end{cases}, i \in \{1, 2, \ldots, n, n+1\}. \tag{3}$$

In Section 3.4 the mathematical description of the planning problem will be continued, but the generic problem will first be applied to a case study in a plant where various types of food additives were made.

3.3.1
Case Study in a Food Additives Plant

In a multiproduct multipurpose batch plant different food additives are produced on two reactors. The present portfolio consists of two product groups A and B. Having strong indications about a growing demand for a new product C, operations management wants to reevaluate the current product portfolio and the production planning for the coming year.

The product groups A and B are manufactured on two exchangeable reactors. Planning management has estimated the production times based on the current *annual operation plan*. The total amount of available operating time for the reactors is determined by the available time in a year minus 15 % down and changeover time, resulting in 4625 hours for reactor 1 and 4390 hours for reactor 2. Together this results in a total production time for the coming year of $T = 9015$ hours.

Table 3.1 shows the estimated values for all parameters in the planning problem. From the figures it is clear that the total production capacity could have been sold out in the annual contracts, since the demand for the product groups A and B is high enough. The new product C however is expected to be very profitable and it would very likely be an unwise decision to sell out the total production capacity. Unfortunately the demand for the new product C is not very certain.

Table 3.1 Estimated values for the planning parameters

	Product group A	Product group B	Product C
Production time in hours per ton	$\tau_A = 0.24$	$\tau_B = 0.47$	$\tau_C = 1.4$
Contract profit in $ per ton	$\sigma_A = 1478$	$\sigma_B = 897$	–
Demand for contracts in ton per year	$\delta_A = 20\,000$	$\delta_B = 11\,000$	–
Profit in production period in $ per ton	$\varrho_A = 1534$	$\varrho_B = 953$	$\varrho_C = 3350$
Minimum demand in ton per year	$\alpha_A = 16\,040$	$\alpha_B = 8350$	$\alpha_C = 0$
Mode of demand in ton per year	$\beta_A = 17\,550$	$\beta_B = 8900$	$\beta_C = 850$
Maximum demand in ton per year	$\gamma_A = 19\,900$	$\gamma_B = 9150$	$\gamma_C = 1600$

This case study will be continued in Section 3.4.1.

3.4
Modeling the Profit of the Production Planning

The criterion on which planning will be assessed is the total profit that is achieved after the production period when the production is executed in accordance with the production planning. For that, not only is the expected profit important, but also the certainty that this profit will be achieved should be taken into account in the final decision. If a small deviation of the expected demand results in a much lower profit than expected, it could be safer to choose a more robust planning with a lower, but more certain profit.

Let z_i, $i \in \{1, 2, \ldots n, n+1\}$ be the sold amount of products when the production period is finished. Together with the products that are sold before the production period in annual contracts, the final profit that will be made in this period equals:

$$P(y_1, \ldots, y_n, z_1, \ldots, z_n, z_{n+1}) = \sum_{i=1}^{n} \sigma_i y_i + \sum_{i=1}^{n+1} \rho_i z_i \qquad (4)$$

The amount of products sold during the production period will depend on the demand for these products and the available production time.

If the demand is lower than the amount that can be produced in the available production time, then the total demand can be satisfied. However, if the demand is larger than the available capacity then only that amount of product can be made and sold. Therefore, the total amount of product sold during the production period will equal $z_i = \min(d_i, x_i)$, $i \in \{1, 2, \ldots n, n+1\}$. The same holds for the amounts sold in annual contracts. The overall profit will then be

$$P(y_1, \ldots, y_n, x_1, \ldots, x_n, x_{n+1}) = \sum_{i=1}^{n} \sigma_i \min(\delta_i, y_i) + \sum_{i=1}^{n+1} \rho_i \min(d_i, x_i). \qquad (5)$$

For fixed values of the decision variables, the maximum and minimum total profit that can be achieved depends on the planned amounts of the products on the production planning, and on the maximum, respectively minimum demand for the products:

$$\max P(y_1, \ldots, y_n, x_1, \ldots, x_n, x_{n+1}) = \sum_{i=1}^{n} \sigma_i \min(y_i, \delta_i) + \sum_{i=1}^{n+1} \rho_i \min(x_i, y_i)$$

$$\min P(y_1, \ldots, y_n, x_1, \ldots, x_n, x_{n+1}) = \sum_{i=1}^{n} \sigma_i \min(y_i, \delta_i) + \sum_{i=1}^{n+1} \rho_i \min(x_i, \alpha_i). \qquad (6)$$

For fixed values of the decision variables the probability density of the final profit can now be derived from the probability density of the demand for the products during the production period, under the assumption that the demands for these products are mutually independent.

In general for a linear combination $w = ax + by$, where the stochastic variables x and y are independent, the density function of w reads (Papoulis 1965):

$$f_W(w) = \frac{1}{|ab|} \int_{-\infty}^{\infty} f_x \left(\frac{w - y_1}{a} \right) f_y \left(\frac{y_1}{b} \right) dy_1 \quad \text{with } y_1 = by. \qquad (7)$$

The final profit was defined by

$$P(y_1, \ldots, y_n, x_1, \ldots, x_n, x_{n+1}) = \sum_{i=1}^{n} \sigma_i \min(\delta_i, y_i) + \sum_{i=1}^{n+1} \varrho_i \min(d_i, x_i).$$

Let p_i, $i \in \{1, 2, \ldots, n, n+1\}$ be the profit made by selling the amount z_i of product i during the production period, then $p_i = \varrho_i z_i$, with a minimum of 0 and a maximum of $\varrho_i x_i$.

Applying the general proposition (Eq. (1)) on the final profit P, the probability density function of P reads:

$$f_P(p) = \frac{1}{\prod\limits_{i=1}^{n+1} \rho_i} \int_0^{\rho_1 x_1} \int_0^{\rho_2 x_2} \cdots \int_0^{\rho_n x_n} \prod_{i=1}^{n+1} f_{z_i}\left(\frac{p_i}{\rho_i}\right) dp_n \ldots dp_2\, dp_1$$

$$\tag{8}$$

with $p_{n+1} = p - \prod\limits_{i=1}^{n} p_i - \sum\limits_{i=1}^{n} \sigma_i y_i$.

If the actual demand d_i, $i \in \{1, 2, \ldots, n, n+1\}$ is smaller than the planned capacity x_i then the sold amount of product z_i will equal the demand d_i. In that case, the probability density of z_i will follow the probability density of the demand d_i. However, if the actual demand d_i is larger than the planned capacity x_i then only the amount $z_i = x_i$ will be produced and sold. The probability that this will happen is the probability of a demand larger than the planned capacity, that is, $d_i \geq x_i$ (see Fig. 3.3).

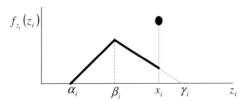

Figure 3.3   The probability density function of the sold amount of product

By this observation, the probability density $f_{z_i}(z_i)$, $i \in \{1, 2, \ldots, n, n+1\}$ of the sold amount of product z_i satisfies:

$$f_{z_i}(z_i) = \begin{cases} f_{D_i}(z_i) & 0 < z_i < x_i \\ 1 - F_{D_i}(z_i) & z_i = x_i \\ 0 & z_i > x_i \end{cases} \quad, i \in \{1, 2, \ldots, n, n+1\} \tag{9}$$

Unfortunately, by the local discontinuities in the probability density function of the final profit, the integrals cannot be solved analytically. The derived theoretical results will now be applied on the case study described in Section 3.3.1.

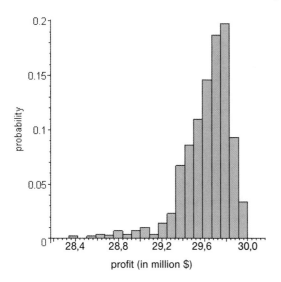

3.4.1
Modeling the Profit for the Food Additives Plant

In the aforementioned case study the production planning should be made for two current product groups A and B and one new product C. For reasons of comprehensibility, the assumption is made that no products were sold before the production period, that is $y_A = y_B = 0$.

The total profit that can be made will now depend on the demand for the products during the production period and on the planned amounts of the different products, that is, $P(x_A, x_B, x_C) = 1534 \min(d_A, x_A) + 953 \min(d_B, x_B) + 3350 \min(d_C, x_C)$,

under the restriction that the total production time will not be exceeded, $0.24 \, x_A + 0.47 \, x_B + 1.4 \, x_C = 9015$.

The probability density function of the profit satisfies:

$$f_P(p) = \frac{1}{1534 \cdot 953 \cdot 3350}$$
$$\cdot \int_0^{1534 x_A} \int_0^{953 x_B} f_{Z_A}\left(\frac{p_A}{1534}\right) \cdot f_{Z_B}\left(\frac{p_B}{953}\right) \cdot f_{Z_C}\left(\frac{p - p_A - p_B}{3350}\right) dp_B \, dp_A \quad (10)$$

Although this probability density cannot be solved analytically, it can be simulated for fixed values of x_A, x_B, x_C by randomly picking a certain demand for the products A, B and C from their individual probability density functions.

Figure 3.4 shows a probability histogram of the simulated profit for a production planning with $x_A = 17\,000$, $x_B = 9588$, $x_C = 950$ ton per year. The sample size taken is 1000. The unequal distribution in the left tail is caused by the sample size and would not be present in the theoretical distribution. This histogram shows a very skewed distribution to the left. This skewness is caused by the discontinuities in the function $P(x_A, x_B, x_C)$. In Section 3.5.1 this case study will be continued.

3.5
Modeling the Objective Functions

The quality of a certain production planning will be assessed on the expected value of the profit that can be achieved with the planning. The expected value of the final profit satisfies:

$$E_P(y_1, \ldots, y_n, x_1, \ldots, x_n, x_{n+1}) = \sum_{i=1}^{n} \sigma_i \min(\delta_i, y_i) + \sum_{i=1}^{n+1} \rho_i E_{Z_i}(x_i) \qquad (11)$$

with $y_i \leq \delta_i$, $i \in \{1, 2, \ldots, n\}$

From the probability density function $f_{Z_i}(z_i)$, $i \in \{1, 2, \ldots, n, n+1\}$ of the amount of sold products, the expected value of the amount z_i can be determined by:

$$E_{Z_i}(x_i) = \int_0^{x_i} z_i f_{d_i}(z_i)\, dz_i + x_i \left(1 - F_{d_i}(x_i)\right), \quad i \in \{1, 2, \ldots, n, n+1\} \qquad (12)$$

There are four possibilities for the planned amount x_i in comparison to the expected demand d_i, $i \in \{1, 2, \ldots, n, n+1\}$. Remember that d_i was expected to lie between α_i and γ_i with a mode β_i. Elaboration of Eq. (12) yields:

1. If $x_i \leq \alpha_i$ then $E_{Z_i}(x_i) = x_i$.

2. If $\alpha_i \leq x_i \leq \beta_i$ then

$$E_{Z_i}(x_i) = \frac{1}{(\beta_i - \alpha_i)(\gamma_i - \alpha_i)} \left[-\frac{1}{3}x_i^3 + \alpha_i x_i^2 + (\beta_i \gamma_i - \alpha_i \beta_i - \alpha_i \gamma_i)x_i + \frac{1}{3}\alpha_i \right].$$

3. If $\beta_i < x_i \leq \gamma_i$ then

$$E_{Z_i}(x_i) = \frac{1}{(\beta_i - \alpha_i)(\gamma_i - \alpha_i)} \left[\frac{2}{3}\beta_i^3 - \alpha_i \beta_i^2 + \frac{1}{3}\alpha_i^3 \right]$$
$$+ \frac{1}{(\gamma_i - \beta_i)(\gamma_i - \alpha_i)} \left[\frac{1}{3}x_i^3 - \gamma_i x_i^2 + \gamma_i^2 x_i + \frac{2}{3}\beta_i^3 - \gamma_i \beta_i^2 \right]$$

4. If $x_i > \gamma_i$ then $E_{Z_i}(x_i) = \dfrac{\alpha + \beta + \gamma}{3}$.

Unfortunately, the expected value of the profit that can be achieved with a certain production planning, does not guarantee that this profit will be achieved in reality.

Due to the skewed density function, for most choices of the production planning, the median of the profit will be higher than the expected value of the profit. This means that with a probability of more than 50% the real profit will be higher than the expected value. And with a probability of less than 50% the real profit will be lower than the expected value. As a consequence, the average deviation below the expected profit will be larger than the average deviation above the expected profit.

As a measure for the robustness of the planning the first quartile $Q_P(y_i, x_i)$ of the profit is chosen. The probability that the real profit will be lower than this first quartile equals 25%. Under the assumption that the demands for the different products are mutually independent, the first quartile of the total profit will be a linear combination of the first quartiles of the sold amounts of products z_i, $i \in \{1, 2, \ldots, n, n+1\}$ and can be written as:

$$Q_P(y_1, \ldots, y_n, x_1, \ldots, x_n, x_{n+1}) = \sum_{i=1}^{n} \sigma_i y_i + \sum_{i=1}^{n+1} p_i Q_{Z_i}(x_i) \tag{13}$$

with $y_i \leq \delta_i$, $i \in \{1, 2, \ldots, n\}$

The first quartile of the sold product, $Q_{Z_i}(x_i)$, will equal the first quartile of the demand d_i if the planned amount x_i is larger than this demand, otherwise it will equal x_i:

$$Q_{Z_i}(x_i) = \min\left(Q_{D_i}(d_i), x_i\right), \ i \in \{1, 2, \ldots, n, n+1\} \tag{14}$$

As long as the first quartile $Q_{D_i}(d_i)$ is smaller than the mode β_i of the demand, i.e.,

$$F_{D_i}(\beta_i) = \frac{(\beta_i - \alpha_i)^2}{(\beta_i - \alpha_i)(\gamma_i - \alpha_i)} \leq 0.25 \Leftrightarrow \beta_i \leq 0.75 \ \alpha_i + 0.25 \ \gamma_i, \text{ the first quartile } Q_{D_i}(d_i) \text{ is}$$
the

solution of $F_{D_i}(d_i) = \dfrac{(d_i - \alpha_i)^2}{(\beta_i - \alpha_i)(\gamma_i - \alpha_i)} = 0.25$ for d_i, i.e.,

$$Q_{D_i}(d_i) = \alpha_i + \sqrt{0.25(\beta_i - \alpha_i)(\gamma_i - \alpha_i)}. \tag{15}$$

If the first quartile $Q_{D_i}(d_i)$ is larger than the mode β_i of the demand, i.e., $\beta_i > 0.75 \ \alpha_i + 0.25 \ \gamma_i$, then

$$Q_{D_i}(d_i) = \gamma_i + \sqrt{0.75(\gamma_i - \beta_i)(\gamma_i - \alpha_i)}. \tag{16}$$

3.5.1
Modeling the Objective Functions of the Food Additives Plant

For the case study described in Section 3.4.1, the objective functions can now be modeled.

The expected value of the final profit satisfies:

$$E_P(x_A, x_B, x_C) = 1534 E_{Z_A}(x_A) + 953 E_{Z_B}(x_B) + 3350 E_{Z_C}(x_C) \tag{17}$$

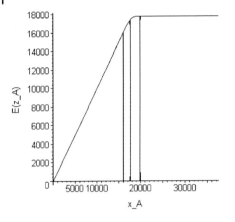

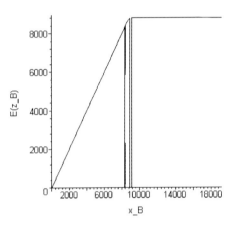

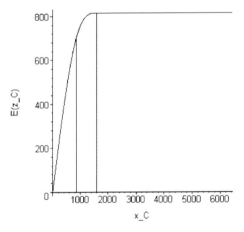

Figure 3.5 The expected amount of sold products A, B and C for a certain production planning

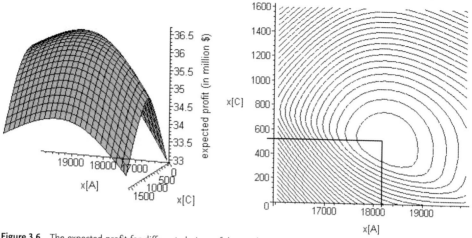

Figure 3.6 The expected profit for different choices of the production planning

Figure 3.5 shows the functions of $E_{Z_A}(x_A)$, $E_{Z_B}(x_B)$ and $E_{Z_C}(x_C)$, respectively. The vertical lines indicate the different parts of the piecewise functions, that is, $x_i < a_i$, $a_i \leq x_i \leq b_i$, $\beta_i < x_i \leq \gamma_i$, $x_i > \gamma_i$, $i \in \{A, B, C\}$.

Figure 3.6 shows the three-dimensional plot and the contour plot of $E_p(x_A, x_B, x_C)$ with the restriction that the total production time is filled, but not exceeded, that is,

$$x_B = \frac{9015 - 0.24x_A - 1.4x_C}{0.47} \tag{18}$$

The figures show that there is one production planning that leads to a maximum value for the expected profit $E_P(x_A, x_B, x_C)$. A rough estimation can already be made from the contour plot that in this production planning around 18,200 tons will be planned of product A, around 500 tons will be planned for product C, which will leave capacity for about 8400 tons of product B.

The second objective function is the first quartile of the total profit, that is,

$$Q_P(x_A, x_B, x_C) = 1534 \min(17247, x_A) + 953 \min(8682, x_B) + 3350 \min(583, x_C).$$

$$\tag{19}$$

Figure 3.7 shows the three-dimensional plot and the contour plot of $Q_P(x_A, x_B, x_C)$ again with the restriction that the total production time is filled, but not exceeded (Eq. (18)).

Also these figures show that there is one production planning that leads to a maximum value for the first quartile $Q_P(x_A, x_B, x_C)$ of the final profit. Again a rough estimation can be made from the contour plot. In this production planning around 17,200 tons will be planned of product A, around 600 tons will be planned for product C, which will leave capacity for about 8600 tons of product B.

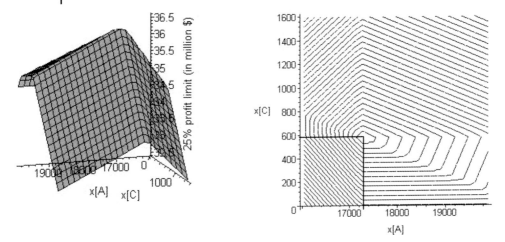

Figure 3.7 The first quartile of the profit for different choices of the production planning

3.6
Solving the Optimization Problem

The production planning problem is translated into a multicriteria piecewise linear optimization problem. The problem, however, will not be solved as a multicriteria problem, since the objectives are the extremes of one scale, from an uncertain but high profit to a more certain but low profit. For planning management willing to run a higher risk for a higher expected profit, the most profitable planning, corresponding with the maximum expected profit, may be the right choice. For planning management not willing to run any risk the most robust planning, that is, the one with the highest first quartile will be a more certain choice, although the expected profit will be much lower in that case. For every nuance of profitableness at a certain risk, a production planning in between those two extremes can be found.

The optimization problem is as follows:

determine $\gamma_1, \ldots, \gamma_n, x_1, \ldots, x_n, x_{n+1}$ for which

$$E_P(\gamma_1, \ldots, \gamma_n, x_1, \ldots, x_n, x_{n+1}) = \sum_{i=1}^{n} \sigma_i \gamma_i + \sum_{i=1}^{n+1} \rho_i E_{Z_i}(x_i)$$

or

$$Q_P(\gamma_1, \ldots, \gamma_n, x_1, \ldots, x_n, x_{n+1}) = \sum_{i=1}^{n} \sigma_i \gamma_i + \sum_{i=1}^{n} \rho_i Q_{Z_i}(x_i)$$

is maximized, subject to

$$T = \sum_{i=1}^{n} \tau_i \gamma_i + \sum_{i=1}^{n+1} \tau_i x_i.$$

Due to the piecewise character of the objective functions, common gradient methods for optimization cannot be used. Therefore a choice is made to use a direct-search method, the simplex method of Nelder and Mead (Nelder 1965). The Nelder-Mead algorithm is mentioned in many textbooks, but very seldom explained in detail. That is why a short description of the working method is given here (Fig. 3.8).

If there are n decision variables in the optimization problem, the Nelder-Mead algorithm will start by choosing $n + 1$ points arbitrarily. For reason of simplicity, assume that there are two decision variables. Then there will be 3 starting points (P1, P2, P3), which together form a triangle. This triangle is called the *simplex*. If the objective function is to be maximised then the point with the smallest value (P1) is reflected into the middle of the opposite side, in the supposition that this will lead to a better value for the objective function.

There are four different possibilities on how the search will continue.

1. If the objective value of the new point (P4) lies between the best and the worst values of the other points (P2, P3), then P4 is accepted as a new starting point and the new simplex is (P2, P3, P4) (Fig. 3.8a).
2. If the objective value of the new point (P4) is better than all others then the point is even further drawn out, twice as far from the reflecting point as P4. Therefore, P5 will form the new simplex with P2 and P3 (Fig. 3.8b).
3. If the objective value of the new point (P4) is worse than all others but better than the original (P1), then a new point P6 is evaluated half as far from the reflecting point as P4 (Fig. 3.8c). Again there are two possibilities:
    a. If P6 is worse than all others then the whole simplex is decreased by half towards the best point in the simplex. The new simplex is then (P2, P3', P6') (Fig. 3.8d).
    b. Otherwise, the new simplex is (P2, P3, P6).
4. If the objective value of the new point (P4) is worse than P1 then a new point P7 is defined halfway between the reflecting point and P1 itself. The new simplex will then be (P2, P3, P7) (Fig. 3.8e).

The new simplex is now used as starting point and the same procedure is performed until the best point and the second best point differ less than a fixed value ε. For both objective functions the Nelder-Mead algorithm could be used to find the production planning with the highest expected profit and the production planning with the highest first quartile, that is, the most robust planning. The planners will be provided with information about the robustness and the expected profit of different possibilities of the planning, on which they can base their final choice.

In the next paragraph, the Nelder-Mead algorithm will be applied on the objective functions in the case study.

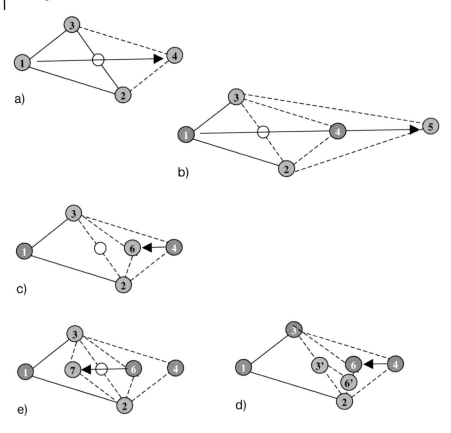

a)

b)

c)

e)

d)

Figure 3.8 Nelder-Mead algorithm

3.6.1
Solving the Optimization Problem in the Case Study

For the case study the production planning problem is translated into the following optimization problem:

Determine x_A, x_B, x_C for which

$$E_P(x_A, x_B, x_C) = 1534\ E_{Z_A}(x_A) + 953\ E_{Z_B}(x_B) + 3350\ E_{Z_C}(x_C)$$

or

$$Q_P(x_A, x_B, x_C) = 1534\ \min(17247, x_A) + 953\ \min(8682, x_B) + 3350\ \min(583, x_C)$$

is maximized, subject to

$$x_B = \frac{9015 - 0.24\ x_A - 1.4\ x_C}{0.47}.$$

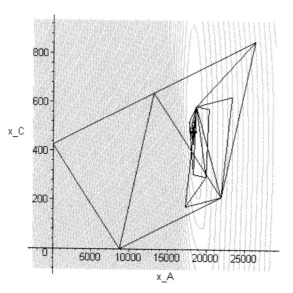

Figure 3.9 The Nelder-Mead algorithm to determine the maximum expected profit

Figure 3.9 shows for the decision variables x_A and x_C, the contour plot of the expected profit $E_P(x_A, x_B, x_C)$ with the simplices resulting from the Nelder-Mead algorithm.

The optimal planning found is the planning in which 18,221 tons for product A, 8445 tons for product B and 481 tons for product C are planned (Table 3.2). The expected profit for this planning equals: $E_P(18\,221, 8445, 481) = 3.67 \cdot 10^7$ dollars. The first quartile for this planning, that is, the robustness of the planning, equals $Q_P (18\,221, 8445, 481) = 3.61 \cdot 10^7$ dollars.

The most robust planning, that is, the one with the highest first quartile, is found by applying the Nelder-Mead algorithm to:

$$Q_P (x_A, x_B, x_C) = 1534 \min(17247, x_A) + 953 \min(8682, x_B) + 3350 \min(583, x_C)$$
$$\text{subject to } x_B = \frac{9015 - 0.24\, x_A - 1.4\, x_C}{0.47}.$$

The company itself should now decide if they will speculate on a higher expected profit or if they will prefer a lower but more certain profit. The diagram in Fig. 3.10 below can serve as an informative tool for the decision making.

Table 3.2 Results of the optimization

Optimization	$E_P(x_A, x_B, x_C)$ (million dollars)	$Q_P(x_A, x_B, x_C)$ (million dollars)	x_A (ton)	x_B (ton)	x_C (ton)
Max. expected profit	36.7	36.1	18 221	8445	481
Max. robustness	36.3	36.6	17 248	8647	579

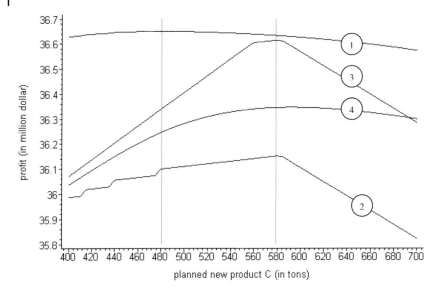

Figure 3.10 The profit levels for planned amounts of product C

In the planning with the maximum expected profit 481 tons were planned for product C. In the planning with the maximum 25 % limit of the profit 579 tons were planned for product C. Figure 3.10 shows for the interesting region for the planned amounts of product C, that is, between 400 and 700 tons, four different profit levels:

- Profit level 1 is the maximum expected profit that can be achieved with the planned amount of product C, assuming that the remaining capacity is optimally divided over the product groups A and B.
- Profit level 2 is the 25 % profit limit if the planning with the maximum expected profit (see profit level 1) is implemented.
- Profit level 3 is the maximum 25 % profit limit that can be achieved with the planned amount of product C, assuming that the remaining capacity is optimally divided over the product groups A and B.
- Profit level 4 is the expected profit if the planning with the maximum 25 % profit limit (see profit level 3) is implemented.

Assume that the company, based on the information from Fig. 3.10, decides to plan 579 tons of product C. This seems to be a profitable but not too risky choice. Compared to the planning with for example $x_C = 481$, the maximum expected profit is a bit lower, but all other profit levels are very high. It will now depend on the choice for the planned amounts of products in the groups A and B, if they can expect a higher profit with more uncertainty or a lower profit with less uncertainty. However, the 25 % profit limit gives no information about how the profit is distributed below this limit. To have more information on how low the profit could be, Fig. 3.11 shows, for 579 tons planned for product C, the probability distribution of the profit for different amounts planned for product groups A and B.

From Fig. 3.11 it is clear that the more product A is planned for, the more profit can be expected, but the more uncertainty exists whether this profit will be achieved. For instance, if the company decides to plan 18,000 tons for products from product group A, then the maximum profit they can achieve is about US $37.5 million and there is a probability of 60 % that this profit will not be achieved. There even is a probability of 25 % that the profit will be lower than US $36.3 million.

3.7
Sensitivity Analysis of the Optimization

The planners settle several parameters on which the determination of the optimal planning is based. Some deviation from the expected values will lead, after completion of the production period, to profit results that differ from what was expected. To be able to make a good evaluation of the risks, the sensitivity of the profit and the optimal planning will be studied for small deviations from the expected values of the following parameters in the objective functions:

1. the profit ϱ_i, $i \in \{1, 2, ..., n, n + 1\}$ made on the sold amounts of the current products and the new one during the production period;
2. the minimum α_i, the mode β_i and the maximum γ_i of the demand d_i, $i \in \{1, 2, ..., n, n + 1\}$ of all products during the production period.

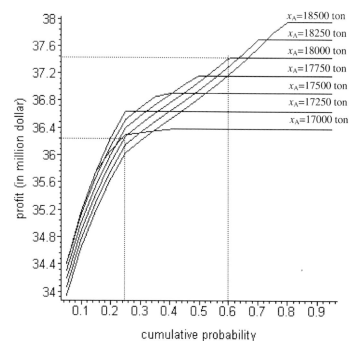

Figure 3.11 Probability distributions of the profit for different amounts of product A

Also the sensitivity of the solution to the parameters in the production time constraint will be studied:

3. the total production time T available per year;
4. the production time τ_i, $i \in \{1, 2, ..., n, n + 1\}$ that is needed to make one ton of product.

Let $(\underline{y}_I^{OPT}, \underline{x}_I^{OPT})$ be the optimal planning with respect to the maximum expected profit. Let $E_P^{OPT} = E_P (\underline{y}_I^{OPT}, \underline{x}_I^{OPT})$ be the maximum value of the expected profit.

The effect on the optimal value E_P^{OPT} of a small change in for example the parameter ϱ_1 is now given by $\dfrac{\partial E_P^{OPT}}{\partial \varrho_1}$, the absolute sensitivity coefficient of ϱ_1. The same can be done for all other parameters and for both objective functions.

The stepwise character of the objective functions is in this case for the differentiation not a problem, since the discontinuities of the function are only found in the decision variables, not in the parameters. For the parameters, the objective functions are continuous and therefore differentiable.

The size of the absolute sensitivity coefficients depends on the scale on which the parameters are measured. To make them comparable they are scaled by $\dfrac{\partial E_P^{OPT}}{\partial \varrho_1} \cdot \dfrac{\varrho_1}{E_P^{OPT}}$, which form the relative sensitivity coefficients, for example, for the parameter ϱ_1.

The other uncertain parameters influence the only constraint in the problem:
$$\sum_{i=1}^{n} \tau_i y_i + \sum_{i=1}^{n+1} \tau_i x_i = T.$$ Changes in the parameters τ_i or in T will cause the optimal planning to be no longer feasible. In practice a decrease in comparison to the expected values of τ_i, $i \in \{1, 2, ..., n, n + 1\}$ or an increase in comparison to the expected value of T will not cause any problem. The planned amounts of products can still be made and it will be possible to make an even higher amount of product than was planned, although it is not guaranteed that this extra amount can be sold as well. Information is needed to determine for which product the extra production time should be used to make as much profit as possible.

An increase, however, of the values of τ_i, $i \in \{1, 2, ..., n, n + 1\}$ or a decrease of the overall production time T will cause the optimal planning to be unachievable, and information is needed at the expense of which product the reduction of production time should be found to keep the profit as high as possible.

In general Lagrange multipliers can be used to investigate the influence of changes in the right hand side of the constraint parameters, but due to the piecewise character of the objective functions the Lagrange multiplier cannot analytically be calculated for an arbitrary T. The sensitivity analysis will be illustrated on the basis of the case study from Section 3.6.1.

3.7.1
Sensitivity Analysis in the Case Study

Assume that the company, based on the information given in Figs. 3.10 and 3.11, has decided to plan 18,000 tons for product group A, 8265 tons for product group B and 579 tons of product C. The expected profit for this planning equals US $36.6 million and there is a probability of 25 % that the real profit is lower than US $36.3 million.

To study the sensitivity of the results for small changes in the profit parameters ϱ_A, ϱ_B, ϱ_C and the demand parameters α_A, β_A, γ_A, α_B, β_B, γ_B, α_C, β_C, γ_C, the relative sensitivity coefficients are calculated in the neighbourhood of the expected values of these parameters, presented in Table 3.3.

Table 3.3 Sensitivity of the profit for different planning parameters

Parameter	Expected value	Expected profit		25 % Probability limit	
		Absolute sensitivity	Relative sensitivity	Absolute sensitivity	Relative sensitivity
ϱ_A	1534 $ per ton	17,578	0.74	17,247	0.73
ϱ_B	953 $ per ton	8265	0.22	8265	0.22
ϱ_C	3350 $ per ton	531	0.05	579	0.05
α_A	16,040 ton	411	0.18	0	0
β_A	17,550 ton	347	0.17	0	0
γ_A	19,900 ton	166	0.09	0	0
α_B	8350 ton	0	0	0	0
β_B	8900 ton	0	0	0	0
γ_B	9150 ton	0	0	0	0
α_C	0 ton	539	0	0	0
β_C	850 ton	188	0.00	0	0
γ_C	1600 ton	100	0.00	0	0

The expected profit and the first quartile of the profit are most sensitive for the profit that can be made with the current product A, due to the high amounts of this product planned to be made and sold, according to expectations. Furthermore, the profit is sensitive to changes with respect to the profit that can be made with one ton of product B, and for changes in the demand of product A. Small changes in the other parameters have hardly any influence on the profit that can be made with the chosen planning.

A change in the total production time will also influence the profit that could be achieved with the implementation of the chosen planning. Figure 3.12 shows the change in expected profit, respectively 25 % profit limit, if the increase or decrease in production time T is totally covered by an increase, i.e., a decrease in planned amounts of product group A, B or product C.

Figure 3.12 shows clearly that a decrease in the total production time should never be covered at the expense of product group A, but should be found in a smaller amount of product group B or product C. On the other hand, when the production time is higher than expected, the extra time should be used to produce more of product C, although the differences are not so large. The robustness of the planning for

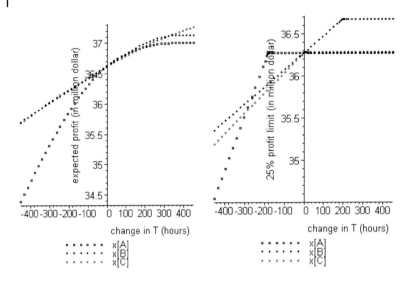

Figure 3.12 The changes in profit for smaller or large production time

a small decrease in total production time will not change if less of product A is made. However, a large decrease should go at the expense of the other products. An increase of the production time should as well be used to make more of the product group B or product C.

Figure 3.13 shows the change in expected profit, respectively 25 % profit limit, if the increase or decrease in the time t_A needed to produce one ton of product A is totally covered by an increase, i.e., a decrease in the planned amounts of product group A, B or product C.

Figure 3.13 shows that an increase of the production time needed to make one ton of product A can best be covered by making less of product group B or product C, although to keep the same robustness it is better to make less of product A. For a decrease, it is best to make more of product C. An increase or decrease of the production time needed to make one ton of product group B or of product C, gives more or less the same results as for product group A.

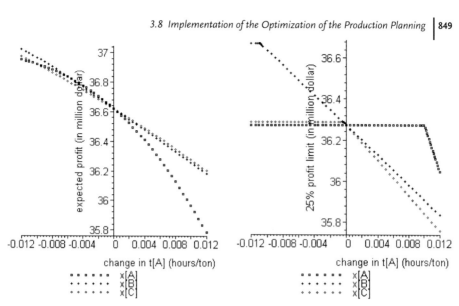

Figure 3.13 The changes in profit for a smaller or larger time/ton for product A

3.8
Implementation of the Optimization of the Production Planning

In the development of the method the focus was on the practical use for the planners in a multiproduct plant. The users of the method should not be bothered with the mathematical background of the method, which from their point of view can be considered as a black box. The emphasis should be on the information that should be acquired from them as an input for the method and on the results obtained from this information to be presented in a comprehensible and useful way. Knowledge of the transformation from the input into the output can increase the confidence in the results, but is not required to be able to use the planning method (see Fig. 3.14).

The *information from the planners*, needed as an input for the method, should consist of:

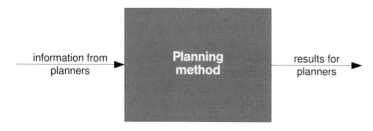

Figure 3.14 Implementation of the planning method

- the profits made on the current products sold in annual contracts in dollars per ton;
- the profits made on sold amounts of the current products and of new ones during the production period in dollars per ton;
- the total production time available in hours per year;
- the production times needed to make the products in hours per ton;
- the demand for the current products that can be sold in the annual contracts;
- the demand for the current products and for the new ones during the production period in ton per year described by:
  - the minimum demand in ton per year;
  - the mode of the demand in ton per year;
  - the maximum demand in ton per year.

Table 3.1 is an example of such input information.

The *results for the planners*, produced by the planning method, consist of:

- results of the optimization showing the optimal planning with respect to the maximum expected profit and the optimal planning with respect to the maximum 25 % profit limit (Table 3.2);
- the profit levels for planned amounts of new products showing the maximum expected profit and its corresponding 25 % profit limit and the maximum 25 % limit and its corresponding expected profit for different choices of free capacity for the new product(s). If more than one new product is taken into consideration, then the aggregate free capacity for these new products will be showed on the *x*-axis (Fig. 3.10).
- probability distributions of the profit for different planned amounts of the products showing the total probability distribution of a certain planning. For practical use, it should be easy to change the chosen amounts for all products to assess the effect of the changes on the profit distribution (Fig. 3.11).
- sensitivity of the profit for different planning parameters showing for a chosen planning which parameters are really influencing the profit, and by that require a good estimation of the expected value (Table 3.3);
- the changes in profit for smaller or larger production time (Fig. 3.12);
- the changes in profit for a smaller or larger time per ton for the products showing which adaptation to the planning should be made if the real values of production times differ from the expected ones (Fig. 3.13).

When the results are presented in such a way that the planners have full insight into the consequences of a chosen planning, the method will serve as a valuable decision support tool.

3.9
Conclusions and Final Remarks

A production planning method has been presented for a multiproduct manufacturing plant, which optimizes the profit under uncertainties in product demands. In the method these uncertainties are modeled by means of a simple triangular probability distribution, which is easy to specify. The optimization goal can be formulated as either a maximum expected profit or a robust profit (first quartile of the profit) to lower the risk. Due to discontinuities in the probabilistic distribution function of sold products a direct search optimization technique, Nelder-Mead, must be applied rather than a gradient-based optimization. The development and the application of the method have been highlighted by means of a case study taken from a food additives plant.

This method is considered practical because the required input data for the demand and process models and the profit function is easy to get by the users of the method, while the output information facilitates the interpretation of sensitivities of the optimized production planning in terms of common economic and product demand specification parameters. The method should be accessible to plant production management rather than to operations research specialized planning experts.

References

1 *Bansal V. Perkins J. D. Pistikopoulos E. N.* Flexibility analysis and design of dynamic processes with stochastic parameters. Comp. Chem. Eng. 22(Suppl.) (1998) p. S817–S820
2 *Bengtsson J.* Manufacturing flexibility and real options: a review. Int. J. Prod. Ec. 74 (2001) p. 213–224
3 *Berry W. L. Cooper M. C.* Manufacturing flexibility: methods for measuring the impact of product variety on performance in process industries. J. Op. Mgmt. 17 (1999) p. 163–178

4 *Corrêa H. L.* Managing unplanned change in the automotive industry. Dissertation University of Sao Paulo Warwick Business School, Avebury, Aldershot 1994
5 *Nelder J. A. Mead R.* A simplex method for function minimization. Comput. J. 7 (1965) p. 308–313
6 *Papoulis A.* Probability, Random Variables, and Stochastic Processes. McGraw-Hill, Boston 1965
7 *Swaney R. E. Grossmann I. E.* An index for operational flexibility in chemical process design. AIChE J. 31(4) (1985) p. 621–641

Indices

Authors' Index

Abildskov, Jens   V-1*
Alva-Argaez, Alberto   II-2
Belaud, Jean-Pierre   IV-5
Bogle, I. David L.   II-4
Braunschweig, Bertrand   IV-5
Cameron, Ian T.   I-6, IV-2
Dua, Vivek   III-5
Engell, Sebastian   III-4
Espuña, Antonio   IV-3
Fernholz, Gregor   III-4
Buzzi-Ferraris, Guido   I-1
Gani, Rafiqul   IV-1, V-1
Gao, Weihua   III-4
Georgiadis, Michael C.   I-4, II-1, III-1
Gerbaud, Vincent   I-3
Gernaey, Krist V.   I-7
Grievink, Johan   V-3
Hangos, Katalin M.   I-5, I-6
Heijnen, Petra   V-3
Heyen, Georges   III-3, V-2
Ingram, Gordon D.   I-6
Jørgensen, Sten Bay   I-2, I-7
Joulia, Xavier   I-3
Kalitventzeff, Boris   II-3, III-3, V-2
Kikkinides, Eustathois S.   I-4

Kokossis, Antonis   II-2
Kostoglou, Margaritis   I-4
Kraslawski, Andrey   II-5
Lakner, Rozalia   I-5
Lim, Young-il   I-2
Lind, Morton   I-7
Linke, Patrick   II-2
Manca, Davide   I-1
Marechal, Francois   II-3
Mateus, Miguel   V-2
Newell, Robert B.   IV-2
Papageorgiou, Lazaros G.   III-7
Perkins, John D.   III-5
Pistikopoulos, Efstratios N.   II-1, III-5
Proios, Petros   II-1
Puigjaner, Luis   III-6, IV-3
Romero, Javier   III-6
Sass, Richard   IV-4
Shah, Nilay   III-2
Toumi, Abdelaziz   III-4
Tsiakis, Panagiotis   III-1
Ydstie, B. Erik   II-4

*(Section-Chapter)

Computer Aided Process and Product Engineering. Edited by Luis Puigjaner and Georges Heyen
Copyright © 2006 WILEY-VCH Verlag GmbH & Co. KGaA, Weinheim
ISBN: 3-527-30804-0

Subject Index

Computer Aided Process and Product Engineering. Edited by Luis Puigjaner and Georges Heyen
Copyright © 2006 WILEY-VCH Verlag GmbH & Co. KGaA, Weinheim
ISBN: 3-527-30804-0

concentration dynamics, regulatory networks 233
concentration profiles
– ammonia formation 815
– model-based control 568
conceptual design
– life cycles (CliP) 669, 676 ff, 691
– multiscale process modeling 203
– shale oil processing 682
condensers
– complex multiphase reactor 394
– data validation 812
– distillation 283, 291
condition number 24
conditioning 519
conductive media simulation 401
configuration
– interaction methods 121 ff
– nonazeotropic mixtures 286
conjoint analysis 424
connection relation 255
conservation balances 176 ff
conservation element/solution element (CE/SE) method 39, 57, 61 ff
conservation laws
– distributed dynamic models 43, 62
– equipment design 384, 388 ff
consistency 183
constitutive equations 173–182
– multiscale process modeling 191 ff, 202 ff
constraint logic programming (CLP)
– product scheduling 500 ff
– resource planning 469
– supply-chain management 627
constraint propagation methods 30
– chemical product-process design 659
– cost-effective manufacturing 831 f
– data validation 802
– educational modules 778
– flexible recipe model 613
– hybrid processes 594, 601 ff
– model-based control 558 ff
– process monitoring 520 ff, 529 ff
– product scheduling 483, 489 f, 496 f
– real-time optimization 581 ff
– resource planning 456
– supply chain management 714
– utility systems 350, 365 f
construction projects 465 ff
consumer requirements 5, 424
contaminant profiles 322 ff, 375 ff
continuation methods 31 f
continuity, boundaries 54
continuous multiscale process modeling 199
continuous time discretization 490, 497
continuous time representation 594

control
– hybrid processes 599
– model-based 541–576
– multiscale process modeling 194
– resource planning 453
– transcriptional regulation 235 f
controllability
– complex multiphase reactor 397
– dynamical properties 185
– life cycles 676
– product development 431
convection
– cell population dynamics 93
– complex multiphase reactor 401
– distributed dynamic models 35, 43, 59, 75
convergence 17 f, 25
– material flows 448
– real-time optimization 589
– separation systems 142
conversion systems 329 ff
convex hull disjunction formulation 277, 284 f
cooling requirements 330, 370
coordinate transformation 177
CORBA-IIOP standards 750, 754 f
Corporate Web site management 764
correction model
– distributed dynamic models 60
– flexible recipes 610
correlations
– molecular modeling 114 ff
– multiscale process modeling 204
– utility systems 335
cosmetics 438
costs
– chemical product-process design 648 f
– correlations 335
– data validation 810
– effective manufacturing 829–854
– flexible recipe model 603 ff
– model-based control 548, 566
– nonlinear functions 361
– process monitoring 537
– real-time optimization 580
– resource planning 450
– supply-chain management 623
– utility systems 334 ff
Coulombic interactions 109, 121 f, 124 f
counter currents
– carbothermic aluminum process 400
– utility systems 330
– intensification 307
– separation systems 142
coupling
– multiscale processmodeling 189
– regulatory networks 244
Courant–Friedrichs–Lewy (CFL) number 59